Physical and Geotechnical Properties of Soils

Physical and Geotechnical Properties of Soils

Joseph E. Bowles

Professor of Civil Engineering, Bradley University

McGraw-Hill Book Company

New York · St. Louis · San Francisco · Auckland · Bogotá · Düsseldorf
Johannesburg · London · Madrid · Mexico · Montreal · New Delhi · Panama · Paris · São Paulo
Singapore · Sydney · Tokyo · Toronto

PHYSICAL AND GEOTECHNICAL PROPERTIES OF SOILS

Copyright © 1979 by McGraw-Hill, Inc. All rights reserved. Printed in the United States of America. No part of this publication may be reproduced, stored in a retrieval system, or transmitted, in any form or by any means, electronic, mechanical, photocopying, recording, or otherwise, without the prior written permission of the publisher.

1234567890 DODO 78321098

This book was set in Times Roman. The editors were B. J. Clark, Frank J. Cerra, and Madelaine Eichberg; the cover was designed by Albert M. Cetta; the production supervisor was Dominick Petrellese. The drawings were done by Arthur E. Dini.
R. R. Donnelley & Sons Company was printer and binder.

Library of Congress Cataloging in Publication Data

Bowles, Joseph E
 Physical and geotechnical properties of soils.

 Includes index.
 1. Soil mechanics. I. Title.
TA710.B684 624′.1513 78-3790
ISBN 0-07-006760-0

Contents

Chapter 6
Soil Structure and Clay Minerals   **149**

Chapter 7
Compaction and Soil Stabilization   **173**

Chapter 8
Soil Hydraulics, Permeability,
Capillarity, and Shrinkage   **201**

Chapter 9
Seepage and Flow Net Theory   233

Chapter 10
Stresses, Strains, and Rheological Concepts   273

Chapter 11
Consolidation and Consolidation Settlements   299

Chapter 12
Rate of Consolidation   320

Chapter 13
Shear Strength of Soils   345

Chapter 14
Stress-Strain Characteristics
and Stresses at a Point   393

Preface

This text is an up-to-date assemblage of material needed for a basic understanding of geotechnical engineering. Geotechnical engineering is defined as the science and practice of that part of civil engineering involving the interrelationships between the geologic environment and the works of people. The scope of this text has been deliberately limited to physical properties of soils, soil origins, and those engineering properties and mechanical behavior of soils needed for flow, strength, and stability analyses. It is the author's opinion that if just these fundamentals can be introduced and reasonably understood in a first course, this is a significant accomplishment.

Foundation engineering makes use of the physical and engineering properties of soils in the design of earth or foundation structures. This is a very important part of geotechnical engineering, but it has been left for other courses and texts to cover. An introduction to Boussinesq stresses, bearing capacity, and lateral earth pressures are included as necessary to develop and reinforce the basic concepts of strength and stability analysis. Some emphasis has been placed on statistics as a tool to aid in quantifying engineering judgment in estimating data reliability.

The text focus on basic fundamentals, but including material on the geological origin of soils, allows a more in-depth presentation without the textbook size becoming excessive. A fairly extensive introduction to geology, with emphasis on geotechnical uses, is included, since this is not a course requirement for many students. The text format allows more discussion, cross-referencing, repetition, and illustrative material, including both figures and examples, than would be possible if the topic coverage were expanded. This format also allows the instructor to emphasize "pet" areas of interest while simultaneously providing the student with additional study material for interest/stimulation whether or not it is assigned.

I did not attempt a literature survey, but a substantial number of references are cited, of which over one-half are from the past 10 years. These, or the bibliographies in the cited references, should be sufficient for any additional study or as a start for any necessary literature search.

A very substantial number of examples, illustrations, and homework problems are included. Many of the homework problems include answers or partial answers. Several of the homework problems refer to alternative solutions to example problems to produce the double benefit of careful study and problem solution. All the photographs, unless noted otherwise, were taken by the author.

I wish to express appreciation to Professor Turgut Demirel of Iowa State University, and Professor Archie M. Richardson of University of Pittsburgh, who carefully reviewed the manuscript and made numerous valuable suggestions.

My wife, Faye J. Bowles, did a considerable amount of the manuscript typing and helped tremendously in cross-checking figures, tables, and references; however, I assume responsibility for any errors or omissions.

Joseph E. Bowles

Physical and Geotechnical Properties of Soils

Chapter 1

Introduction to Soil Mechanics, Statistics, and SI

1-1 INTRODUCTION

Soil mechanics is concerned with the behavior and performance of soil as a construction material or as a support for engineered construction. The geotechnical engineer is concerned with obtaining representative soil samples to test for identification and to determine suitability as a construction material, e.g., for roadway fill, airfields, dams, and other works. Since sampling and testing is of necessity on small sample populations, some kind of statistical method is required to estimate the reliability of the results. Until very recently this reliability has been estimated more by a "feel" for the data than by any type of quantifiable analysis. Basic elements of statistics are introduced later in this chapter for use in making quantifiable test reliability estimates in the later subject matter.

The engineer is also concerned with prediction of the load-deformation characteristics of natural, as well as compacted, fill underlying any structure, or as a soil structure. Other important concerns include effects of water on soil behavior and the movement of water through the soil mass.

Geologically, soil mechanics is concerned with the unconsolidated mantle of weathered rock material overlying solid rock. A distinction is generally made between soil and rock in that soil is a particulate mass forming a skeleton structure, whereas rock is a dense structure with the constituent particles firmly bonded together.

Summarizing, the science of soil mechanics is concerned with soil stability, deformation, and water flow into, out of, and through soil masses, and with the associated risk which is economically tolerable.

1

1-2 THE STUDY OF SOIL MECHANICS

Virtually all civil, environmental, transportation, structural, and geotechnical engineers are intimately concerned with soil mechanics concepts. This is because almost all the construction endeavors of these individuals are concerned with soil behavior, since either the soil is used as a construction material or a structure is placed upon it. Foundation engineering is the design specialty that is specifically concerned with soil behavior and performance and the interface of either a superstructure (above ground) or substructure (the foundation) with the ground.

The study of soil mechanics is of considerable economic importance, since soil is the most readily available construction material at any site. All above-ground structures are supported by either soil or rock, and much of the public water supply moves through soil (to wells) or is retained by it in reservoirs.

A study of soil mechanics, and of the various standardized testing procedures available for identification and performance, enables the engineer to rapidly gain experience and obtain a "feel" for soil behavior. Of course, no matter how standardized the test, or how carefully it is performed, if the soil sample is not representative, the test results will not be of much value.

Soils engineering is more "state-of-the-art" than many of the more scientifically oriented disciplines because of the heterogeneity of natural soil deposits; hence, experience is a very important factor. Careful study of the literature together with appropriate soil tests, however, does allow an engineer with less experience to compete with the more experienced engineer.

All the topics of soil mechanics and foundation engineering and many aspects of geological engineering can be grouped under the term *geotechnical engineering*. In the broadest sense this can be described as the "science and practice of that part of civil engineering involving the interrelationship between the geological environment and the works of man." The engineer doing this specific type of engineering work is a geotechnical engineer.

A problem encountered by many students just starting the study of geotechnical engineering is the introduction of a new soil mechanics language, or terminology, and exposure to a series of seemingly unrelated and diverse topics. Generally some relationship between the topic and engineering practice is made as the subjects are presented, but the varied nature of geotechnical engineering is such that this must of necessity be brief in a textbook. A student who has mastered the material in this text should be able to integrate the various seemingly unrelated topics into a unified body of knowledge which can be used as a tool for any future geotechnical work.

1-3 TEXT OBJECTIVES

This text introduces the reader to:

1. Soil mechanics terminology
2. The physical, index, and engineering properties of soils and some of the methods of measurement

3. Classification of soils in the several widely used systems and in terms of geological formation
4. Methods of determining suitability of soils for various types of engineered construction
5. Evaluation of soil response to changes in loading and soil moisture
6. Effects of water on soil properties and movement of water through soil
7. Basic elements of statistics and probability as a tool to evaluate a soil testing program or data reliability
8. Methods of solution of certain soil mechanics problems

This list indicates that there is considerable diversity in the subject matter. The student should continually strive to integrate the material, since the real problem often requires meshing the effects of several factors to develop an overall solution.

1-4 SOME TYPICAL SOIL PROBLEMS

To put the discussion into perspective, some typical soil problems with which a geotechnical engineer might be involved include:

1. In a soil exploration program to investigate site conditions, how many borings are necessary, and how deep? How many samples are required? What soil tests will need to be performed?
2. What is the stress in the soil at a given depth from the imposed superstructure or fill load? Can the soil carry this stress without a shear failure?
3. How much settlement can be expected for a structure as a result of the increase in soil stress? How long will it take for this settlement to occur?
4. Is this soil suitable for a highway or railroad fill? For use as a dam where water will be retained? For an embankment to retain industrial waste without leaking environmental pollutants?
5. Can this soil be used directly in fills, or will it require admixtures to modify certain undesirable index properties prior to use? What additives can be used? Can we use additives which alone are environmental pollutants, such as fly ash (power plant byproduct of coal burning), paper mill wastes, or mine wastes?
6. What happens to the soil structure if the ground water table fluctuates? Will pumping to dewater an excavation cause environmental problems?
7. What is the effect of frost or ice formation? Can the effects be avoided or reduced?
8. What is the effect of soil moisture change on the volume of the soil mass? How can volume change be controlled for pavements? For other structures, including residential construction?
9. What is the rate of water movement through a soil mass, i.e., can it be easily drained? Will a well provide an adequate supply? Will a dam built over this soil hold water?

(a)

(c)

(e)

(b)

(d)

(f)

(g)

Figure 1-1 Some soil engineering projects. (*a*) Upstream face (riprapped to control erosion) of Garrison Dam near Pick City, North Dakota, and part of Missouri River Basin development. (*b*) Downstream face of Garrison Dam (powerhouse in distance). Dam is rockfill, but downstream face has soil and grass cover for erosion protection. (*c*) Portion of Mississippi levee in Missouri. River is several hundred meters behind trees on right. Levee is built from soil obtained between levee location and river rather than from farm land on land side. (*d*) Irrigation canal in Montana. (*e*) Small dam in a residential land development project near Peoria, Ill. Concrete overflow weir can be seen at far shore line. (*f*) Large above ground sanitary landfill in Michigan. Fill is more than 30 m high, with an anticipated future height of about 50 m. (*g*) Rockfill dam being used for a roadway. Dam is a reservoir for Rapid City, South Dakota.

4

10. What kind of excavation slope can be cut in the soil—1 : 2 (1 vertical on 2 horizontal), $1 : \frac{1}{2}$, 1 : 1 or what? This can cause serious economic problems for highway and railroad work in particular because of the additional cut volume and additional right-of-way required.
11. Can a site be used as a landfill or for impondment of industrial wastes without pollution of the groundwater?
12. How can an environmentally safe disposal be made for mine wastes, industrial waste solids, sewage sludge, etc.?
13. Is a site safe for a radiation-producing plant? Can settlements be controlled so that no leakage occurs? Will an earthquake produce a disaster?

Working solutions for some of these problems are shown in Fig. 1-1. It should be obvious that if some of these questions are inadequately answered or if the risk factor is too high, a failure may occur. Failure may take the form of:

1. Structural damage to buildings from excessive settlements or differential settlements
2. Bumpy roads resulting from differential settlements within a fill or at the junction of a cut and fill
3. Embankment failures, which may be slope (landslide) failures or excessive settlements or the underlying foundation or within the fill itself
4. Dam failures of various types, including "embankment" as in (3) as well as excessive leakage through the embankment or through the underlying soil

Not all the problems cited will be considered in this text, since neither time nor space is available. The soil properties, both physical and engineering, which will be needed to solve these problems will, however, be considered in some detail. Topics beyond the scope of this textbook can be found in texts on foundation engineering (e.g., Bowles, 1977).

1-5 HISTORICAL DEVELOPMENT OF SOIL MECHANICS

Most authorities date the beginning of soil mechanics as an engineering science to the publication of *Erdbaumechanik auf bodenphysikalischer Grundlage* by Karl Terzaghi (the first textbook in soil mechanics) in Germany in 1925. Because of this publication, Terzaghi's early work in Europe, Asia, and the United States, and his over 250 technical papers, Terzaghi is often called the "father of soil mechanics." The Terzaghi Lecture is given annually at the American Society of Civil Engineers in honor of Terzaghi. A historical perspective of Terzaghi as a soils engineer and a bibliography of most of his publications, including those considered most noteworthy, are given in *From Theory to Practice in Soil Mechanics*, by Bjerrum et al. (1960).

Soil construction and the associated problems have been with people ever since they began digging caves and building mud huts for shelter. The Bible makes reference to a preference for building on a rock foundation as opposed to sand (Matt. 7:24–27). The Egyptians were aware of soil problems and even used caissons in pre-Biblical times to sink shafts through very soft Nile River sediments. Later the Romans were involved with building roads, parts of which are presently in use as subgrades, and foundations for aqueducts, some of which carried water for many kilometers. The Romans also constructed many large buildings, some of which are still in existence, such as the Forum and Colosseum. These and later medieval constructions, including the notable St. Peter's Church, were built by utilizing principles which are as valid today as they were in historical times, solid, well-drained foundations. In Asia the Chinese made considerable use of soil from early times. A notable construction is the "Great Wall of China," first built during the Ch'in Dynasty (221–207 B.C.).

During medieval times many huge religious structures such as churches and bell towers were constructed in Europe. One which became famous because of an unsolved soil settlement problem is the Leaning Tower of Pisa in the city of that name in central Italy. This tower was begun in 1174 and discontinued after uneven settlement began to occur. Construction was later resumed, and the tower was finished in 1350. It is now some 5 m out of plumb in a 60-m height and is expected to turn over in about 200 years as a result of the continuing differential settlement. This settlement has now been continuing for more than 600 years. Recently the city of Pisa offered a prize (money) to anyone who could design a feasible method of halting the settlement without damage to the tower. As of this writing no satisfactory scheme seems to have been proposed. A brief engineering analysis of this project is given by Mitchell et al. (1977).

Some soil problems of more recent date follow.

A General Failures

Mexico City settlements. These settlements range from 1 to 4 m and are primarily due to massive pumping of underground water from an extremely porous subsoil aquifer (void ratio e up to 14 and natural water contents up to 650 percent) as reported by Hiriart and Marsal (1969). The Palace of Fine Arts in Mexico City, begun in 1904 and completed in 1944, has settled some $3\frac{1}{2}$ m, according to Leggett (1962).

Houston, Texas, settlements. These are areal (the whole area) settlements underlying Houston and adjacent towns. Parts of the area have settled as much as 3 m, and the current settlement rate is some 150 mm/yr. Recently in a bayou location a bridge redesign was necessary because of the 2.4-m settlement at the site between the time the bridge was designed in 1970 and the beginning of construction in 1977 (ENR† 6/9/77, p. 11). These settlements are attributed to pumping of under-

---

† ENR, 6/9/77, p. 11 = Engineering News-Record, June 9, 1977, page 11.

ground water for local water supplies. Some also believe that pumping of oil from deeper strata is contributing to the areal subsidence.

Transcona elevator. In 1914 a 1-million-bushel grain elevator at Winnipeg, Canada, consisting of 5 rows of 13 bins each and 30 m high suddenly (in approximately 12 h) tilted to about 30° from the vertical after filling. The structure was later tilted back to the vertical. The tilt was attributed to a shear failure caused by overloading the foundation soil [Peck and Bryant (1952), White (1952)].

B Settlements Due to Lateral Flow of Soil from Beneath Foundation

Vertical movement of soil resulting from loss of lateral support, such as that caused by an adjacent excavation, is still a common problem. For example:

1. Construction of a depressed section of Interstate Highway in California caused a lateral flow of soil into the cut and subsidence of adjacent ground and buildings along the cut (ENR, 10/10/68, p. 22).
2. During construction of a 32-story office building in Los Angeles, California, the bulkheads retaining the excavation slipped laterally on the order of 75 mm because of excessive lateral pressure buildup from soil/rainfall conditions not properly accounted for in design. This much slip can cause large settlements called *ground loss* around the perimeter, which, in an urban area, means ruined pavements and cracks in the closer buildings (ENR, 9/26/68).
3. During construction of a nine-story office building in Osaka, Japan, a loss of soil beneath footings on one side resulted in the building tilting about 5° out of plumb. The cause was believed to be excavation for a foundation on the adjoining lot (ENR, 10/17/68, p. 30).

C Differential Settlements

These are unequal settlements beneath different areas of a building or embankment; they result in cracking when they are sufficiently large. The notable exception to differential settlements being unwanted is the Leaning Tower of Pisa, which has become a tourist attraction—but even here the owner wants the settlement to stop! Other examples of unwanted differential settlements include:

1. The Charity Hospital in New Orleans, Louisiana, where the addition of a new wing to the original hospital resulted in differential settlements between the two parts of some 380 mm (a ramp was finally used to bridge the discontinuity).
2. Library-office building on the Cleveland State University campus in Cleveland, Ohio, had differential settlements of some 25 mm between the interior high-rise portion and the attached perimeter low-rise part. While this amount of settlement does not seem large, it was sufficient to cause detrimental cracking in the masonry walls and required remedial measures (ENR, 2/18/71, p. 12).

3. A building collapse in Akron, Ohio, was due to an isolated sinkhole formation missed during boring operations. The failure was initiated by the buildup of a roof water load which caused settlement in the sinkhole area. This caused the roof water to "pond" at that point and collapse the roof. The building was about 6 years old at the time of the roof collapse (ENR, 12/11/69, p. 23).

D Slope Failures

Slope failures are extremely numerous—especially small ones involving 5 to 50 m³ of earth—along the sides of road cuts. Other failures, including dam embankment failures, are far less numerous but often cause considerable property damage and loss of life. A few slope failures to indicate the scope of the problem include:

1. *Panama Canal slides.* These slides are still occurring, but were particularly troublesome during construction of the canal. The original earthwork estimate increased from 79 to 177 million m³ of earth. Fortunately increased efficiency in earth moving kept the final cost within the original appropriation of $375 million, according to Mills (1913).
2. *Fort Peck Dam.* This dam failed during construction in 1938 by the U.S. Bureau of Reclamation; the failure involved approximately 5 million m³ of fill material. It was apparently caused by an excess pore pressure buildup in the fill and a resulting loss of shear strength [Casagrande (1965), Middlebrooks (1942)].
3. A landslide occurred in 1806 at Goldau, Switzerland, and killed 457 people.
4. A small landslide in Virginia in 1933 involved only some 380 m³ of soil but was sufficient to wreck two trains.
5. Landslides occurring on Interstate I-40 in Tennessee resulted in some $10 million in cost overruns (ENR, 5/31/73, p. 15).

Leggett (1962) describes numerous landslides in various parts of the world. The reader can often observe landslides along highway cuts and on the sides of hills when traveling along highways in more rugged terrain in many areas of the world.

E Dam Failures

Dam failures may occur because of slope failures, but more often the failure is caused by overtopping, which erodes a channel (the velocity increases across the dam crest), or by seepage forming a channel which gradually tunnels (or forms a pipe) back, via progressive velocity increase, to the upstream face. The tunnel then usually collapses, and water rushes through the channel thus formed and erodes a cut to the dam base. Later inspection may never reveal the exact cause, since the remaining evidence is simply a channel through the embankment. Several dam failures include:

1. Malpasset Dam in southern France failed in 1959 and killed some 344 people. The failure was reportedly caused by failure to detect a clay-filled rock seam, which was only 10 to 20 mm thick, during the rock exploration phase of the design work. The seam became soft when saturated under the head of water retained and caused the abutment to slip (Civil Engineering, 1/60, p. 91).
2. St. Francis Dam at Saguache, California, provided water for the city of Los Angeles. It failed in 1928, killing 426 persons, and was assumed to have failed in a situation similar to the Malpasset dam, i.e., a weakened stratum of foundation material permitted abutment movement after being saturated under pressure from the reservoir.
3. Baldwin Hills reservoir, located near Los Angeles, failed in 1963, killing five people and causing some $50 million property damage. This was a 1.1-million-m^3-reservoir dam some 40 m high. The cause of failure was considered to be the development of tensile forces between the compacted fill and the abutment due to foundation settlement, forming a crack which enlarged through erosion (piping) to a failure (Casagrande, 1965). Some were of the opinion that pumping oil from the substrata caused ground subsidence, which in turn produced the tension crack (ENR, 10/31/68, p. 12).
4. Teton Dam failure. This recent failure involved a 93-m-high × 930-m-long dam with approximately 7.2 million m^3 of earth fill located in eastern Idaho. The dam failed June 5, 1976, resulting in 11 deaths and approximately $1 billion in property damage. This dam failed by piping (it was observed and took about 6 h to fail) believed to be due to inadequate grouting of the badly cracked and shattered rocks in the abutments (ENR, 6/15/76).

Other types of foundation failures include excessive settlement of pile foundations [see those cited by Blessey (1970) and Miller (1938)]. Another problem of widespread occurrence is caused by expansive soils. Soils which expand on wetting and shrink on drying are "expansive." The amount of volume change is very difficult to evaluate; however, methods to be presented later will alert one to the problem. This problem involves residential construction as well as larger structures and pavements. Gromko (1974) gives a review of this problem which should be read since the problem is so widespread.

Many soil failures are of a much lesser order of magnitude than those just cited; several of these are shown in Fig. 1-2. Small failures may be just as damaging to a small client as a major failure is for a governmental agency. Also, small failures such as those caused by settled fills, the maintenance of highway slopes for slide removal, and repair of cracked masonry in buildings are expensive.

This list of failures and photographs showing "failures" is not to emphasize failures but rather to show that:

1. Failures do occur in spite of the considerable advances in soil mechanics technology over the past 20 years.

Figure 1-2 Some soil failures. (*a*) Large landslide near Vicksburg, Miss., which completely closed both lanes of U.S. 61 south of Vicksburg. (*b*) Large piping failure which temporarily closed Route 9 near Pekin, Ill. The piping was caused during the flood stage of the Illinois River, which is about 300 m to the right, by an inadequately plugged drill hole. Piling is around initial piping; it was driven in a futile attempt to halt the piping. Piping enlarged the hole and carried material from beneath Route 9, causing the pavement to settle about 0.8 m. (*c*) Large pavement settlement in Kentucky. (*d*) Teton dam. Light material is dam fill. Failure caused erosion of a channel through the dam at far end. Close inspection can reveal the poor quality of the rock in the far abutment. (*e–h*) Typical local slope failures in highway cuts. These vary in size as shown and represent considerable maintenance expense. A large number are shown to give an indication of the magnitude of the problem. (*e*) Illinois; (*f*) Kentucky; (*g*) West Virginia; (*h*) Ohio.

2. Soil is an uncertain material with which to work or to take any excessive risk. While some risk is nearly always inherent in geotechnical engineering work, the risk factor should be assessed, and unnecessarily high risks must be avoided.

Actually, for the quantity of geotechnical work which is, or has been, done, the percentage of failures is quite small. Note that failures are "newsy" and tend to be reported, whereas the successes are generally hidden from view and almost never reported. When failures do occur, they are almost always due (in spite of disclaimers) to carelessness or to taking an excessive risk.

1-6 SOURCES OF INFORMATION FOR GEOTECHNICAL ENGINEERS

The following are readily available sources of information on geotechnical engineering (in English):

1. *Journal of the Geotechnical Engineering Division, ASCE.* The Geotechnical Engineering Division sponsors periodic specialty conferences and publishes the conference proceedings. The eleventh specialty conference was held at Pasadena, California, in June 1978.
2. *Canadian Geotechnical Journal (Canada).* A publication similar to the *ASCE Geotechnical Journal;* it began publication in 1963.
3. *Geotechnique (United Kingdom).* Journal which in 1948 originated the idea of a separate publication in geotechnical engineering. Papers are primarily British (although authors from other countries are published) and often highly theoretical.
4. *Soils and Foundations (Japan).* A publication similar to *Geotechnique* with the authors primarily Japanese, although authors from other countries are published. This journal began publication in 1960.
5. *Proceedings of the International Conference on Soil Mechanics and Foundation Engineering (ISSMFE).* The proceedings of international conferences held approximately every four years (1936, 1948, 1953, 1957, 1961, 1965, 1969, 1973, 1977). The ninth conference was held in Tokyo in 1977. Papers are published in the conference proceedings from every country which has members in the international society.
6. *Proceedings of Regional Soil Conferences.* Some of these include:
   European—sixth held in Vienna, Austria, in 1976
   African—sixth held in Durban, Natal, South Africa, in 1975
   Asian—fifth held in Bangalore, India, in 1975
   Pan-American—fifth held in Buenos Aires, Argentina, in 1975
   Australia–New Zealand—seventh held in Brisbane, Australia, in 1975
   Southeast Asian—fifth held in Bangkok, Thailand, in 1977
7. *University Engineering Experiment Station Bulletins.*
8. *American Society for Testing and Materials (ASTM).* Periodic conference

proceedings are published in Special Technical Publications (STPs). Occasional papers on soil testing are published in the journals.

9. *Transportation Research Board (formerly Highway Research Board) publications.* Areas coded by TRB which pertain to soil mechanics are nos. 61, 62, 63, and 64.
10. *Civil Engineering.* A monthly journal published by ASCE. It often contains notes on other publications of interest to geotechnical engineers.
11. *Proceedings, Institution of Civil Engineers of London, Australia, and India.*

Publications of the state geological survey are often of interest to geotechnical engineers. Sometimes local conferences sponsored by the state departments of transportation or under the auspices of the local sections of ASCE include papers of interest to geotechnical engineers.

1-7 ELEMENTS OF STATISTICS AND PROBABILITY

Statistics is the collection, tabulation, and analysis of data so that intelligent decisions may be made. Because of the inevitable uncertainties in every undertaking, statistics is also concerned with the probable accuracy of the data analysis and/or with whether the sample (data) size is sufficiently large. For example, if 100 manufactured devices represents the total production and we test each device, the results are said to be 100 percent reliable. If we test only four of the devices, the reliability is something less than 100 percent; depending on how the test samples were obtained, or how representative the four are of the 100 produced, the testing program may approach a prediction reliability of nearly zero (precisely the minimum prediction reliability is $4/100 = 0.04$ or 4 percent). If we select the four devices to be tested in a random manner, as we will find later, the prediction reliability may approach 60+ percent, depending on the allowable deviation from perfection termed "standard deviation" that we allow for a successful test.

The statistics and probability concepts introduced in this text will be very elementary, but will be generally sufficient for most soil testing programs. This is because soils generally obey random laws of distribution in terms of:

1. Grain size, type of soil, and horizontal and vertical variations.
2. Physical and engineering properties—that is, the cohesion, angle of internal friction, or modulus of elasticity should vary from sample to sample in a random manner. The properties listed here will be taken up in later chapters, since the reader is presumed not to know what they are at this point.

If soil is naturally altered in any manner, such as drying, wetting, or increasing or decreasing in density, the process should be random. Note carefully, however, that:

1. Testing techniques can introduce nonrandom errors (termed *bias*).

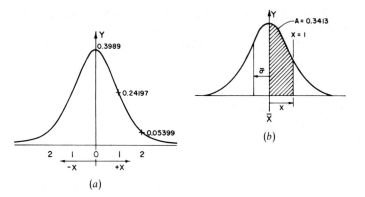

Figure 1-3 Normal distribution curves. Curve (*b*) shows the area given in Table 1-1. (*a*) Normal distribution curve; (*b*) normal distribution curve mean and standard deviation.

2. Random distribution of effects pertains to a particular soil. That is, where a soil mass consists of several layers of different soils, as clay layers, sand layers, silty-sand layers, etc., we must consider each layer separately in applying statistical concepts of the mean, standard deviation, and numbers of tests as introduced in the following paragraphs. If all the layers are combined, erroneous statistical conclusions will be obtained—from incorrectly applying the statistical methods.

   With random laws of distribution assumed to apply, the results produce a normal distribution curve which is symmetrical about the centroid, or mean, as in Fig. 1-3*a*.
   In statistics we will be concerned with the following terms (some of which are identified in Fig. 1-3):

N = number of tests or the size of the sample population
V = test value of interest (such as unit weight, water content, or any other numerical quantity)
$\bar{X}$ = sample mean computed as

$$\bar{X} = \frac{\sum V}{N}$$

Note that this differs from:

Median = midvalue so chosen that half of the tests are above and half are below it
   Mode = maximum value from the peak of the distribution curve (is not maximum value of all numbers in the test series of N values)

$\bar{\sigma}^2$ = sample variance computed as

$$\bar{\sigma}^2 = \frac{1}{N}\left(\sum_1^N V^2 - N\bar{X}^2\right)$$

Table 1-1 Areas under the standard distribution curve for χ as indicated. Double table values for $\pm\chi$ to get total area

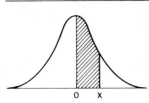

X	0.00	0.01	0.02	0.03	0.04	0.05	0.06	0.07	0.08	0.09
0.00	0.0000	0.0040	0.0080	0.0120	0.0160	0.0199	0.0239	0.0279	0.0319	0.0359
0.10	0.0398	0.0438	0.0478	0.0517	0.0557	0.0596	0.0636	0.0675	0.0714	0.0753
0.20	0.0793	0.0832	0.0871	0.0910	0.0948	0.0987	0.1026	0.1064	0.1103	0.1141
0.30	0.1179	0.1217	0.1255	0.1293	0.1331	0.1368	0.1406	0.1443	0.1480	0.1517
0.40	0.1554	0.1591	0.1628	0.1664	0.1700	0.1736	0.1772	0.1808	0.1844	0.1879
0.50	0.1915	0.1950	0.1985	0.2019	0.2054	0.2088	0.2123	0.2157	0.2190	0.2224
0.60	0.2257	0.2291	0.2324	0.2357	0.2389	0.2422	0.2454	0.2486	0.2517	0.2549
0.70	0.2580	0.2611	0.2642	0.2673	0.2703	0.2734	0.2764	0.2793	0.2823	0.2852
0.80	0.2881	0.2910	0.2939	0.2967	0.2995	0.3023	0.3051	0.3078	0.3106	0.3133
0.90	0.3159	0.3186	0.3212	0.3238	0.3264	0.3289	0.3315	0.3340	0.3365	0.3389
1.00	0.3413	0.3437	0.3461	0.3485	0.3508	0.3531	0.3554	0.3577	0.3599	0.3621
1.10	0.3643	0.3665	0.3686	0.3708	0.3729	0.3749	0.3770	0.3790	0.3810	0.3830
1.20	0.3849	0.3869	0.3888	0.3906	0.3925	0.3943	0.3962	0.3980	0.3997	0.4015
1.30	0.4032	0.4049	0.4066	0.4082	0.4099	0.4115	0.4131	0.4147	0.4162	0.4177
1.40	0.4192	0.4207	0.4222	0.4236	0.4251	0.4265	0.4279	0.4292	0.4306	0.4319
1.50	0.4332	0.4345	0.4357	0.4370	0.4382	0.4394	0.4406	0.4418	0.4429	0.4441
1.60	0.4452	0.4463	0.4474	0.4484	0.4495	0.4505	0.4515	0.4525	0.4535	0.4545
1.70	0.4554	0.4564	0.4573	0.4582	0.4591	0.4599	0.4608	0.4616	0.4625	0.4633
1.80	0.4641	0.4648	0.4656	0.4664	0.4671	0.4678	0.4686	0.4693	0.4699	0.4706
1.90	0.4713	0.4719	0.4726	0.4732	0.4738	0.4744	0.4750	0.4756	0.4761	0.4767
2.00	0.4772	0.4778	0.4783	0.4788	0.4793	0.4798	0.4803	0.4808	0.4812	0.4817
2.10	0.4821	0.4826	0.4830	0.4834	0.4838	0.4842	0.4846	0.4850	0.4854	0.4857
2.20	0.4861	0.4864	0.4868	0.4871	0.4874	0.4878	0.4881	0.4884	0.4887	0.4890
2.30	0.4893	0.4896	0.4898	0.4901	0.4904	0.4906	0.4909	0.4911	0.4913	0.4916
2.40	0.4918	0.4920	0.4922	0.4924	0.4927	0.4929	0.4930	0.4932	0.4934	0.4936
2.50	0.4938	0.4940	0.4941	0.4943	0.4945	0.4946	0.4948	0.4949	0.4951	0.4952
2.60	0.4953	0.4955	0.4956	0.4957	0.4958	0.4960	0.4961	0.4962	0.4963	0.4964
2.70	0.4965	0.4966	0.4967	0.4968	0.4969	0.4970	0.4971	0.4972	0.4973	0.4974
2.80	0.4974	0.4975	0.4976	0.4977	0.4977	0.4978	0.4979	0.4979	0.4980	0.4981
2.90	0.4981	0.4982	0.4982	0.4983	0.4984	0.4984	0.4985	0.4985	0.4986	0.4986
3.00	0.4986	0.4987	0.4987	0.4988	0.4988	0.4988	0.4989	0.4989	0.4990	0.4990
3.10	0.4990	0.4991	0.4991	0.4991	0.4991	0.4992	0.4992	0.4992	0.4993	0.4993
3.20	0.4993	0.4993	0.4994	0.4994	0.4994	0.4994	0.4994	0.4995	0.4995	0.4995
3.30	0.4995	0.4995	0.4995	0.4996	0.4996	0.4996	0.4996	0.4996	0.4996	0.4996

$\bar{\sigma}$ = standard deviation = $\sqrt{\bar{\sigma}^2}$

The mean deviation is computed simply as $(1/N) \sum (V - \bar{X})$ and is generally less than the standard deviation $\bar{\sigma}$.

$\bar{C}$ = coefficient of variance computed as

$$\bar{C} = \frac{\bar{\sigma}}{\bar{X}}$$

Y = ordinate of the normal distribution curve (may use symmetrical with the origin to simplify the mathematics for drawing the curve)

$$Y = \frac{1}{\bar{\sigma}\sqrt{2\pi}} \exp\left[-\frac{1}{2}\left(\frac{X - \bar{X}}{\bar{\sigma}}\right)^2 \right]$$

where X = any abcissa value such as 0, ± 1, ± 2, ± 3, etc.

$Y_0 = 0.39894$
$Y_1 = 0.24197$
$Y_2 = 0.05399$ when $\bar{\sigma} = 1.0$ and $\bar{X} = 0.00$ (origin)

This equation may be used to plot a normal distribution curve (symmetrical) as shown in Fig. 1-3a. The total area of this curve is 1.00, and the area under the curve bounded by any two ordinates represents the probability that X is between these two points. Table 1-1 gives selected values of curve areas between $X = 0$ and $X = 3.9$ which can be doubled to give the total area on both sides. A table such as this is most readily obtained by integrating the equation above for Y and programming on the computer. Note that it is a simple matter to transform the curve of Fig. 1-3b to put the origin to the left so that the curve peak is at a normal distribution for the actual numerical values of V, such as 10.0, 100.0, etc., depending on what the curve represents, such as water contents, densities, specific gravities, or whatever.

From the table, we find that at $X = 1.0$ the area is 0.3413; thus, doubling for symmetry, we have 0.6826, or the percent of the curve included between ± 1.0 is

$$\text{Percent} = \frac{0.6826}{1}(100) = 68.26 \text{ percent of total curve area}$$

and the area excluded is $1.0000 - 0.6826 = 0.3174$ or 31.74 percent. For a standard deviation $\bar{\sigma} = 1.0$, the probability of any test N_i being between the values of

$$\bar{X} - \bar{\sigma} \le V \le \bar{X} + \bar{\sigma}$$

is 68.26 percent.

t = Student's t-distribution values. This is a widely used series of numbers for normal random distributions. The equation can be found in most textbooks on statistics and requires a computer to obtain tables such as Table 1-2. From these tables, how many tests N would be required to obtain a reliability of 68.26 percent? Solution: Inspecting the tables at $N = 1$, we find that a single test will produce a condition of 75 percent reliability that the value V will not be different from $\bar{X}$ by more than a maximum of $\bar{\sigma} = 1$, or

$$\bar{X} - 1 \le V \le \bar{X} + 1$$

Example 1-1 Given are the following numbers from a series of plastic limit tests (see Chap. 3 for definition):

18.2	16.0
19.1	17.3
20.1	17.9
20.3	21.4
20.7	20.5
21.5	20.6
24.6	

REQUIRED

(a) Compute the mean, median, and mode values.
(b) Compute the standard deviation and coefficient of variance.
(c) Estimate the number of tests needed to produce a 75 percent reliability that a test will be within the standard deviation.

Table 1-2 Values of Student's t distribution for several percentages indicated and for numbers of test values N as shown

R / N	0.999	0.995	0.975	0.950	0.900	0.800	0.700	0.600
1	318.309	63.657	12.706	6.314	3.078	1.376	0.726	0.325
2	22.327	9.925	4.303	2.920	1.886	1.061	0.617	0.289
3	10.215	5.841	3.182	2.353	1.638	0.979	0.584	0.277
4	7.173	4.604	2.776	2.132	1.533	0.941	0.569	0.271
5	5.893	4.032	2.571	2.015	1.476	0.919	0.559	0.267
5	0.893	4.032	2.571	2.015	1.476	0.919	0.559	0.267
7	4.785	3.500	2.365	1.895	1.415	0.896	0.549	0.263
8	4.501	3.355	2.306	1.859	1.397	0.889	0.546	0.262
9	4.297	3.250	2.262	1.833	1.383	0.883	0.544	0.261
10	4.144	3.169	2.228	1.813	1.372	0.879	0.541	0.260
12	3.930	3.054	2.179	1.782	1.356	0.873	0.539	0.259
14	3.787	2.977	2.145	1.761	1.345	0.868	0.537	0.258
16	3.686	2.921	2.120	1.746	1.337	0.865	0.535	0.258
18	3.611	2.878	2.101	1.734	1.330	0.862	0.534	0.257
20	3.552	2.845	2.086	1.725	1.325	0.860	0.533	0.257
22	3.505	2.819	2.074	1.717	1.321	0.858	0.532	0.256
24	3.467	2.797	2.064	1.711	1.318	0.857	0.531	0.256
26	3.435	2.779	2.056	1.706	1.315	0.856	0.531	0.256
28	3.408	2.763	2.048	1.701	1.313	0.855	0.530	0.256
30	3.385	2.750	2.042	1.697	1.310	0.854	0.530	0.256
35	3.340	2.724	2.030	1.690	1.306	0.852	0.529	0.255
40	3.307	2.705	2.021	1.684	1.303	0.851	0.529	0.255
45	3.281	2.690	2.014	1.679	1.301	0.850	0.528	0.255
50	3.261	2.678	2.009	1.676	1.299	0.849	0.528	0.255
55	3.245	2.668	2.004	1.673	1.297	0.848	0.527	0.255
60	3.232	2.660	2.000	1.671	1.296	0.848	0.527	0.254
65	3.220	2.654	1.997	1.669	1.295	0.847	0.527	0.254
70	3.211	2.648	1.994	1.667	1.294	0.847	0.527	0.254
75	3.203	2.643	1.992	1.665	1.293	0.846	0.527	0.254
80	3.195	2.639	1.990	1.664	1.292	0.846	0.526	0.254
85	3.189	2.635	1.988	1.663	1.292	0.846	0.526	0.254
90	3.183	2.632	1.987	1.662	1.291	0.846	0.526	0.254
95	3.178	2.629	1.985	1.661	1.290	0.845	0.526	0.254
100	3.174	2.626	1.984	1.660	1.290	0.845	0.526	0.254
105	3.170	2.623	1.983	1.660	1.290	0.845	0.526	0.254

(d) What is the percent probability that any test would be within the standard deviation?

SOLUTION
Compute values of (a):
The mean is

$$\bar{X} = \frac{\sum V}{N} = \frac{258.2}{13} = 19.86 \text{ percent}$$

The median is obtained by ranking the 13 tests in order to obtain

16.0   17.3   17.9   18.2   19.1   20.1   20.3

20.5   20.6   20.7   21.4   21.5   24.6

to obtain 20.3 percent by inspection.

The mode value will be approximately $(20.1 + 20.3 + 20.5 + 20.6 + 20.7)/5 = 20.44$, without plotting a distribution curve.

(b) The standard deviation and coefficient of variance are now computed.

$$\bar{\sigma}^2 = \frac{1}{N}\left(\sum_1^N V^2 - N\bar{X}^2\right) = \frac{1}{13}\left[\sum_1^{13} V^2 - 13(19.86)^2\right]$$

$$= \frac{1}{13}(5186.32 - 5127.45) = 4.528$$

and the standard deviation $\bar{\sigma}$:

$$\bar{\sigma} = \sqrt{\bar{\sigma}^2} = \sqrt{4.528} = 2.128$$

The coefficient of variance:

$$\bar{C} = \frac{2.128}{19.86} = 0.1072$$

(c) Estimate the number of tests needed to produce 75 percent reliability. The range of values for 75 percent reliability should be within

$$X \pm n\bar{\sigma}$$

where n = the value of the coefficient of the standard deviation to produce the necessary range of variance. If $n = 0$ we would have an "exact" value. Obviously this is not physically possible (although it is often assumed as a convenience). The value of $n \to 1$ is often of some interest.

From the table of Student's t numbers (Table 1-2) under the heading of 75 percent (requiring interpolating), we note that more than 105 samples are necessary to produce $n = 0$, since at $N = 105$ samples the value of n is computed as

$$n = 0.526 + 0.5(0.845 - 0.526) = 0.686$$

Let us decide that $n \cong 1$ is sufficiently accurate for this work. We will investigate at our sample population of $N = 13$ tests, which requires double interpolation, to obtain

$$n = 0.873 + 0.5(1.3505 - 0.873) = 0.968 \qquad \text{say, } 1.0$$

In other words, 13 tests gives an approximate 75 percent reliability that any single test will be 19.86 ± 2.128 percent for our example.

(d) The probability that any test will be within one standard deviation can always be obtained from Table 1-2 at coordinates of percent reliability versus the number of samples (or tests) n to obtain $n = 1$.

In this example we have established that there is a 75 percent probability that any single plastic limit test will be within the value 19.86 ± 2.13 percent. This means one test out of four will be unsatisfactory in the sense that the value will fall out of this range. Let us check this: We had 13 tests of which one is greater than 21.99 and two are less than 17.73 $(19.86 - 2.13 = 17.73)$, which means three tests fell out of the limits and the percentage of satisfactory tests was

$$\text{Percent good tests} = \frac{10(100)}{13} = 77 \text{ percent}$$

This is only slightly more than the theoretical value of 75 percent, and the difference is due to a combination of rounding 1.051 and the necessity of describing the test program using integers. The agreement between test and theory tends to rapidly deteriorate as the number of tests decreases.

1-8 SI UNITS

This textbook uses SI units and/or preferred usage metric units throughout. Most soil laboratory equipment will last for many years if properly maintained; thus, mass units of grams and kilograms will continue to be used to obtain the mass (formerly—and likely to continue to be commonly—called " weight ") of a sample. Grams and kilograms are valid SI mass units, so no problems will arise in making mass measurements. A problem will arise with the unit weight term due to the common practice of using the terms weight and mass interchangeably. In this text the term *unit weight* will be used to define a body force unit (force per unit volume) of kilonewtons per cubic meter. The kilonewton will be used exclusively since the newton unit is too small. The abbreviation for kilonewton per cubic meter will be, kN/m^3. Density will be taken as mass per unit volume kilograms per cubic meter or grams per cubic centimeter; this is the definition presently used by the scientific community. Unfortunately, in soil mechanics, not only have mass and weight been used interchangeably, but density and unit weight have been also. It is the author's opinion that if the reader carefully adheres to using density for mass/volume and unit weight for force/volume and recognizes that laboratory " weights " are in mass units, no problems will arise. The use of unit weight as a force unit is necessary in order to be able to compute lateral and vertical pressures in a soil mass due to gravitational body forces. Pressure is defined to be a force per unit area (kilonewtons per square meter = kilopascals = kPa in this text).

Most soil samples will be on the order of 50 to 75 mm diameter × 100 to 200 mm in length. An area using 50 to 75 mm diameter is not practical in the standard SI units of millimeters (it results in too large a number) or meters (too small); thus square centimeters is the only practical unit and is used by the author in this text.

Weighing measurements will be most convenient in grams (g) or kilograms (kg), and for density the best laboratory measuring unit is grams per cubic centimeter. Note that grams is not strictly SI, but most laboratory scales measure in either grams or kilograms and are not likely to be changed in the near future. Volumetric devices not calibrated in cubic centimeters can be converted (soft conversion) and marked, once and for all. The use of grams per cubic centimeter is particularly advantageous since:

In SI: $\qquad$ $1 \text{ g/cm}^3 \times 9.807 = 9.807 \text{ kilonewtons/m}^3 \text{ (kN/m}^3)$

In fps: $\qquad$ $1 \text{ g/cm}^3 \times 62.4 = 62.4 \text{ pounds/ft}^3 \text{ (pcf)}$

both of which are the *unit weight* of water. This conversion factor(s) and the length units of:

$$3.2808 \text{ ft} = 1 \text{ m}$$

$$1 \text{ in} = 25.4 \text{ mm} = 2.54 \text{ cm}$$

and force units of

$$1 \text{ g} = 980.7 \text{ dynes} \quad \text{and} \quad 1 \text{ newton (N)} = 1 \times 10^5 \text{ dynes}$$

and the gravitational constant

$$g = 980.7 \text{ cm/s}^2$$

will enable the reader to make any conversions needed for derived SI units. Note that these conversion factors have been located at the end of this chapter for rapid reference.

HOMEWORK PROBLEMS

1-1 A material has a density of 1.76 g/cm^3. What is the unit weight in fps and SI units?

1-2 A soil sample has a diameter of 62.3 mm. What is the area in fps and SI?

1-3 Given are the following soil compaction unit weights from a laboratory section:

1 17.3 kN/m^3	5 17.1	9 17.7
2 18.4	6 16.9	10 18.1
3 17.9	7 17.2	11 18.0
4 18.3	8 17.0	12 17.9

Compute the mean, standard deviation, and coefficient of variance for these 12 tests. Using Table 1-2, estimate the percent chance that any single test will be within $\bar{X} \pm \bar{\sigma}$ and compare this estimate to what actually happens in the data above.

1-4 Given are the following plastic limit water content values from a laboratory section:

1 24.3 percent	4 24.1	7 23.8
2 26.2	5 19.3	8 25.1
3 18.1	6 24.2	9 24.9

Compute the mean, standard deviation, and coefficient of variance. Using Table 1-2, estimate the percent chance that a single test will be within $\bar{X} \pm \bar{\sigma}$ and compare this to the data. What happens if the most extreme value(s) is discarded?

1-5 Estimate the minimum number of tests needed to give an 85 percent reliability that any given test value is within $\bar{X} \pm \bar{\sigma}$.

1-6 What is the best (minimum) possible value of k to give $\bar{X} \pm k\bar{\sigma}$ to obtain 92 percent reliability and not perform more than ten tests?

Chapter 2
Soil Properties –
Physical and Index

2-1 INTRODUCTION

This chapter contains numerous terms and definitions that are used by geotechnical engineers in describing the physical and index properties of soils. Many of these terms will be used throughout the text. Those equations used as basic definitions in the following sections should be memorized for convenience.

2-2 BASIC DEFINITIONS AND MASS-VOLUME RELATIONSHIPS

In soil mechanics, as in any other area of engineering, certain relationships are so fundamental that they become definitions even though the relationships are expressed as mathematical formulas. In other cases, the definitions are used in the usual context. Several of the more fundamental soil mechanics definitions will be introduced or developed in this and following sections.

If one were to go into the field, clear off the ground surface, and (if it were physically possible) extract a cube of soil of unit dimensions (say, $1 \times 1 \times 1$ cm), as shown in Fig. 2-1, a visual inspection would reveal the block to be made up of

1. *Pores* or *voids*, which are the open spaces between soil grains, of various sizes.
2. *Soil grains*, which may be macroscopic† or microscopic in size. Obviously

---

† *Macroscopic* refers to particles which are visible to the eye, whereas *microscopic* refers to particles which can only be seen with the aid of a microscope or other magnifying device.

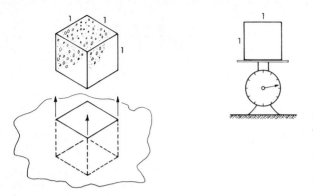

Figure 2-1 Weight-volume relationship for soil. Note that one does not require a regular shaped cube; the conditions shown are valid as long as one can obtain the volume of the hole and the weight of soil removed from the hole.

microscopic sizes cannot be observed with the unaided eye, but it is not unreasonable to expect that they would be present in varying amounts.

3. *Soil moisture*, which may cause the soil to appear wet to damp to somewhat dry. The water in the pores or voids, or pore water, may be of sufficient quantity to completely fill the voids (a saturated soil) or may only surround the soil grains to some extent.

The soil pores which do not contain water are, of course, full of air and/or water vapor. Referring again to Fig. 2-1, if we place the cube (of unit volume) of soil on a scale before any water drains from the block, the resulting weight may be taken as the wet unit weight γ_m of the soil. A special case exists when *all* the voids are filled with water; the resulting weight is the *saturated unit weight* γ_{sat} of the soil. If the cube of soil is placed in an oven and dried to a constant weight and reweighed, the *dry unit weight* γ_d of the soil is obtained. These, and other symbols used in this text, are consistent with those recommended by the Geotechnical Committee on Nomenclature for Soil Mechanics. It might be observed that even if the cube crumbles in the oven as it dries, the weights obtained are still unit weights. For the case shown in Fig. 2-1, the unit weight will be X grams per cubic centimeter (g/cm³). If the foot-pound-second (fps) system of units had been used, the unit weight would have been given in pounds per cubic foot (pcf). The correct SI unit is kilonewtons per cubic meter (kilonewtons since newtons is too large a number) (kN/m³). From the preceding discussion, it follows that the unit weight of soil (or any material) is

$$\gamma = \frac{\text{weight of material}}{\text{volume of material}} \tag{2-1}$$

Since water at 4°C has a mass of 1 g/cm³ (or approximately 62.38 pcf), the unit weight of water at 4°C is

$$\gamma_w = \frac{62.4}{1} = 62.4 \text{ pcf}$$

or in SI units

$$\gamma_w = 1\,\frac{g}{cm^3} \times 980.7\,\frac{dynes}{g} \times \left(1 \times 10^{-5}\,\frac{N}{dyne}\right)\left(1 \times 10^6\,\frac{cm^3}{m^3}\right) \times \frac{1\,kN}{1000\,N}$$

$$= 1 \times 9.807 = 9.807\ kN/m^3$$

from which, the conversion factor to convert the common laboratory mass unit of g/cm^3 to the correct SI unit of kN/m^3 is:

$$g/cm^3 \times 9.807 = kN/m^3$$

This factor should be memorized, as nearly all the derived SI units can be obtained from this and the relationship between feet and meters:

$$1\ ft = 0.3048 \qquad or \qquad 1\ m = 3.2808\ ft$$

The conversion factor from g/cm^3 to pcf is

$$g/cm^3 \times 62.4 = pcf$$

The user may wish to memorize this conversion factor also.

The density (and corresponding unit weight) of water varies slightly depending on the temperature, but for all practical engineering purposes, γ_w may be taken as 9.807 kN/m^3 (or a mass of 1 g/cm^3) for a temperature range of 0 to about 20°C unless the problem context indicates otherwise.

It should be observed at this point that the 1-cm^3 block being used for illustration in Fig. 2-1 represents only a minuscule portion of the total volume of soil at the location from which it was extracted. Whether the unit weight obtained is valid for the entire mass will depend on how *representative* the cube of soil is of the mass.

If the water is collected and the soil grains are melted down so that they fuse into a nonporous solid, one can depict the cube of Fig. 2-1 in the form of a block diagram (Fig. 2-2a) which shows the same unit of bulk soil volume.

From Fig. 2-2 we will define several useful terms.

1. Void ratio e is defined as

$$e = \frac{V_v}{V_s} \tag{2-2}$$

where in this and following equations the terms are defined on Fig. 2-2. The void ratio e is normally expressed, and used, as a decimal. This definition is widely used in geotechnical engineering. Inspection of Fig. 2-2 and Eq. (2-2) indicates that the possible range of e is

$$0 < e \ll \infty$$

Typical values of void ratios for natural sands may range from 0.5 to 0.8, typical values for cohesive soils from 0.7 to 1.1.

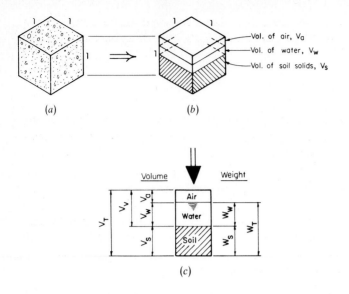

Figure 2-2 Volumetric and weight relationships for a soil mass. (*a*) Volume of soil removed from the ground; it may be any size or shape, but is shown here as a unit cube. (*b*) If constituent parts of the soil sample are separated and full recovery of air and water in soil pores is possible, the soil sample may be shown as above. (*c*) It is usual practice (and more convenient) to use a line diagram with the relationships as shown rather than a three-dimensional block as in (*b*); it is understood that the line diagram is one unit in the third dimension. The left side of the line diagram displays volume relationships; the right side of the line diagram displays weight relationships. Note that the volume of air is significant but contributes no weight.

2. Porosity n is defined as

$$n = \frac{V_v}{V_T} \times 100 \qquad (2\text{-}3)$$

Porosity is often used by agricultural and to a lesser extent by geotechnical engineers. It is expressed most often as a percentage, although it is used in engineering computations as a decimal. Inspection of Fig. 2-2 and Eq. (2-3) indicates that the range of n is

$$0 \le n \le 1$$

A combination of Eqs. (2-2) and (2-3) results in

$$e = \frac{n}{1-n} \qquad (2\text{-}4)$$

3. Water content w is defined as

$$w = \frac{W_w}{W_s} \times 100 \qquad (2\text{-}5)$$

This equation gives the water content as an independent variable, since W_s is a constant for a constant state soil condition. A few authorities have used a water content definition

$$w' = \frac{W_w}{W_T} = \frac{W_w}{W_s + W_w} \times 100$$

but this is a dependent equation with the weight of water present in both the numerator and denominator. This equation is not generally used by geotechnical engineers for this reason.

The range of water content is

$$0 \leq w, \text{ percent} \ll \infty$$

It is not unusual for marine and organic lake soils to have water content values up to 300–400 percent, but the natural water content for most soils is under 60 percent.

4. Degree of saturation S is defined as

$$S = \frac{V_w}{V_v} \times 100 \qquad (2\text{-}6)$$

This equation expresses the ratio of water present in the soil pores to the total amount which could be present if all the pores were full of water. The degree of saturation is the percentage of total void volume that contains water. Inspection of Eq. (2-6) indicates that if the soil is dry (no water), $S = 0$ percent, and if the pores are full of water, the soil is *saturated* and $S = 100$ percent; therefore, the range of S is

$$0 \leq S, \text{ percent} \leq 100$$

5. Specific gravity G_i. This uses some of the terms from Fig. 2-2. There are two definitions of specific gravity which may be used. The basic definition can be found in any physics text and computed according to the following equation:

$$G = \frac{\text{weight of unit volume of any material}}{\text{weight of unit volume of water at 4°C}} \qquad (2\text{-}7)$$

Generally, geotechnical engineers need the specific gravity of the soil grains (or solids) G_s, and this should be assumed as the value under consideration when no qualification is used. The specific gravity of the soil solids G_s is computed as

$$G_s = \frac{\gamma_s}{\gamma_w} = \frac{W_s}{V_s \gamma_w} \qquad (2\text{-}8)$$

where γ_s = unit weight of the soil solids (no pores). Typical values of G_s for soil solids are 2.65 to 2.72. The specific gravity of mercury is 13.6; of gold, 19.3, or 19.3 times heavier than water.

The bulk specific gravity is rarely used, but it can be computed as

$$G_m = \frac{\gamma_m}{\gamma_w}$$

From inspection of this equation, it is evident that the bulk specific gravity is numerically equal to the mass density when the mass density is in grams per cubic centimeter.

Obtaining the specific gravity of the soil grains G_s requires finding the weight and volume of a representative portion of the soil grains and using Eq. (2-8). The exact method is in any soil mechanics laboratory manual.†

From the preceding basic definitions, many other useful relationships and/or quantities can be computed, as the following examples illustrate.

Example 2-1

GIVEN 1870 g of wet soil is compacted into a mold with a volume of 1000 cm³. The soil is put into the oven and dried to a constant weight of 1677 g. The specific gravity G_s is assumed to be 2.66.

REQUIRED Compute the following quantities:
(a) Water content
(b) Dry unit weight γ_d
(c) Porosity n
(d) Degree of saturation S
(e) Saturated unit weight γ_{sat}

SOLUTION It is suggested that the student draw a block diagram when working problems such as this as an aid in noting what is given and what is required. Initially the block diagram is as shown in Fig. E2-1a.

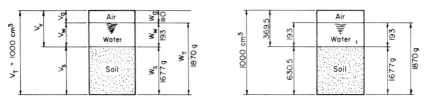

Figure E2-1a   Figure E2-1b

Step 1 Compute the water content w:
When the soil is dried in the oven, the loss in weight is only the loss of the

---

† See *Engineering Properties of Soils and Their Measurement*, 2nd ed. by J. E. Bowles, McGraw-Hill Book Co., New York, 1978.

water which evaporates. The oven temperature is not high enough to evaporate any soil; therefore,

$$W_w = 1870 - 1677 = 193 \text{ g}$$

$$w = \frac{W_w}{W_s} = \frac{193}{1677} \times 100 = 11.5 \text{ percent}$$

Step 2 Compute the dry unit weight:

$$\gamma_d = \frac{W_s}{V_T} = \frac{1677}{1000} \times 9.807 = 16.45 \text{ kN/m}^3$$

Note that generally the unit weight in SI should be reported to the nearest 0.01 kN/m^3, which is equivalent to 0.1 pcf used in fps, and water content to no more than the nearest 0.1 percent.

Step 3 Compute the porosity n:
   Rearrange Eq. (2-8) to obtain

$$V_s = \frac{W_s}{G_s \gamma_w} = \frac{1677}{2.66(1)} = 630.5 \text{ cm}^3$$

The total volume was 1000; therefore, the volume of voids is

$$V_v = 1000 - 630.5 = 369.5 \text{ cm}^3$$

$$n = \frac{V_v}{V_T} = \frac{369.5}{1000} \times 100 = 36.95 \text{ percent}$$

A similar computation gives the void ratio as 0.586 (the reader should verify this value). At this point redraw the block diagram with new entries (Fig. 2-1b).

Step 4 Compute the degree of saturation S:
   The weight of water from Step 1 was 193 g; using Eq. (2-8) rearranged,

$$V_w = \frac{W_w}{G_w \gamma_w} = \frac{193}{1(1)} = 193 \text{ cm}^3$$

From this, the reader should note that when using grams and centimeters, $V_w = W_w$.

$$S = \frac{V_w}{V_v} = \frac{193}{369.5} \times 100 = 52.2 \text{ percent}$$

Step 5 Compute the saturated unit weight γ_{sat}:
   The saturated unit weight is obtained when the voids are completely full of water; thus,

$$\gamma_{sat} = \gamma_d + \text{weight of water in } V_v$$

$$\gamma_{sat} = 16.45 + \frac{(369.5)(1)(9.807)}{1000} = 20.07 \text{ kN/m}^3$$

Example 2-2 Redo Example 2-1 using fps units.

SOLUTION We will work the problem slightly differently to illustrate alternative methods of accomplishing the same result.

Step 1 Find the water content:

This can be worked only one way; therefore, the answer of $w = 11.5$ percent is independent of units but would most likely be as computed in Example 2-1 since most laboratories use scales calibrated in grams.

Step 2 Compute the dry unit weight:

We will use the water content and wet unit weight as an alternative computation procedure.

$$\gamma_{wet} = \gamma_d + w\gamma_d$$

$$\gamma_{wet} = \frac{W_T}{V_T} = \frac{1870}{1000} = 1.87 \text{ g/cm}^3$$

$$\gamma_d + w\gamma_d = 1.87$$

Solving for γ_d, obtain

$$\gamma_d = \frac{1.87}{1 + 0.115}(62.4) = 104.7 \text{ pcf}$$

Step 3 The porosity would be computed exactly as in Ex. 2-1.

Step 4 The degree of saturation would be computed as in Ex. 2-1.

Step 5 Compute the saturated unit weight.

$$\gamma_{sat} = \gamma_d + \text{weight of water in filled voids}$$

$$\gamma_{sat} = 104.7 + 0.3695(62.4) = 127.7 \text{ pcf}$$

Example 2-3 Express the porosity in terms of the void ratio $[n = f(e)]$.

SOLUTION Referring to the block diagram of Fig. E2-3, and since V_s is not defined, let us arbitrarily assume $V_s = 1$ (as shown on the figure).

With $V_s = 1$, the void ratio is

$$e = \frac{V_v}{V_s} = \frac{V_v}{1} = V_v$$

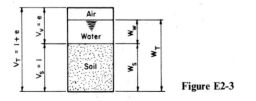

Figure E2-3

With $V_s = 1$ and $e = V_v$, the total volume is

$$V_T = V_s + V_v = 1 + e$$

From Eq. (2-3) we have by substitution

$$n = \frac{V_v}{V_T} = \frac{e}{1+e} \qquad (2\text{-}9)$$

Example 2-4 Given are the following quantities:

  w, water content *when $S = 100$ percent* (i.e., saturated condition)

  G_s, specific gravity of soil solids

  γ_w, unit weight of water

REQUIRED Express the dry unit weight γ_d in terms of w, G_s, and γ_w, or

$$\gamma_d = f(w, G_s, \gamma_w)$$

SOLUTION Construct a block diagram as Fig. E2-4 for $S = 100$ percent and note that

$$\gamma_d = \frac{W_s}{V_T}$$

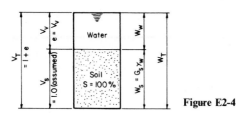

Figure E2-4

The following steps are used to derive the required formula:

$$V_w = V_v \qquad \text{when } S = 100 \text{ percent}$$

also,

$$V_w = \frac{W_w}{G_s \gamma_w} = W_w \qquad (a)$$

Assume $V_s = 1.0$, from which

$$e = \frac{V_v}{V_s} = \frac{V_v}{1} = V_v \qquad (b)$$

From the definition of water content,

$$W_w = wW_s \qquad (c)$$

From the definition of specific gravity,

$$G_s = \frac{W_s}{V_s \gamma_w}$$

The weight of solids, with $V_s = 1.0$, is

$$W_s = G_s \gamma_w \qquad (d)$$

Now, substituting for W_s and V_T, obtain

$$\gamma_d = \frac{W_s}{V_T} = \frac{G_s \gamma_w}{V_T} = \frac{G_s \gamma_w}{1 + V_w} \qquad (e)$$

The volume of water is

$$V_w = \frac{W_w}{G_w \gamma_w} = \frac{w G_s \gamma_w}{1(1)} = w G_s \qquad (f)$$

Now, substituting Eq. (f) into Eq. (e), we obtain the desired equation as

$$\gamma_d = \frac{G_s \gamma_w}{1 + w G_s} \qquad (2\text{-}10)$$

Note from the definition of void ratio in Eq. (b) that

$$e = w G_s \qquad \text{when } S = 100 \text{ percent}$$

and substitution into Eq. (2-10) gives

$$\gamma_d = \frac{G_s \gamma_w}{1 + e} \qquad (2\text{-}11)$$

Example 2-5 Given w for any S, G_s, γ_w.

REQUIRED
(a) $e = f(w, S, G_s)$
(b) $\gamma_d = f(w, S, G_s, \gamma_w)$

SOLUTION Draw the block diagram of Fig. E2-5. The following steps are used in the derivations required.

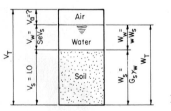

Figure E2-5

The dry unit weight is defined as

$$\gamma_d = \frac{W_s}{V_T} \qquad (a)$$

From the definition of degree of saturation, obtain

$$V_w = SV_v \qquad (b)$$

From the definition of void ratio, obtain

$$V_v = eV_s \qquad (c)$$

Substituting (c) into (b), obtain

$$V_w = SeV_s \qquad (d)$$

From the definition of water content, obtain

$$W_w = wW_s \qquad (e)$$

Again assume $V_s = 1$ as on the block diagram and

$$W_s = 1(G_s)\gamma_w$$

Also,

$$V_w = \frac{W_w}{\gamma_w} = \frac{wW_s}{\gamma_w} = \frac{wG_s\gamma_w}{\gamma_w} \qquad (f)$$

Now, equate (d) and (f) with $V_s = 1$ to obtain

$$Se = wG_s \qquad (2\text{-}12)$$

which is the desired equation for part (a) of this example. The reader should verify that although S and w are reported as percents, they are used in decimal form in this and other equations.

Now, to find the equation for part (b), rearrange Eq. (2-12) to obtain

$$e = \frac{w}{S}G_s$$

and substitute into Eq. (2-11) for e to obtain

$$\gamma_d = \frac{G_s\gamma_w}{1 + (w/S)G_s} \qquad (2\text{-}13)$$

Example 2-6

GIVEN A *saturated* sample of soil has a specific gravity $G_s = 2.67$ and water content data as follows:

$$\text{Weight of can + wet soil} = 150.63 \text{ g}$$

$$\text{Weight of can + dry soil} = 131.58 \text{ g}$$

$$\text{Weight of can} = 26.48 \text{ g}$$

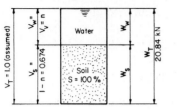

Figure E2-6

REQUIRED
(a) Water content of soil
(b) Dry and saturated unit weights of soil (both SI and fps)

SOLUTION   Construct a block diagram (Fig. E2-6) and place selected data and results on the diagram as computed.

Step 1   Compute water content from given water content data:
   Weight of dry soil $= 131.58 -$ weight of can
$$= 131.58 - 26.48 = 105.10 \text{ g}$$
   Weight of water $= 150.63 - 131.58 = 19.05 \text{ g}$

from which the water content is computed as

$$w = \frac{19.05}{105.10} \times 100 = 18.1 \text{ percent}$$

Step 2   Find dry and saturated unit weights:
   Assume $V_T = 1.0$. (It may be easier to assume $V_s = 1.0$ for given data, but this is to illustrate that either volume may be assumed.) With $V_T = 1.0$, the porosity n is

$$n = \frac{V_v}{V_T} = V_v$$

and the volume of solids is

$$V_s = V_T - V_v = 1 - n \qquad \text{values shown on Fig. E2-6}$$

From Fig. E2-6, the definition of w, and taking $\gamma_w = 1.0$,

$$V_v = wW_s = wV_sG_s = w(1 - n)G_s$$
$$= 0.181(1 - n)2.67 = 0.483 - 0.483n$$

Since $V_T = 1 = V_s + V_v$, we have on substitution of values

$$0.483 - 0.483n + 1 - n = 1.0$$
$$n = 0.326 \text{ (units)}$$

and the volume of solids is

$$V_s = 1 - n = 1 - 0.326 = 0.674$$

from which the dry unit weight is readily computed as

$$\gamma_d = \frac{W_s}{V_T} = 0.674(2.67)(1)(9.807) = 17.65 \text{ kN/m}^3$$

$$= 0.674(2.67)(1)(62.4) = 112.3 \text{ pcf}$$

The saturated unit weight is computed as (using the definition of water content)

$$\gamma_{\text{sat}} = \gamma_d + W_w$$

$$= 17.65 + 0.181(17.65) = 20.84 \text{ kN/m}^3$$

and in fps units the saturated unit weight is

$$\gamma_{\text{sat}} = \frac{20.84}{9.807}(62.4) = 132.6 \text{ pcf}$$

These six examples cover several types and methods of computing volumetric-gravimetric relationships. Combinations of these methods should enable the reader to compute any other needed relationship.

2-3 NONCOHESIVE AND COHESIVE SOILS

If an inherent physical characteristic of the mass of soil grains is that upon wetting and/or any subsequent drying, the soil grains stick together so that some force is required to separate them in the dry state, the soil is *cohesive*. If the soil grains fall apart after drying and only stick together when wet due to surface tension forces in the water, the soil is *cohesionless*, or noncohesive. A cohesive soil may be nonplastic, plastic, or a viscous fluid, depending on the instantaneous water content value. A cohesionless soil exhibits no demarcation line between the plastic and nonplastic states, as this type of soil is nonplastic for all ranges of water content. Under certain conditions, however, a *cohesionless* soil with a sufficiently high water content may act as a viscous fluid.

2-4 ATTERBERG (OR CONSISTENCY) LIMITS

A Swedish soil scientist, A. Atterberg, engaged in ceramics and agriculture work proposed (ca. 1911) five states of soil consistency. These limits of soil consistency are based on water content and are:

1. *Liquid limit* w_L. The water content above which the soil behaves as a viscous liquid (a soil-water mixture with no measurable shear strength). In soils engineering the liquid limit is rather arbitrarily defined as the water content at which 25 blows of the liquid limit machine closes a standard groove cut in the soil pat

for a distance of 12.7 cm. Casagrande (1958) and others have modified the test as initially proposed by Atterberg so that it is less operator-subjective and more reproducible. With standard equipment various operators can reproduce liquid limit values to within 1 to 2 percent (i.e., say $w_L = 39 \pm 2$ percent and not 39×0.02).

2. *Plastic limit w_P.* The water content below which the soil no longer behaves as a plastic material. It is in the range of water contents between w_L and the plastic limit w_P that the soil behaves as a plastic material. This range is termed the *plasticity* (or *plastic*) *index* and is computed as

$$I_P = w_L - w_P \tag{2-14}$$

By definition of I_P it is impossible to obtain a negative value.

   The plastic limit is also arbitrarily defined as the water content at which a thread of soil, when rolled down to a diameter of 3 mm, will just crumble. This test is more operator-subjective than the liquid limit test, but experience indicates that results are reproducible within 1 to 3 percent by more experienced technicians.

3. *Shrinkage limit w_S.* That water content, defined at degree of saturation = 100 percent, below which no further soil volume change occurs with further drying. This limit is of considerable importance in arid areas and for certain soil types which undergo considerable volume change with changes in water content.

   One should note that the smaller the shrinkage limit, the more susceptible a soil is to volume change—that is, the smaller the w_S, the less water is required to start the soil to change in volume. If the shrinkage limit is 5 percent, then when the in situ water content exceeds this value, the soil will begin to expand. The relative locations of w_L, w_P, and w_S on a water content scale are shown in Fig. 2-3.

4. *Sticky limit.* That water content at which a soil loses its adhesive property and ceases to stick to other objects such as the hands or the smooth metal surface of a spatula blade. This limit is of some significance in agriculture and to earth-work contractors, as the drag on the plow moldboard is increased if the soil is wet enough to stick to it.

5. *Cohesion limit.* That water content at which the soil grains just cease to stick together, e.g., at which cultivation of the soil does not result in clods or lumps

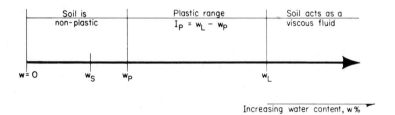

Figure 2-3 Relative locations of the plastic and liquid limits of a soil.

forming. This limit also has more significance for the agriculturist than for the soils engineer.

The liquid, plastic, and shrinkage limits are well known worldwide. The sticky limit has been used in Europe, but by and large, the sticky and cohesion limits have not been used by geotechnical engineers.

From this discussion it is evident that the Atterberg limits exist only for *cohesive* soils. Surface tension gives cohesionless soils an *apparent cohesion*, so called because the cohesion disappears when the soil is either completely dry or saturated. From a practical standpoint, a cohesionless soil with apparent cohesion (damp to wet but not saturated) may be excavated on a vertical cut for shallow depths, or bore holes where the hole will stand open may be made. When the soil dries or becomes saturated, the apparent cohesion disappears and the sides of the cut or bore holes will collapse.

The liquid and plastic limits, together with the plasticity index, are required for the several systems of soil classification considered in Chap. 4.

2-5 SOIL MOISTURE

Moisture or water content of a soil has previously been defined as the ratio of the weight of water in the pores of a soil to the weight of soil solids (grains). A distinction has been made between the water content determination made in the laboratory on laboratory specimens and the water content which indicates the instantaneous field value. This latter value is termed the *natural moisture* or water content of the soil and given the symbol w_N. The value of natural field moisture w_N will vary depending on the location of the soil sample, i.e., at or near the ground surface, deep in the ground, beneath a lake; recency of rainfall; etc. It should be evident that samples obtained from beneath a permanent ground water level will probably not change in water content from day to day or year to year. On the other hand, samples of soil near the ground surface or above some permanent water level will have a varying natural moisture content due to climatic factors such as temperature; amount, duration, and recency of rainfall; and length of dry periods. The actual water content depends on the void ratio, the location of the sample, and the climatic factors as cited above.

2-6 INDICES OF SOIL CONSISTENCY

The state or potential state of consistency of a natural soil can be established through a relationship termed the *liquidity index* I_L,

$$I_L = \frac{w_N - w_P}{I_P} \tag{2-15}$$

where w_N is the natural moisture or water content of the soil at the field site or in

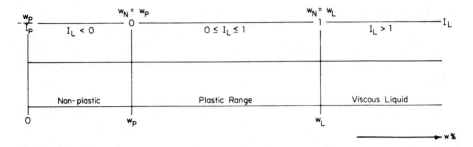

Figure 2-4 Relationship between w_P, w_L, and w_N in computing I_L.

situ. The relationship between water content and I_L is illustrated in Fig. 2-4.
From this expression it can be seen that if

$$0 < I_L < 1$$

the soil is in the plastic range. If

$$I_L \geq 1.0$$

the soil is in a liquid state or potentially liquid. While it may be currently stable, a sudden shock or remolding may transform the mass to a liquid. Such a soil would be known as a sensitive clay. Such clay deposits exist in southwest Canada and in Scandinavia.

Another relationship occasionally used is the *index of consistency*, defined as

$$I_C = \frac{w_L - w_N}{I_P} \qquad (2\text{-}16)$$

with all terms as previously defined. Both Eqs. (2-15) and (2-16) give an index value between 0 and 1 when the field moisture is between w_P and w_L. The essential differences are in the numerical values when the field soil has a natural moisture content greater than the liquid or less than the plastic limit.

The most important consistency index is the plastic index (or plasticity index) I_P, previously defined by Eq. (2-14). In general, the larger the plasticity index, the greater will be the engineering problems associated with using the soil as an engineering material, such as foundation support for residential building, road subgrades, etc.

There are many engineering correlations of soil properties and behavior associated with the plasticity index. These include soil strength, horizontal in situ or in place earth pressure, and shrinkage-swell potential. Most of these correlations should be used cautiously; some are reasonably reliable, but many are little better than guesses.

2-7 SPECIFIC SURFACE

Specific surface relates the surface area of a material to either weight or volume of the material, with volume being generally preferred. Using this latter definition, specific surface is:

$$\text{Specific surface} = \frac{\text{surface area}}{\text{volume}} \qquad (2\text{-}17)$$

Physically the significance of specific surface can be demonstrated using a 1 cm × 1 cm × 1 cm cube as follows:

$$\text{Specific surface} = \frac{\text{surface area}}{\text{volume}} = \frac{6}{1} = 6$$

Now let us subdivide the cube so that each side is 0.5 cm.

$$\text{Number of cubes} = 2 \times 2 \times 2 = 8$$

$$\text{Surface area} = (0.5)^2(6)(8) = 12 \text{ cm}^2$$

and

$$\text{Specific surface} = \frac{12}{1} = 12$$

Now divide the sides by 10:

$$\text{Number of cubes} = 10 \times 10 \times 10 = 1{,}000$$

$$\text{Surface area} = (0.1)^2(6)(1{,}000) = 60 \text{ cm}^2$$

and

$$\text{Specific surface} = \frac{60}{1} = 60$$

This illustrates that large particles, whether cubes or soil particles, have smaller surface areas per unit of volume and thus smaller specific surfaces than small soil grains.

Now if sufficient water were present to just dampen the surface area in the preceding example, it would take 10 times as much water to wet the surface of all the grains when the cubes were 0.1 cm × 0.1 cm × 0.1 cm as when the same volume occupied a single cube of 1 cm³. Note also that if one were trying to remove water from the surface-wet soil, there would be 10 times as much water to remove from the smaller grains.

This discussion illustrates that specific surface is inversely related to the grain size of a soil. We generally do not compute the value for practical cases, since naturally occurring soil grains are too irregular in shape, but one can extrapolate from the preceding paragraphs that a soil mass made up of many small particles will have a larger specific surface than the same mass made up of large particles.

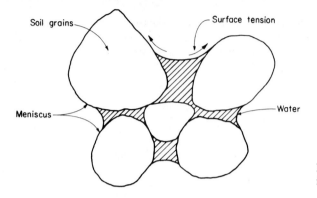

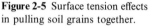

Figure 2-5 Surface tension effects in pulling soil grains together.

From the concept of specific surface, one could expect larger moisture contents for small-grained soils than for coarse-grained soils, other things being equal; however, one must also consider the effect of grain size on the void ratios of the soils and exterior factors such as climate and location of the sample.

Specific surface is a primary factor in concrete and asphalt mix design, as in both types of design it is necessary to provide sufficient cement paste or asphalt to coat the particle surfaces.

Specific surface is a primary factor in soils subject to volume change and to the surface tension effects of water at particle interfaces, as shown in Fig. 2-5. With large numbers of particles and small sizes, the cumulative effect of surface tension of the water film in holding or pulling the particles together is very large (even though the actual amount of surface tension per unit of area is very small).

2-8 SPECIFIC GRAVITY

Specific gravity was defined by Eq. (2-8) as the ratio of a unit weight of material to the unit weight of water. The laboratory test to determine the specific gravity of soil (G_s of the soil grains) is briefly described in Chap. 3.

Table 2-1 lists the specific gravity of a number of minerals which are common in soils. Most soils (individual grains in aggregate) contain large amounts of quartz and feldspars and to a lesser extent mica and iron-based minerals.

Results of many specific gravity determinations on large numbers of soils indicate that values of 2.55 to 2.80 will bracket nearly all soils, with values for most soils being between 2.60 and 2.75. As a matter of fact, the specific gravity test is not often performed, and values are arbitrarily taken as follows:

Sands, gravels, coarse-grained materials	$G_s = 2.65$–2.67
Cohesive soils, as mixtures of clay, silt, sand, etc.	$G_s = 2.68$–2.72
Clay	Use values in Table 2-1 for specific type

Table 2-1 Typical values of specific gravity for soil minerals

Mineral	Specific gravity	Mineral	Specific gravity
Bentonite	2.13–2.18	Muscovite (mica)	2.80–2.90
Gypsum	2.30	Dolomite	2.87
Gibbsite	2.30–2.40	Aragonite	2.94
Montmorillonite	2.40	Anhydrite	3.00
Orthoclase feldspar	2.56	Biotite (mica)	3.00–3.1
Illite	2.60	Hornblende	3.00–3.47
Quartz	2.60	Augite	3.20–3.40
Kaolinite	2.60–2.63	Olivine	3.27–3.37
Chlorite	2.6 –3.0	Limonite	3.8
Plagioclase feldspar	2.62–2.76	Siderite	3.83–3.88
Talc	2.70–2.80	Hematite	4.90–5.30
Calcite	2.80–2.90	Magnetite	5.17–5.18

An indication of the computation error in the void ratio if a G_s of 2.65 is used when a true value of 2.60 would have been obtained from a laboratory test is computed as follows:

Given a soil of $\gamma_d = 1.80$ g/cm³ (should be representative, but implies error of ± 0.05 g/cm³).

From the definition of specific gravity,

$$G_s = \frac{W_s}{V_s \gamma_w}$$

Therefore the volume of solids is

$$V_s = \frac{1.80}{2.60(1)} = 0.692 \text{ cm}^3 \text{ (true)}$$

$$V_s' = \frac{1.80}{2.65(1)} = 0.679 \text{ cm}^3 \text{ (assumed)}$$

The volume of voids V_v for the two cases is

$$V_v = 1 - 0.692 = 0.308 \rightarrow e = \frac{0.308}{0.692} = 0.445$$

and

$$V_v' = 1 - 0.679 = 0.321 \rightarrow e' = \frac{0.321}{0.679} = 0.473$$

The percent increase in void ratio due to the use of the erroneous value of G_s is

$$\text{Percent} = \frac{e'}{e} \times 100 = \frac{0.473}{0.445} \times 100 = 106 \text{ percent}$$

The void ratio is 6 percent too large due to using the erroneous value 2.65. Since this solution also depends on $\gamma_d = 1.80$ being statistically correct, there may be some compensating error. If similar calculations are performed for $\gamma_d = 2.0$ and 1.5 g/cm^3, the percent increase is found to be 8.2 and 4.5 percent, respectively. Also the " true " value of 2.60 will be true only if the small quantity of soil (up to 150 g) used to determine G_s is truly representative of the soil mass.

2-9 SOIL TEXTURE

Soil texture may be defined as the visual appearance of a soil based on a qualitative composition of soil grain sizes in a given soil mass. Large soil particles with some small particles will give a coarse-appearing or *coarse-textured* soil. A conglomeration of smaller particles will give a *medium-textured* material, and a conglomeration of fine-grained particles yields a *fine-textured* soil. It can be observed, however, that lumps of fine-grained materials will give a *coarse* texture, so we must also relate texture to the state of elemental soil particles. Figure 2-6 indicates the textural classification of several soils.

Texture based on visual appearance is often used in soil classification of cohesionless materials such as coarse sand, medium coarse sand and gravel, fine sand, etc. Texture is not used for cohesive soils, since the soil state is a factor in the texture (i.e., lumps can be pulverized), as illustrated with soils 8 and 9 of Fig. 2-6.

2-10 SOIL PHASES

The definition of phase as used by the chemist is "any homogeneous part of the material system separated from other parts by physical boundaries, as water in ice, with water vapor above." In this context it is evident that a soil mass may be a:

1. Two-phase system consisting of soil and air ($S = 0$ percent), soil and water ($S = 100$ percent), or soil and ice ($S = 100$ percent)
2. Three-phase system (refer to Fig. 2-1) consisting of soil, water, and air ($0 < S < 100$ percent); soil, ice, and air ($0 < S < 100$ percent); or soil, water, and ice ($S = 100$ percent)
3. Four-phase system consisting of soil, water, ice, and air
   Of course the "air" may contain appreciable amounts of water vapor.

2-11 GRAIN SIZE

Grain size of a soil refers to the diameters of the soil particles making up the soil mass. Since a macroscopic examination of a mass of soil grains indicates that few, if any, of the particles are round and thus possess a diameter, one can conclude that this is a rather loose description of the soil.

Figure 2-6 Soil texture of several soils.

Grain size is determined by sieving a quantity of soil through a stack of sieves of progressively smaller mesh openings from top to bottom of the stack. The quantity of soil retained on a given sieve in the stack is termed one of the grain sizes of the soil sample. Actually this operation only brackets a portion of the soil as being between two sizes, as shown in Fig. 2-7—the size indicated by the particular sieve under consideration and the size of the one immediately above it in the stack.

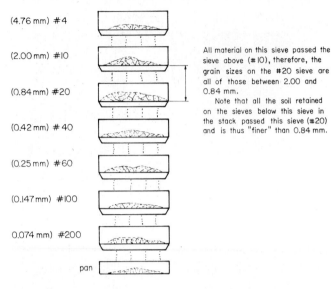

(4.76 mm) #4

(2.00 mm) #10

All material on this sieve passed the sieve above (#10), therefore, the grain sizes on the #20 sieve are all of those between 2.00 and 0.84 mm.

(0.84 mm) #20

Note that all the soil retained on the sieves below this sieve in the stack passed this sieve (#20) and is thus "finer" than 0.84 mm.

(0.42 mm) #40

(0.25 mm) #60

(0.147 mm) #100

0.074 mm) #200

pan

Figure 2-7 Sieve analysis of a soil. Note that the sieve arrangement is such that sieve openings decrease in size from top to bottom of the stack.

Figure 2-8 illustrates both the shape and size of particle ranges bracketed. All the soil particles except soil No. 2 were sieved through the No. 10 and retained on the No. 20 sieve, and they clearly show a range of particle sizes. Soil 2 is a fine beach sand from Atlantic City, New Jersey, and was retained on the No. 60 sieve. Soils 1 through 3 are transported soils and are well rounded, soil 1 being Ottawa sand and soil 3 being a sand from a river near Peoria, Illinois. Soils 4 and 5 are what is called residual (refer to Chap. 5) soil and are typified by angular particles. Soil 4 is from Kentucky and soil 5 from Georgia. Soil 6 is crushed rock and is shown for comparison.

Sieve sizes range from openings of 101.6 mm (4 in) down to 0.037 mm (No. 400). All the openings are square; thus, what constitutes the diameter of a soil particle is somewhat academic since the likelihood of a particle passing a given sieve opening depends on both its size and its orientation with respect to the sieve opening.

The sieve sizes are referred to mesh openings from 101.6 mm down to 6.35 mm; then the sieves are designated by numbers (see Table 3-2). Using sieves of numbers larger than No. 200 (0.074 mm)† is largely impractical, as soil can be sieved through this size mesh only with some difficulty. This mesh is fine enough to just begin to provide resistance to water flow, and soil provides considerably more resistance to passage through the mesh than does water.

† Or closest corresponding sieve opening for other standards, as shown in Table 3-2.

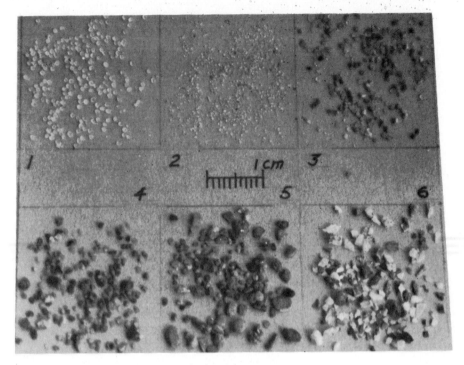

Figure 2-8 Several soils to illustrate grain size and range.

To determine approximately particle diameters smaller than 0.07 mm, an analysis based on the velocity of fall of spheres through a viscous fluid (Stokes' law) is used. One method of using Stokes' law uses a hydrometer to measure the specific gravity of a soil-water suspension and is called hydrometer analysis (see Chap. 3).

The grain-size analysis is useful, as it helps identify such soil properties as

1. Whether a given soil can be easily drained
2. Whether the soil is suitable for use in construction projects such as dams, levees, and roads
3. Potential frost heave
4. Estimated height of capillary rise (Chap. 8)
5. Whether the soil can be used in asphalt or concrete mixes (It should be understood that the definition of "soil" includes the sand and gravel used in the manufacture of concrete.)
6. Filter design, to prevent fine-grained material from being washed out of a soil mass and lost (see Sec. 9-15)

The grain-size analysis is an important part of most of the engineering classifications of soil to be considered in Chap. 4.

2-12 UNIT WEIGHT OF SOILS (DRY, WET AND SUBMERGED)

The basic definition of soil unit weight was given in Sec. 2-2 as weight per unit volume of material. From an inspection of Fig. 2-1, one arrives at the conclusion that, in general, unit weight is

$$\gamma = f(G_s, e, w)$$

That is, the unit weight of a soil can only depend on the weight of the individual soil grains $[f(G_s)]$, the total number of soil particles present $[f(e)]$, and the amount of water present in the voids $[f(w)]$. It should be kept in mind that unit weight can be altered only by changing the void ratio and/or the water content of the soil mass (since G_s = constant for a given soil mass). Strictly, the unit weight is a state vector and should include the void ratio and water content in the description; however, except for such qualifying terms as "wet" or "dry" unit weight, this is seldom done, it being commonly understood that the unit weight is for the correct soil state; i.e., the state (condition) is assumed to be a constant.

A special unit weight of considerable interest to the soils engineer is the *buoyant* (or *submerged*) *unit weight*, given the symbol γ'. Referring to Fig. 2-9, take a cube of saturated soil 1 cm $\times$ 1 cm $\times$ 1 cm and weigh it. This weight is, of course, γ_{sat}, since it is the weight of a unit volume of saturated material. Now suspend the cube under water, as shown in Fig. 2-9b. The question is, What is the weight showing on the scales?

From a free-body analysis,

$$\sum F_v = 0$$

$$P_s + P_{up} - \gamma_{sat} = 0$$

but

$$P_{up} = \sigma A = \gamma_w(h)(A) = \gamma_w(1)(1) = \gamma_w$$

Therefore,

$$P_s + \gamma_w - \gamma_{sat} = 0$$

or

$$P_s = \gamma_{sat} - \gamma_w$$

but with a unit of volume, P_s is the unit weight of the submerged soil; therefore,

$$P_s = \gamma'$$

and the *submerged* (also called *buoyant*) *unit weight* is

$$\gamma' = \gamma_{sat} - \gamma_w \qquad (2\text{-}18)$$

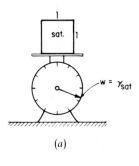

(a)

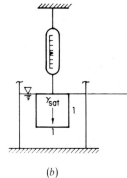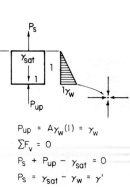

(b)

$$P_{up} = A\gamma_w(1) = \gamma_w$$
$$\Sigma F_v = 0$$
$$P_s + P_{up} - \gamma_{sat} = 0$$
$$P_s = \gamma_{sat} - \gamma_w = \gamma'$$

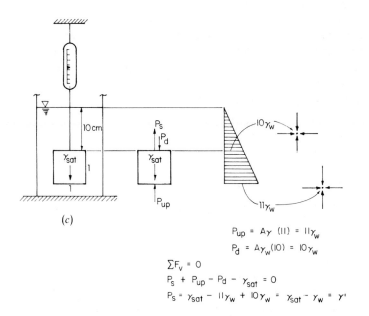

(c)

$$P_{up} = A\gamma\ (11) = 11\gamma_w$$
$$P_d = A\gamma_w(10) = 10\gamma_w$$

$$\Sigma F_v = 0$$
$$P_s + P_{up} - P_d - \gamma_{sat} = 0$$
$$P_s = \gamma_{sat} - 11\gamma_w + 10\gamma_w = \gamma_{sat} - \gamma_w = \gamma'$$

Figure 2-9 Development of the concept of buoyant weight of a soil. From (b) and (c) it can be seen that depth of submergence does not affect the submerged (buoyant) unit weight of a soil. (a) Unit cube of saturated soil placed on scales. (b) Unit cube of soil submerged from scales. Note that the soil will saturate if it is not already saturated. (c) Unit cube of soil submerged 10 cm to illustrate that the effect of buoyancy is independent of depth of submergence.

One might ask what the scale would show the cube of soil to weigh if the cube were 10 cm beneath the water surface, as in Fig. 2-9c. Again

$$\sum F_v = 0$$

$$P_s + P_{up} - \gamma_{sat} - 10\gamma_w = 0$$

but

$$P_{up} = \sigma A = \gamma_w hA = \gamma_w(1)(10 + 1) = 10\gamma_w + \gamma_w$$

therefore,

$$P_s + 10\gamma_w + \gamma_w - \gamma_{sat} - 10\gamma_w = 0$$

or the scale reads

$$P_s = \gamma_{sat} - \gamma_w = \gamma'$$

as before, indicating the submergence effect is independent of water surface location above the soil.

The submerged unit weight of a soil has particular importance in soil mechanics, as the next section and later chapters indicate.

2-13 INTERGRANULAR PRESSURES—SATURATED SOILS

Figure 2-10a illustrates a soil mass to a large scale as it might exist in situ (in the field or in place). As there is no water present, there is obviously nothing supporting the soil above plane A-A except the contacts between the grains on plane A-A. It should be evident that the air in the pore spaces cut by plane A-A is not carrying any of the weight of the soil above this plane. The grain-to-grain contact pressure is termed the *intergranular* (or *effective*) *pressure*. It is this pressure which is used in the usual friction equations of physics to evaluate friction resistance (in units of N) as

$$F_f = vN$$

where v = coefficient of friction
N = normal force or stress

In the case of a steel block on top of a second block, the normal force (or stress) N is distributed uniformly over the base of the top block, and the pore space is negligible. For soils, referring again to Fig. 2-10, the pore space is not negligible, and the normal force N is the intergranular pressure or force. This does not mean to imply that the contact pressure is not computed as

$$\sigma_n = \frac{P}{A_{nominal}}$$

for both materials, but rather that the contact pressure is the contributing factor in the friction equation. It is convenient to compute σ_n in soil work using the nom-

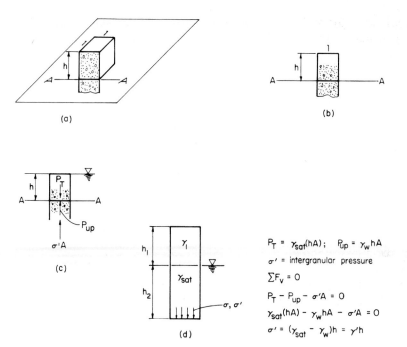

$$P_T = \gamma_{sat}(hA); \quad P_{up} = \gamma_w hA$$

$$\sigma' = \text{intergranular pressure}$$

$$\Sigma F_v = 0$$

$$P_T - P_{up} - \sigma'A = 0$$

$$\gamma_{sat}(hA) - \gamma_w hA - \sigma'A = 0$$

$$\sigma' = (\gamma_{sat} - \gamma_w)h = \gamma'h$$

Figure 2-10 Diagram illustrating total, neutral (or pore water), and intergranular pressures.

inal cross section being stressed rather than the area of grain-to-grain contact points, since one does not know this latter area and there is no way to compute it. What is of interest is the frictional stress $\sigma_f = v\sigma_n$, and if we devise a means to measure σ_f and σ_n, then effectively, v takes care of providing a compatible equation.

If the soil system of Fig. 2-10b is submerged as in Fig. 2-10c, what is the effective pressure? From reasoning alone we can deduce that the contact pressure will be less than before by the amount of buoyancy of the water. One can mathematically compute the effective pressure as follows:

$$\sum F_v = 0 \qquad \text{on plane } A\text{–}A$$

which gives

$$\gamma_{sat} h(A) - \gamma_w h(A) - \sigma' A = 0$$

Solving for the effective pressure, obtain

$$\sigma' = (\gamma_{sat} - \gamma_w)h = \gamma'h \tag{2-19}$$

where the symbol γ' is always used for the buoyant unit weight defined in Eq. (2-18).

The total pressure σ_t on plane A–A is computed as

$$\sigma_t = \gamma_{sat} h \tag{2-20}$$

Both of these latter equations neglect any resistance to vertical movement along the sides of the column of soil (i.e., side friction); however, considering that this is a one-unit plan projection out of a very large plan-area in situ, the error is negligible.

Equation (2-20) is based on the concept of stacking a series of h unit cubes of some weight into a column of soil; thus the total stress σ_t is

$$\frac{W}{A}(\text{total}) = \sigma_t = \sum h\gamma_{\text{sat}} \qquad \text{kPa}$$

and the effective stress σ' is (for water to top of soil column of h cubes, in meters)

$$\sigma' = \sum \gamma'h \qquad \text{kPa}$$

These equations must be adjusted for the weight of the column of soil above the plane of interest when the water level is only part of the column height, as illustrated in Fig. 2-10d and Ex. 2-7. In general, the total and effective stresses are

$$\sigma = \gamma_1 h_1 + \gamma_{\text{sat}} h_2$$
$$\sigma' = \gamma_1 h_1 + \gamma' h_2$$

Equation (2-19) can also be written as

$$\sigma' = \sigma - u \qquad\qquad\qquad\qquad (2\text{-}21)$$

where $\sigma' =$ intergranular or effective pressure
  $\sigma =$ total pressure as obtained from Fig. 2-10d
  $u = \gamma_w h =$ pore water, or neutral, pressure

Equation (2-21) can also be written to include the effect of change in pore water pressure, Δu, from some initial (usually static) pore water pressure as

$$\sigma' = \sigma - (u + \Delta u) \qquad\qquad\qquad (2\text{-}21a)$$

Raising the ground water level is a $+\Delta u$, and lowering it produces a $-\Delta u$. If we insert a vertical tube into the soil and allow the water to rise in it to the static water level, the height at any time is the pore pressure at the tip of the tube. A tube such as this is called a *piezometer* and is widely used in the field to measure the pore water pressure. Since we observe that water entering a high-pressure water transmission line does not instantly travel from point A to point B some distance away even though the line provides a nearly ideal passage, it would be logical to assume that if we pour water into, and fill the piezometer tube, it will take some time for the additional water to drain and the water level to return to the static level. During the time that the water level in the piezometer tube is at a height Δh above the static water table, the pore pressure at the tip of the piezometer is

$$u + \Delta u = \gamma_w h + \gamma_w \Delta h = (h + \Delta h)\gamma_w$$

If the excess (amount above static water level) pore pressure Δu is sufficiently large, it is evident that we could have a condition where the effective pressure approaches or even becomes (a negative value has no significance) zero as

$$\sigma' = \sigma - (u + \Delta u) \rightarrow 0$$

which means that in the equation $F_f = vN$ and with the substitution of σ' for N, the friction resistance $F_f \rightarrow 0$. Part of the strength of cohesive soils and all the strength of cohesionless soils is due to this friction resistance. Physically, when $\sigma' = 0$, the soil grains are just touching, or essentially "floating" in the pore water.

An alternative approach to the intergranular stress problem is to consider that the total pressure minus the actual water pressure which acts on the pore space defines the effective stress as

$$\sigma' = \sigma - (1 - a)u_w \qquad (2\text{-}22)$$

where a = interparticle contact area and other terms have been previously defined. This equation becomes Eq. (2-21) if $a = 0$. There can be no doubt that a is very small in soils, but it is surely not zero. Unfortunately there is no ready means of measuring it, but little error seems to be associated with computations where it is arbitrarily taken as zero. In this text all effective stress computations will use some form of Eq. (2-21).

Example 2-7

GIVEN The soil profile shown in Fig. E2-7.

REQUIRED What are the total and effective pressures at point A?

SOLUTION

Step 1 Find γ_{dry} and γ_{sat} of the sand. Let $V_T = 1.0$, from which

$$n = V_v$$

The volume of solids is

$$V_s = 1 - V_v = 1 - n$$

The weight of solids is

$$W_s = G_s V_s \gamma_w$$

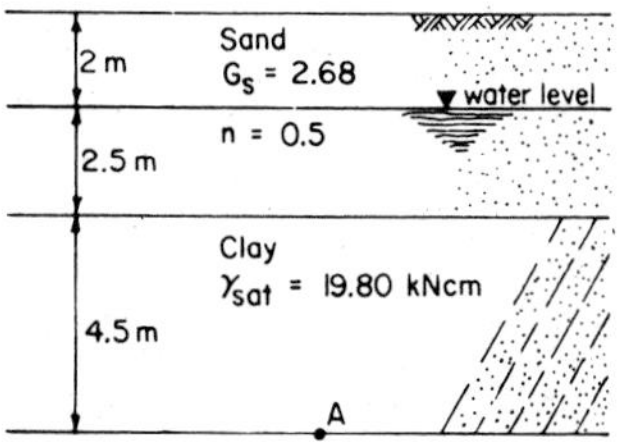

2 m	Sand $G_s = 2.68$
2.5 m	$n = 0.5$
4.5 m	Clay $\gamma_{\text{sat}} = 19.80$ kN/cm

water level

A

Figure E2-7

Substituting $1 - n$ for V_s, obtain

$$W_s = 2.68(1 - 0.5)(9.807) = 13.14 \text{ kN/m}^3$$

But $\gamma_{dry} = W_s$ since $V_T = 1$. The saturated unit weight is

$$\gamma_{sat} = \gamma_{dry} + W_w$$

$$\gamma_{sat} = 13.14 + 0.5(9.807) = 18.04 \text{ kN/m}^3$$

The total pressure σ is

$$
\begin{aligned}
2(13.14) = \quad & 26.28 \\
+2.5(18.04) = \quad & 45.10 \\
+4.5(19.80) = \quad & 89.10 \\
\hline
\sigma = \; & 160.48 \text{ kPa}
\end{aligned}
$$

The effective pressure σ' is

$$\sigma' = 160.48 - 9.807(7) = 91.83 \text{ kPa}$$

The effective pressure can also be computed as follows:

$$
\begin{aligned}
2(13.14) = \; & 26.28 \\
+2.5(18.04 - 9.807) = \; & 20.58 \\
+4.5(19.80 - 9.807) = \; & 44.97 \\
\hline
\sigma' = \; & 91.83 \text{ kPa}
\end{aligned}
$$

(which checks the previous value, as it should) ///

2-14 INTERGRANULAR PRESSURES IN PARTIALLY SATURATED SOILS

The preceding section has shown that intergranular pressures for saturated soils can be computed using Eq. (2-21). Considerable engineering experience to date indicates that this equation is sufficiently accurate for most geotechnical work. When the soil is partially saturated, laboratory tests indicate that Eq. (2-21) can be considerably in error (see Skempton, 1961). A more general expression can be written as

$$\sigma' = \sigma - [u_a - \psi(u_a - u_w)] \tag{2-23}$$

where u_a = pore air pressure

ψ = parameter relating to degree of saturation; it is 1 when $S = 100$ percent and must be experimentally determined for $S < 100$ percent

Other terms have been previously defined.

Equation (2-23) is little better than an estimate at the present level of technology. This is because at $S < 100$ percent, the actual pore water distribution and the resulting pore pressures are indeterminate. In small laboratory samples the distribution may be less complex than in situ. For this reason the only reliable means of obtaining in situ pore pressures is by use of a piezometer system.

2-15 SUMMARY

This chapter has introduced several fundamental definitions of soil mechanics which the student should memorize: *void ratio, porosity, water content*, and *degree of saturation*. The student should already be familiar with the concept of *unit weight* and *specific gravity* relationships from physics courses. All the weight-volume relationships needed in soil mechanics can be derived from appropriate combinations of these six quantities. The assumption of a volume = 1.0 for either the volume of solids V_s or total volume V_T of a soil mass, as used in Exs. 2-3 and 2-7, is a convenience, and the user should verify that this is not necessary.

The reader should form the habit of deriving any needed weight-volume relationships using the fundamental definitions rather than searching the literature (or this text) for a "formula" which fits the given data.

The physical significance of the liquid and plastic limits, as well as the plastic index relationship, should be studied to obtain a careful understanding. Note particularly that there are both a definition of the liquid and plastic limits and the arbitrary means of obtaining them. Be especially sure you understand the difference.

The factor 9.807 should be memorized since it is so commonly used in converting mass density to unit weight. Note the number of significant digits used in reporting unit weight (0.01) and water content (0.1).

The concept of effective pressure is one of the most important factors in stability analysis in geotechnical work. The role of pore water, or excess pore water, pressures in developing effective pressure should be thoroughly understood. A very large number of soil failures have been, and still are, caused by buildup of excess pore pressures. This chapter has considered excess pore pressure in its simplest mode—a column of water in a piezometer above the point of interest. Later chapters will examine the mechanisms which might produce this column of water other than physically pouring water down the piezometer tube.

HOMEWORK PROBLEMS

2-1 Define the following terms:
  (*a*) Void ratio
  (*b*) Porosity
  (*c*) Water content
  (*d*) Specific gravity of a soil
  (*e*) Unit weight of a soil

(f) Wet unit weight of a soil

(g) Dry unit weight of a soil

2-2 What is soil *texture* and what is its significance?

2-3 Redo Exs. 2-1 and 2-2 without assuming $V_s = 1$.

2-4 A 105.0-cm³ volume of clay weighs 143.0 g in its undisturbed state. When the clay specimen is dried it weighs 111.3 g. What is the natural water content of the clay, and what is the degree of saturation? Assume $G_s = 2.70$.

Partial Ans: S = 49.7 percent.

2-5 The dry density of a soil is 1.73 g/cm³. It is determined that the void ratio is 0.55. What is the wet unit weight if $S = 50$ percent? If $S = 100$ percent? What would the unit weight be if the voids were filled with oil of $G = 0.9$?

Partial Ans: $\gamma_{\text{wet}} = 18.73 \text{ kN/m}^3$ at $S = 50$ percent.

2-6 The moisture content of a *saturated* clay is 160.0 percent. The specific gravity of the soil solids is 2.40. What are the wet and dry unit weights of the saturated clay?

2-7 Derive an expression for $\gamma_{\text{dry}} = f(G_s, n, \gamma_w)$.

Ans: $\gamma_{\text{dry}} = G_s \gamma_w (1 - n)$.

2-8 Derive an expression for $\gamma_{\text{sat}} = f(G_s, e, \gamma_w)$.

2-9 Derive an expression for $e = f(\gamma_{\text{sat}}, \gamma_w, w)$.

2-10 Plot a curve of $n = f(e)$. Comment on the curve and establish the valid range of the curve.

2-11 Prove that $e = wG_s$ when and only when $S = 100$ percent.

2-12 The following data were obtained from a liquid limit (w_L) test:

No. of blows	20	28	34
Water content w, percent	68.0	60.1	54.3

Two plastic limit (w_p) tests gave values of 28.6 percent and 29.1 percent, respectively. The natural water content of the soil is 78 percent. Required are:

(a) Liquid and plastic limits.

(b) Liquidity index I_L and consistency index I_c.

(c) Make appropriate comments on the soil in the natural state.

2-13 A saturated sample of inorganic clay has a volume of 21.4 cm³ and weighs 36.7 g. After drying at 105°C to constant weight, the volume is found to be 13.7 cm³. The weight of dry soil is 23.2 g. For the soil in its *natural* state, find:

(a) Water content w, percent

(b) Specific gravity G_s

(c) Void ratio e

(d) Saturated unit weight γ_{sat}

(e) Dry unit weight γ_d

(f) Shrinkage limit w_S

2-14 A soil specimen with a volume of 60.0 cm³ weighs 105 g. Its dry weight is 80.2 g, and G_s is 2.65. Compute:

(a) Water content w, percent

(b) Void ratio e and porosity n

(c) *Mass* specific gravity G_m

(d) Degree of saturation S, percent

2-15 A soil (medium sand) has a damp density of 1.77 g/cm³. The moisture content was found to be 15.1 percent. Find:

(a) Void ratio e

(b) Dry density γ_d

(c) Degree of saturation S

2-16 Given the soil profile shown in Fig. P2-16, find:

(a) Intergranular pressure at point A.

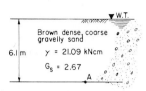

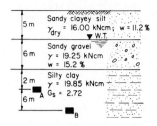

Figure P2-16　　　　　　　　　**Figure P2-17**

(b) Effective and total pressure at point A if the water table drops to A and the water content averages 10 percent.

(c) Values found in (b) in fps units.

Ans: 2.53 ksf.

2-17 Given the soil profile of Fig. P2-17, find:

(a) Total and effective soil pressures at point A for the water table shown.

(b) Total and effective soil pressures at point B if the water table drops to point A with an average water content of 15 percent for the silty clay above the water table.

2-18 Given the soil profile of Fig. P2-18, compute the total and effective soil pressures at the clay-rock interface.

Figure P2-18　　　　　　　　　**Figure P2-19**

2-19 Given the soil profile of Fig. P2-19.

REQUIRED:

(a) Effective pressure at A.

Ans: 215.98 kPa.

(b) Effective pressure at A if the piezometric head drops 0.5 m.

(c) Piezometric head needed to make the effective pressure at $A = 0.0$.

Ans: (c) 26.02 m.

2-20 A saturated soil has $w = 45$ percent and $e = 1.21$. Find:

(a) Specific gravity G_s

(b) Saturated unit weight γ

Ans: $G_s = 2.69$; $\gamma = 17.30$ kN/m³.

2-21 A sand has a porosity of 50 percent and $G_s = 2.65$. Find:

(a) e

(b) γ_d and γ_{sat}

(c) γ_{wet} if $S = 30$ percent

Ans: (a) $e = 1.00$; (c) $\gamma_{30} = 14.47$ kN/m³.

2-22 A 1-m³ volume of dry soil has a mass of 1.7 tons (metric). The water content is 20 percent, and $G_s = 2.70$. Find:

(a) e and n

(b) S

(c) γ in both kN/m^3 and lb/ft^3

Ans: $e = 0.588$; $n = 0.37$.

2-23 A 560-cm^3 volume of soil weighs 1100 g. Its dry weight is 980 g. $G_s = 2.67$. Find:

(a) e, n, w;

(b) γ both wet and dry (in kilonewtons per cubic meter and pounds per cubic foot) (kN/m^3 and lb/ft^3)

Ans: (a) 0.526, 34.5 percent, 12.2 percent; (b) $\gamma_{wet} = 19.22$ kN/m^3 or 122.5 lb/ft^3.

2-24 Given G, w, $S = 100$ percent, derive:

(a) $n = f(w, G)$

(b) $e = f(w, G)$

(c) $\gamma_{sat} = f(G, w, \gamma_w)$

Chapter 3
Soil Tests for Properties/ Index Values and Classification

3-1 INTRODUCTION

Anyone who has driven along a highway in an area where the roadway has cut through hills has observed the many types, or at least colors, of soils which are exposed. One who has visited building sites where basements have been excavated has probably observed several types (or colors) of soils exposed. It may be inferred from even these limited observations that soil occurs in nature in a highly variable manner. On one hand the soil may be relatively homogeneous over an area several hundred meters in extent and for several meters vertically; alternatively, the soil may vary considerably within distances of 1 to 2 m or less, both horizontally and vertically.

Limitations of testing equipment size, methods of obtaining soil samples to test (Sec. 3-9), and economics impose a practical limit on the quantity, and often the quality, of testing that is performed to identify and/or classify a soil mass. Testing is necessary to enable the geotechnical engineer to obtain the index properties (Chap. 2), classify the soil (Chap. 4), determine the seepage and hydraulic properties (Chaps. 8 and 9), estimate the soil strength and settlement characteristics (Chaps. 10 through 15), and predict the stability of soil masses on slopes (Chap. 16) or retained by walls.

It is readily apparent that when a small mass quantity of soil, say a few kilograms, is tested and the results are extrapolated to describe the properties of tons† of material in the field, great care must be taken to obtain a sample which is

---

† Ton is a mass unit that is also used in the SI metric system. One metric ton (sometimes spelled tonne) = 1000 kg = 1.1 tons in the fps system.

representative of the soil to be described. All possible means must be used to reduce and/or eliminate test errors. Probably one of the greatest sources of errors in extrapolating test results to the field is the failure to use a representative soil sample.† This error can occur at two points: first, in collecting the sample from the field, and second, in obtaining a representative test quantity from the field sample. Methods of obtaining a representative test quantity are outlined in both ASTM (1978) (American Society for Testing and Materials) and AASHTO (1971) (American Association of State Highway and Transportation Officials) publications. In any case when one obtains a test sample from a bag of soil collected in the field, the sample should be representative of the soil in the bag. This can be accomplished by thoroughly mixing the soil prior to removing any to test. For other samples where the soil cannot be blended, such as tube samples, portions should be taken from several points in the tube or levels in the strata (stratum) if more than one tube sample is collected. Also it is extremely risky in soil testing to rely on the results of a single test to obtain the desired soil data. Three and preferably more tests are much more desirable. Two tests may be disconcerting unless the results are close together. Results far apart may require that a choice be made, since averaging the results may be overly conservative or unsafe. Three tests allow the rejection of any one test which is out of line.

3-2 TESTS FOR SOIL PROPERTIES

In order to establish a common base for comparison of soils and to interchange ideas between different locales, it is necessary in the field of soil mechanics to use standard methods for determining the soil properties, just as other disciplines use standard test procedures. Where general agreement has been reached on a soil test, ASTM has given the test a standards number designation; for example, the liquid limit determination test discussed in Sec. 3-4 is designated D423-66 (66 is the year of official adoption—1966). AASHTO also has standardized many of the soil tests used by geotechnical engineers engaged in highway work. Often the AASHTO test is simply a copy of the ASTM test method with an AASHTO specification number. In the AASHTO standards the liquid limit test is covered in specification T89-60 (1960 is the year of official adoption). The more common soil tests are discussed in the following sections focusing on their limitations rather than on how to perform the tests. The interested reader may consult ASTM (1978 or later) or AASHTO (1971 or later) for the details of tests which have been standardized. The test procedures can also be found in published soil testing texts such as Bowles (1978). These latter sources may be consulted for nonstandard soil tests. It should be noted in passing that one should not blindly use a standard soil test (except for the index and other simple tests) to obtain data. In many cases,

---

† There are other major errors, such as test equipment limitations and inability to form a mathematical model, but these will be discussed later.

particularly in strength testing, a nonstandard-type test produces more realistic data than a standard test. Depending on the results desired it may be necessary to design test equipment so that realistic data may be obtained.

For a large class of work, namely, foundation engineering for buildings, soil tests to determine soil properties are not relied on as much as is visual observation of the soil, coupled with field or laboratory shear strength testing. These simple identification procedures will be considered in some detail under soil classification in Chap. 4. The remainder of this chapter will describe most of the common soil tests the geotechnical engineer will encounter for classifying soils.

Soil tests for permeability, consolidation, and strength properties and parameters will be described in later chapters.

3-3 WATER CONTENT DETERMINATION

This is a routine test performed by placing a quantity of wet soil in a metal cup and obtaining the wet weight. The soil is then oven-dried at a temperature between 105 and 110°C to a constant weight (say 8 to 12 h), and the loss in weight due to evaporation of water is determined. Equation (2-5) can be used to compute the water content, since both W_w and W_s are obtained from the wet and dry weights of soil.

If the water content determination is made on a soil as obtained from the field, it is termed *natural water content* and given the symbol w_N (for natural moisture). For certain tests such as the Atterberg limits it may be identified as the plastic limit (w_P) or liquid limit (w_L). In other tests it may simply be termed the water content. The soil sample for obtaining w should be *representative* and as large as practical.

Experience has shown that the water content of certain soils (especially clays) is sensitive to the temperature of oven drying; therefore, care should be taken to ensure that the oven temperature is set to the standard value of between 105 and 110°C.

3-4 THE ATTERBERG LIMITS

Terzaghi (1925) is generally credited with recognizing the use of the liquid and plastic limits as consistency index values which could be useful in soil identification and/or classification. A. Casagrande (Casagrande, 1932) modified Atterberg's original method for determining the liquid limit to improve the reproducibility of the test.

The shrinkage limit (modified from the original Atterberg method by Terzaghi, 1925) is also used, especially in arid areas, but is applicable in any area where soils may undergo large volume changes with change in moisture content.

These tests were developed for, and are performed on, cohesive soils. The soil is air-dried and pulverized into elemental particles, and the $(-)$ No. 40 sieve fraction is used for the tests.

Oven-drying (and often air-drying) the soil tends to reduce the liquid limit (often 4 to 6 percent) unless the soil is sieved, then prewetted for 24 to 48 h prior to the test. Even with the use of prewetting, the w_L may be reduced up to 2 to 4 percent. The plastic limit appears to be not much affected by drying. When the w_L may change on drying and for soils for which visual examination indicates more than 90 percent will pass the No. 40 sieve, the sample is separated visually by hand removal of the large particles, then the test is performed on the remainder without drying and sieving.

These tests are reasonably reproducible even with inexperienced operators, as shown by two laboratory sections of one of the author's classes in Table 3-1. The tests in Table 3-1 were on two different soils, but it can be seen that while there is unacceptable scatter, for commercial purposes, in the plastic limit test values, there is very little scatter in the liquid limit values. The plastic limit test is, however, considerably more operator-controlled than is the liquid limit test.

The principal sources of errors which affect the reproducibility of the liquid and plastic limit tests are:

1. Care in drying the soil to obtain the $(-)$ No. 40 sieve material and/or prewetting it prior to performing the tests.
2. Careful adjustment of the liquid limit machine. A fall of less than 1 cm will increase the liquid limit, and a fall of more than 1 cm decreases the liquid limit; it is easy for the machine fall to become out of adjustment.

Table 3-1 Distribution of liquid and plastic limits for two student laboratory sections

Student no.	Section 1		Section 2	
	w_L	w_P	w_L	w_P
1	32.8	21.0	34.0	19.8
2	32.0	21.0	35.8	23.5
3	30.0	22.4	29.8	21.6
4	30.9	19.1	32.5	19.9
5	29.6	22.2	29.4	19.4
6	32.6	21.5	35.2	19.2
7	32.5	21.0	36.3	20.6
8	31.8	22.3	35.5	22.8
9	32.1	21.8	33.4	16.1
10	31.8	22.2	37.1	16.4
11	—	—	32.8	22.7

3. Careful attention to rolling the plastic limit thread to 3 mm and to a water content state of just crumbling. Use of a 3 mm rod for usual comparison in the laboratory can be of considerable help.
4. Controlling the quantity of soil used in the liquid limit cup for a test. This can be done by careful observance of the soil depth relative to the shoulder of the grooving tool. Obviously too much or too little soil in the cup will decrease or increase the limit value.

3-5 SPECIFIC GRAVITY TEST

The specific gravity test generally uses the displacement of water to measure the total volume of the irregular shapes making up the test sample. Any valid method of soil volume determination may be used, but the water displacement method is the most popular. The test as described in laboratory soil manuals is somewhat indirect because of the very major problem of removing any air which may become entrapped in the volume of soil. Since air will occupy volume with negligible weight, entrapped air will markedly reduce the computed specific gravity value of a test sample unless it is removed. For this reason the specific gravity test is usually performed using a closed container to which a vacuum can be applied to remove any entrapped air.

Referring to Fig. 3-1, let

$$W_{bw} = \text{weight of bottle + water}$$

$$W_{bws} = \text{weight of bottle + water + soil grains } (W_s)$$

Note that in both cases the weight of water is to the volume mark. From Chap. 2 we have

$$G_s = \frac{W_s}{V_s \gamma_w} \tag{2-8}$$

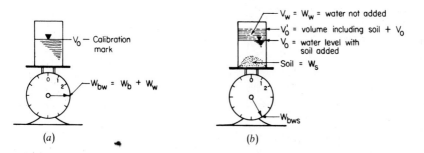

Figure 3-1 Method of determining the specific gravity of a soil using the displacement method. (a) Weight of a clean volumetric flask filled with water to the calibration mark. (b) Weight of a volumetric flask with a known weight of soil W_s added, then filled to the calibration mark with water.

Now if we simply poured soil of weight W_s into the flask already filled to a known volume V_0, the water would rise to a new volume mark V'_0 (after any air removal). The change in volume marks is the volume of soil. In our case we will simply *not pour in* the amount of water necessary to go from V_0 to V'_0 but will instead calculate this volume (or weight) not poured, indirectly, as

$$W_w = W_{bw} + W_s - W_{bws}$$

but

$$W_w = V_s \quad \text{(when using centimeters and grams)}$$

from which G_s is readily computed.

The accuracy of the G_s determination is affected by the size of the sample, error in flask volume (assumed correct volume at 20°C) calibration, density of water, and weighing errors. Strictly, distilled water should be used; however, ordinary tap water has been found by the author to be satisfactory. Computations to three decimal digits are the usual practice, which implies ± 0.5 accuracy in the third decimal digit and final roundoff to two decimals. For this reason it is hardly necessary to calibrate the volumetric flask. In fact, size (and representativeness) of the sample and the presence of air in the flask are the largest sources of error and are the test features which should receive the most attention.

In the absence of test data, typical values of specific gravity of soil particles[†] may be taken as follows:

Soil	G_s
Sand	2.65–2.68
Gravel	2.65–2.68
Clay (inorganic)	2.68–2.72
Clay (organic)	2.62–2.66
Silt	2.65–2.68

3-6 GRAIN-SIZE ANALYSIS

The grain-size analysis test for soil grains larger than the No. 200 sieve (0.074 mm) has been described in Chap. 2. Basically the test consists of:

1. Obtaining a representative sample[‡] of the soil mass.
2. Reducing the lumps in the sample to elemental particle sizes.[§]

[†] Note the distinction between soil particles and the constituent minerals making up the soil particles. Mineral G values are given in Table 2-2.

[‡] Research by the author indicates that this is very difficult to impossible; however, the test has value in that "exact" knowledge of the grain-size distribution is seldom required.

[§] "Washing" the sample on a No. 200 sieve appears to be the most reliable means of separating the (−) No. 200 fraction as well as eliminating the lumps by slaking them.

3. Sieving the resulting soil sample through a representative stack of sieves and weighing the amount of soil retained on each sieve. Refer to Table 3-2 for sieve sizes commercially available, although few laboratories will have the full range of sizes. The sieve stack should contain only enough sieves to adequately define a curve. The arrangement should use a set such that each successive lower sieve is approximately half the opening of the upper sieve; five to eight sieves are usually sufficient.

4. Computing the percent passing each sieve in the stack and plotting a curve of percent passing versus grain diameter. The percent passing is plotted as the ordinate, and grain diameter (also sieve size) is the abscissa. This usually is a logarithmic plot (using semilog paper).

3-7 THE GRAIN-SIZE DISTRIBUTION CURVE

The semilog† plotting of the grain size data has been found most advantageous because:

1. It extends the scale, thus giving all the grain sizes an approximately equal amount of scale separation. Obviously, a grain-size range of 4.76 (No. 4) to 0.074 mm (No. 200) on an arithmetic scale will plot the 0.074 (No. 200), 0.105 (No. 140), and 0.149 (No. 100) (see Figs. 3-2 and 3-3) nearly on each other.
2. Separating the grain sizes somewhat makes it easier to compare soils.

The shape of the grain-size curve is somewhat indicative of the grain distribution. For example:

1. A smooth curve covering a wide range of sizes represents a *well-graded* or *nonuniform* soil.
2. A soil whose curve contains a vertical or nearly vertical portion is deficient in certain grain sizes in the region of the vertical slope. A soil consisting of few grain sizes (or deficient in certain sizes) is a *poorly graded* or *uniform soil.* Typical curves are shown in Fig. 3-2.

An indication of the gradation may be numerically computed from the grain size curve for ($+$) No. 200 sieve sizes using the *coefficient of uniformity,* defined as

$$C_U = \frac{D_{60}}{D_{10}} \tag{3-1}$$

---

† There is not total agreement on how the logarithmic plotting of grain diameters is to be accomplished. Some persons plot increasing sizes from left to right. This discrepancy does not affect either the basic curve shape or the data obtained from the curve.

Table 3-2 Standard U.S., British, French, and German sieves. U.S. sieves are all available in 20-cm diameters and most are available in 30.5-cm diameters

U.S.†		British standard‡		French§		German DIN¶	
Size or no.	Opening, mm	No.	Opening, mm	No.	Opening, mm	Designation, μm	Opening, mm
Size 4″	101.6						
3″	76.1						
$2\frac{1}{2}$″	64.0						
2″	50.8						
$1\frac{3}{4}$″	45.3						
$1\frac{1}{2}$″	38.1						
$1\frac{1}{4}$″	32.						
1″	25.4						25.0
$\frac{3}{4}$″	19.0						20.0
							18.0
$\frac{5}{8}$″	16.0						16.0
$\frac{1}{2}$″	12.7						12.5
$\frac{3}{8}$″	9.51						10.0
$\frac{5}{16}$″	8.00						8.0
$\frac{1}{4}$″ No. 3	6.35						6.3
No. 4**	4.76			38**	5.000		5.0
5	4.00			37	4.000		4.0
6	3.36	5**	3.353				
7	2.83	6	2.812	36	3.150		3.15
8	2.38	7	2.411	35	2.500		2.5
10	2.00	8	2.057	34	2.000		2.0
12	1.68	10	1.676	33	1.600		1.6
14	1.41	12	1.405	32	1.250		1.25
16	1.19	14	1.204				
18	1.00	16	1.003	31	1.000		1.0
20	0.841	18	.853				
25	0.707	22	.699	30	.800	800	.800
30	0.595	25	.599	29	.630	630	.630
35	0.500	30	.500	28	.500	500	.500
40††	0.420	36††	.422	27††	.400	400††	.400
45	0.354	44	.353	26	.315	315	.315
50	0.297	52	.295				

Table 3-2—*Continued*

U.S.[†]		British standard[‡]		French[§]		German DIN[¶]	
Size or no.	Opening, mm	No.	Opening, mm	No.	Opening, mm	Designation, μm	Opening, mm
60	0.250	60	.251	25	.250	250	.250
70	0.210	72	.211	24	.200	200	.200
80	0.177	85	.78	23	.160	160	.160
100	0.149	100	.152				
120	0.125	120	.124	22	.125	125	.125
140	0.105	150	.104	21	.100	100	.100
170	0.088	170	.089			90	.090
				20	.080	80	.080
200	0.074	200	.076			71	.071
230	0.063	240	.066	19	.063	63	.063
						56	.056
270	0.053	300	.053	18	.050	50	.050
325	0.044			17	.040	45	.045
400	0.037					40	.040

† ASTM E-11-70 (Part 41).
‡ British Standards Institution, London BS-410.
§ French Standard Specifications, AFNOR X-11-501.
¶ German Standard Specification, DIN 4188.
** For standard compaction test.
†† For Atterberg limits.

and sometimes the *coefficient of concavity*, defined as

$$C_C = \frac{D_{30}^2}{D_{10} \times D_{60}} \tag{3-2}$$

A large value of C_U indicates a wide grain-size spread between D_{60} and D_{10} (grain size of 60 percent passing and size of 10 percent passing) grain sizes. A value of C_C of approximately 1.00 indicates nearly linear variation of the grain-size distribution curve from the D_{60} to the D_{10} size when C_U is from 4 to 6. If the D_{30} and D_{10} size were the same, C_C would be much smaller than 1.00, and if the D_{30} and D_{60} sizes were the same, C_C would equal C_U (larger than 1); therefore, we can say that if:

1. C_C is close to 1.0, the soil is well graded.
2. C_C is much less or much larger than 1.0, the soil is poorly graded.

The coefficient of uniformity and coefficient of concavity have no meaning when more than about 10 percent of the soil passes the No. 200 sieve.

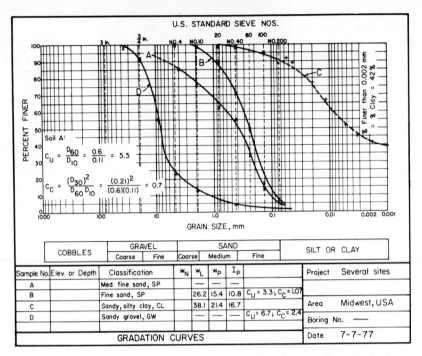

Figure 3-2 Grain-size distribution curves for several soils using a semilogarithmic plot.

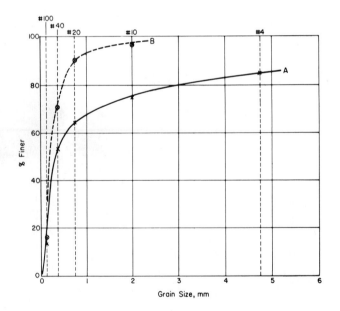

Figure 3-3 Replot of soils *A* and *B* of Fig. 3-2 using an arithmetic scale. Note how much detail is lost on the distribution of soil grain sizes less than 1 mm.

3-8 HYDROMETER ANALYSIS

A *hydrometer* analysis may be used to extend the grain-size distribution curve for sizes smaller than the No. 200 sieve if desired (but this is not often necessary). The hydrometer analysis was apparently first developed by G. J. Bouyoucos, an agricultural scientist, and further improved through use of a hydrometer scaled to read g/cm^3 in suspension (designated the 152H by ASTM) directly (Gilboy, 1933).

The hydrometer test involves dispersing a small quantity of soil in water to form a 1-L suspension (usually with an agent to neutralize soil particle charges to inhibit flocculation) and measuring the specific gravity of the suspension at timed intervals. Assuming Stokes' law (ca. 1850) to be applicable,† the particles will be settling out of the top soil-water zone where the hydrometer is being inserted at a velocity of

$$v = \frac{L}{t}$$

where L = depth of settling zone or distance the particle has fallen in time t (Fig. 3-4)

t = elapsed time

---

† Since soil particles are, in general, anything but spheres, one may certainly question the validity of using Stokes' law.

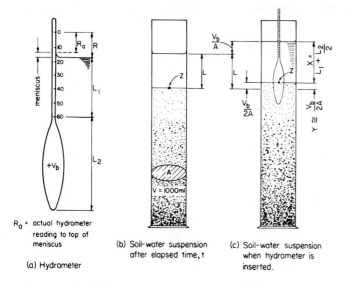

R_a = actual hydrometer reading to top of meniscus

(a) Hydrometer

(b) Soil-water suspension after elapsed time, t

(c) Soil-water suspension when hydrometer is inserted.

Figure 3-4 Hydrometer configuration and concepts used in converting hydrometer reading to length of fall of soil particles L in elapsed time t.

We can now use Stokes' equation to find the equivalent grain size of the settled particles:

$$v = \frac{2}{9} \frac{\gamma_s - \gamma_w}{\eta} \left(\frac{D}{2}\right)^2 \tag{3-3}$$

and rearranging,

$$D = \sqrt{\frac{18\eta v}{\gamma_s - \gamma_w}} \tag{3-4}$$

where γ_s, γ_w = density of soil and water, g/cm^3
$\quad\quad\quad\eta$ = absolute viscosity of water, dyne $\cdot$ s/cm^2 *or* g/s $\cdot$ cm
$\quad\quad\quad D$ = diameter of sphere, cm

Stokes' law is approximately valid over a particle diameter range of

$$0.0002 < D < 0.2 \text{ mm}$$

The distance of particle fall L in elapsed time t is found indirectly from the hydrometer reading corrected for meniscus and considering that as the hydrometer is inserted in the soil-water suspension, the surface rises. In Fig. 3-4b we have a standard jar of soil-water suspension after some elapsed time t. As the hydrometer is inserted and stabilized, the water rises in the test container (1-L glass cylinder) an amount

$$\text{Height of rise} = \frac{V_b}{A}$$

where V_b = volume of hydrometer bulb (*and immersed stem*)
$\quad\quad\quad A$ = area of test cylinder

From Fig. 3-4, and since the soil-water suspension is muddy (turbid), the hydrometer readings R_a are taken to the top of the meniscus:

$$R_a + \text{meniscus} + L_1 = L' \tag{a}$$

and
$$L_1 = L' - R_a - \text{meniscus} \tag{b}$$

By inspection of Fig. 3-4c we can write the following equation:

$$L + \frac{V_b}{A} = X + Y \tag{c}$$

Substituting for X and Y the values shown on Fig. 3-4c, we obtain

$$L = L_1 + \frac{1}{2}\left(L_2 - \frac{V_b}{A}\right) \tag{d}$$

Now substituting Eq. (b) into Eq. (d), we obtain

$$L = L' + \frac{1}{2}\left(L_2 - \frac{V_b}{A}\right) - R \tag{3-5}$$

which can be written also as

$$L = K - R \qquad (3\text{-}6)$$

since the distances L', L_2, and V_b can be measured once and for all for a 152H hydrometer and are essentially (or within test precision) constant for this hydrometer, and the test is performed in a standard 1000-mL sedimentation cylinder. As a convenience tables are generally available which give values of L directly for the hydrometer reading $R = R_a +$ meniscus (see Bowles, 1978).

The 152H hydrometer reads directly grams of soil in suspension based on $G_s = 2.65$; thus, it is directly related to the percent passing by proportion:

$$\text{Percent passing} = \frac{aR_c}{W_s}(100) \qquad (3\text{-}7)$$

where $R_c =$ corrected hydrometer reading
$\quad W_s =$ weight of soil used originally to make the 1000-mL soil-water suspension, g
$\quad a =$ factor when G_s is not 2.65 as $a = [G_s(2.65 - 1)]/[(G_s - 1)2.65]$

The corrected hydrometer reading for this computation is

$$R_c = R_a - \text{zero correction} + \text{temperature correction} \qquad (3\text{-}8)$$

where a reading less than zero (hydrometer sinks below the 0 graduation on stem) is a $(-)$ zero correction and one greater than zero is a $(+)$ value. The zero correction includes the meniscus effect, since it is obtained by using a second hydrometer jar of clear water at the temperature of the test jar and the zero reading is *to the top of the meniscus*, not to the water surface.

The temperature correction is a $(+)$ value when larger than 20°C, since the density of the water is less and the constant weight hydrometer sinks deeper into the suspension that it should (and R_a is too small, since the readings increase down the hydrometer stem).

Example 3-1 Hydrometer Test

GIVEN $W_s = 50$ g
$\quad\quad\quad G_s = 2.70$
$\quad\quad\quad$ Zero correction $= +1.0$ (read $+1$ down hydrometer stem)
$\quad\quad\quad$ Meniscus $= 1.0$

t, min	T	R_{act}
2.0	25°C	40
80.0	15°C	28

FIND D and percent finer for these two test readings.

SOLUTION Tables are available (Bowles, 1978, chap. 6) to solve Eq. (3-5) and to obtain temperature and G_s correction values. Equation (3-5) can be rearranged to give

$$D = K \sqrt{\frac{L}{t}} \quad \text{mm}$$

To obtain L, the actual readings at 2 min and 80 min are corrected for meniscus only:

$$R_1 = 40 + 1 = 41 \qquad \text{reading increased down stem}$$

$$R_2 = 28 + 1 = 29$$

and for Eq. (3-8),

$$R'_1 = 40 - 1.0 + 1.30 = 40.3 \to 40$$

$$R'_2 = 28 - 1.0 - 1.10 = 25.9 \to 26$$

From tables:

$L_1 = 9.6$	(at $R = 41$)	$K_1 = 0.0127$	at 25°C
$L_2 = 11.5$	(at $R = 29$)	$K_2 = 0.0141$	use 16°C

The K values include the temperature effect on the water density and the specific gravity of the soil, 2.70.
   Solving,

$$D_1 = 0.0127 \sqrt{\frac{9.6}{2}} = 0.0277 \text{ mm}$$

$$D_2 = 0.0141 \sqrt{\frac{11.5}{80}} = 0.0052 \text{ mm}$$

The percent finer values, including a correction of 0.99 for $G_s = 2.70$ instead of 2.65 used to calibrate the 152H hydrometer, are as follows:

$$F_1 = \frac{40(0.99) \times 100}{50} = 79.2 \text{ percent}$$

$$F_2 = \frac{26(0.99) \times 100}{50} = 51.5 \text{ percent}$$

Observe that fractional hydrometer readings and corrections give a fictitiously high precision and should not be used.

3-9 FIELD EXPLORATION AND SOIL SAMPLING

Field exploration provides data on underground conditions at the site as well as obtaining soil samples for inspection, for classification, and for strength and/or deformation testing. There are very few situations where site exploration is not necessary in engineered design. Due to the variable nature of natural soils, exploration on one building lot does not ensure the adequacy of the soil on even the adjoining lot perhaps only a meter or so away.

The field exploration program will include inspection of any available topographic maps of the area and viewing of aerial photographs if available, and may include an on-site inspection to observe surface features and conditions of adjoining construction. Discussion of the site and area with local residents may be valuable in determining locations of filled-in areas or recent topographical changes.

The field exploration will ultimately involve drilling holes into the ground to determine the stratigraphy and to obtain disturbed and/or undisturbed soil samples for testing. This operation, coupled with an on-site inspection, study of maps, etc., should enable the geotechnical engineer to:

1. Determine the source and nature of the deposits (recent history of filling or excavation, flooding, and local geology).
2. Determine the depth, thickness, and, with testing, composition of the several strata making up the soil profile.
3. Determine the location of bedrock. The quality of bedrock may also be determined, but this is done only when necessary due to the excessive costs of rock, compared with soil, drilling.
4. Determine the location and variation in the ground water table or determine that the water table is not in the zone of design interest.
5. Obtain the quantity and types of soil samples needed to determine the engineering properties for design. Samples may range from bags of disturbed soil used for index properties to undisturbed samples obtained in thin-wall tubes for strength and settlement tests.

In some cases the exploration program is broken into phases as follows:

Phase 1   Initial reconnaissance (field trip and/or office study of available maps and other information)
Phase 2   Preliminary borings—just sufficient to obtain the general character of the subsoil, obtain index properties, and possibly perform some strength/deformation tests
Phase 3   Detailed program with additional borings and/or more carefully performed strength/deformation tests

For smaller routine projects, phases 1 and 2, or many times just phase 2, complete the exploration. In some cases phase 2 determines the feasibility of using the site. The time lag between phases 2 and 3 may be several years as the client uses the initial data in a preliminary design to run project cost estimates so that money can be obtained for doing the project.

The cost of an adequate site exploration, including laboratory testing and submitting a report to the client which outlines what was found and the recommendations of the geotechnical engineer, runs from about 0.1 to 0.3 percent of the total cost for most structures except bridge and dam sites, where the exploration costs may exceed 1 percent. This cost must be weighed against overdesign (for the designer's protection) costs plus the additional contract bid cost to cover either contractor uncertainty or the cost of hiring a geotechnical expert to check the site. There are numerous cases where overdesign did not work, i.e., where site conditions discovered during construction required a redesign.

A Drilling

As stated earlier, most soil exploration consists of drilling holes to ascertain subsurface conditions. Sometimes a test pit may be excavated using a backhoe or front-end loader, but this is generally limited in depth, requires a large site, and is rather costly. Any type of well-drilling equipment can be used to make the boring, but the most common method, and the only one considered here, is to use a rotary drill rig with continuous flight augers. These devices are commercially available and mount on trucks. Capacities range to about 40 m of depth using a 190- to 200-mm-OD auger.

The auger is usually "hollow stem" with a stem cavity on the order of 75 to 90 mm. The operator can observe the soil and note the depth when stratum changes occur. Bags of auger cuttings may be obtained if desired.

B Testing and Sampling

On nearly all borings a standard penetration test (SPT) is performed at depth increments of 0.75 to 1.5 m (the auger is in 1.5-m segments). The standard penetration test consists of:

1. Augering to the desired level below ground surface, disconnecting the drive unit, then, while the auger prevents the hole from caving, inserting a tube device called a standard split spoon (Fig. 3-5)
2. Driving the split spoon 150 mm to seat it, then driving it an additional distance of 305 mm.
3. Using a 63.5-kg mass falling a height of 0.762 m.
4. Recording the number of blows N required.

Obviously if N is small, the tube easily penetrates the soil, and if large, the soil is of better quality. A rough correlation between N and D_r [see Eq. (6-2)] is as follows:

D_r	0	0.15	0.35	0.65	0.85	1.00
N		4	10	30	50	

Generally a blow count exceeding 25 to 30 per centimeter is considered "refusal."

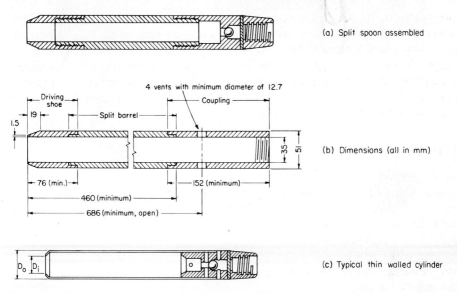

Figure 3-5 Dimensions of the standard split spoon used for penetration testing.

A drilling premium may be charged when $N > 50$ for 30.5 cm due to excessive equipment wear.

Penetration tests are widely used in cohesionless materials (for which the test was originated), but they are also used for cohesive soils with a considerably reduced reliability. When the test is completed, the split spoon is withdrawn with the soil collected during driving. The ends are unscrewed and the barrel separated and the sample removed. Index and/or strength (if the soil is cohesive) tests may be performed on the sample thus obtained.

The SPT recovers a very disturbed soil sample due to the thick walls of the sampler, as shown in Fig. 3-5. For this reason, where better-quality samples are needed, thin-walled tubes are attached to the drill rods, inserted into the hollow stem, and pushed (not driven) to fill the tube. The quality of any tube sample is related to the area and recovery ratios. The area ratio compares the area of the tube wall to the sample area (see Fig. 3-5c) as

$$A_r = \frac{D_{\text{outside}}^2 - D_{\text{inside}}^2}{D_{\text{inside}}^2} \tag{3-9}$$

where $D =$ diameter. Sample tubes should have an $A_r < 0.1$ for minimum disturbance. The recovery ratio compares the distance pushed to the sample length recovered:

$$L_r = \frac{L_{\text{recovered}}}{L_{\text{pushed}}} \tag{3-10}$$

If $L_r > 1$, the sample has expanded during the pushing operation; if $L_r < 1$, the

Figure 3-6 Soil boring operation using a continuous flight auger and drill rig mounted on an "all-terrain" vehicle. (*a*) Operator observing auger cuttings. Top half of 1.5-m auger flight is being advanced to take an SPT test. (*b*) Taking a blow count for an SPT test. (*c*) Split spoon separated, showing part of sample recovered. Driving shoe is in foreground with unconfined compression sampler (Fig. 13-25). (*d*) Using pocket a penetrometer (Fig. 13-24) to obtain unconfined compression strength of soil from SPT test. (*e*) SPT recovery of watery sand. Pronged round device at the left of the compression tester is a split spoon insert to hold sands; representative soil samples from the SPT are stored in the small glass jars shown. Lid markings indicate job, boring no., depth, and blow count. (*Photographs courtesy of A and H Engineering and Testing Corporation.*)

sample has compressed (probably due to tube friction). Seldom will $L_r = 1$, but values of 0.98 to 1.02 would produce samples of good quality.

Figure 3-6 gives a series of photographs of a typical drilling operation using a continuous flight auger to make the boring.

This section has only briefly considered the field of soil exploration. The interested reader may consult a reference such as Bowles (1977) or Hvorslev (1949) for additional details and in-depth discussion.

3-10 SUMMARY

This chapter has presented the tests commonly used to obtain the physical and index properties of a soil, including:

Water content determination
Atterberg limits (liquid, plastic, and shrinkage)
Specific gravity, including comments on the test's computational accuracy
Grain-size distribution, including both sieve and hydrometer analysis, hydrometer
    analysis computations, and the method of presenting the data on a semilog
    graph

We have also looked briefly at the problem of field exploration and the standard penetration test (SPT) commonly used for field data and to obtain laboratory samples.

Tests for obtaining an estimate of the flow of water through soil, time-dependent settlements, and shear strength tests are presented in later chapters as appropriate to the discussion.

HOMEWORK PROBLEMS

3-1 For the assigned set of liquid and plastic limit data in Table 3-1,
    (a) Compute the arithmetic mean for w_L and w_P.
    (b) Compute the standard deviation for w_L and w_P.
    (c) Draw a frequency histogram for both sets of data. A histogram is a bar graph showing the relative occurrence of a data value.
    (d) Using the above information, compute I_P and give an estimate of the reliability of this computation. Be sure to comment on significance of values.

3-2 A sieve analysis performed on two soils produced the following data:

Sieve no.	Percent passing	
	Soil A	Soil B
4	72.0	100.0
10	68.7	96.0
30	49.3	78.1
60	32.4	41.8
100	21.3	23.7
200	5.1	8.8

REQUIRED: Plot the grain-size curves for both soils on the same graph. Find D_{10} and D_{85} and compute C_C and C_U.

Ans: Soil A: $D_{10} = 0.09$ mm; $C_C = 0.68$; $C_U = 12$ (student answers may differ slightly due to plotting accuracy).

3-3 The development of the hydrometer analysis theory requires simplifying assumptions. List these and briefly discuss them.

3-4 A 1000-cm³ soil-water suspension contains 55 g of soil of $G_s = 2.70$. The temperature of the suspension is 20°C. Find the corrected hydrometer reading at $t = 0$. Also:

(a) Compute the specific gravity of the soil-water suspension at $t = 0$ for temperatures of 4 and 20°C.

Ans: $G_{4°} = 1.035$.

(b) Compute the specific gravity if oil instead of water is used to make a soil-oil suspension. The oil has $G = 0.90$ at 4°C.

Ans: $G_{20°} = 0.935$.

3-5 A hydrometer test was performed using $W_s = 50$ g. Other data include zero correction $= +1.5$; meniscus $= 1.0$ unit; $G_s = 2.72$.

Elapsed time t, min	T, °C	Hydrometer reading R
1.0	18	48
5.0	20	38
100.0	24	25
1000.0	24	15

COMPUTE: Percent finer and D for these four readings.

Ans: $t = 5$ min: percent finer $= 72.1$ percent, $D = 0.0189$ mm.

3-6 A grain-size analysis was performed on two soils as follows:

Sieve no.	Percentage passing, soil A	Percentage passing, soil B	
4	98.0		100
10	86.5		100.0
20	71.9		82.5
40	55.9		62.3
60	34.7		51.5
100	18.3		40.1
200	8.7		36.7
		0.05 mm	23.2 Hydrometer
		0.01	15.8
		0.005	8.7
		0.001	3.4

REQUIRED: Plot the two grain-size curves on the same graph sheet, find D_{10} and D_{85} for both soils, and compute C_C and C_U as appropriate.

Ans: Soil B: C_C and C_U, no meaning; $D_{10} = 0.007$ mm; $D_{85} = 0.9$ mm.

3-7 What is the area ratio of a thin-wall tube with OD $= 51$ mm and ID $= 47$ mm?

3-8 Referring to Fig. 3-5, what is the area ratio of the standard split spoon? Comment on the sample quality based on this value.

3-9 What is the recovery ratio of a sample pushed 610 mm with a recovered length of 590 mm? What happened to the sample? How could this be avoided?

Chapter 4

Soil Classification

4-1 GENERAL

Soils may be classified in a general way as *cohesionless* or *cohesive* or as coarse- or fine-grained (see Sec. 2-9). As these terms are too general and cover too wide a range of physical and engineering properties, additional refinement or means of classification is necessary to determine the suitability of a soil for specific engineering purposes and to be able to convey this information to others in an understandable way. Numerous classification systems have been proposed in the past several decades, and occasionally someone still proposes a new classification system. Of the large number of proposed systems, the Unified Classification System based on Casagrande's (1948) work on military airfields and the American Association of State Highway and Transportation Officials (AASHTO, formerly the Bureau of Public Roads) system are widely used worldwide. Nearly all the state departments of transportation (highway departments) in the United States currently use the AASHTO system for soil classification.

The several classification systems which have been proposed use particle sizes to differentiate general group classifications as gravel, sand, silt, and clay as illustrated in Fig. 4-1. Atterberg limits are usually also a part of the final classification process—especially for engineering work.

The Unified Soil Classification System is widely used by the engineering agencies of the U.S. government, including the Bureau of Reclamation and the Corps of Engineers, city building codes, and many commercial soil testing laboratories and geotechnical engineering consulting firms. With slight to no

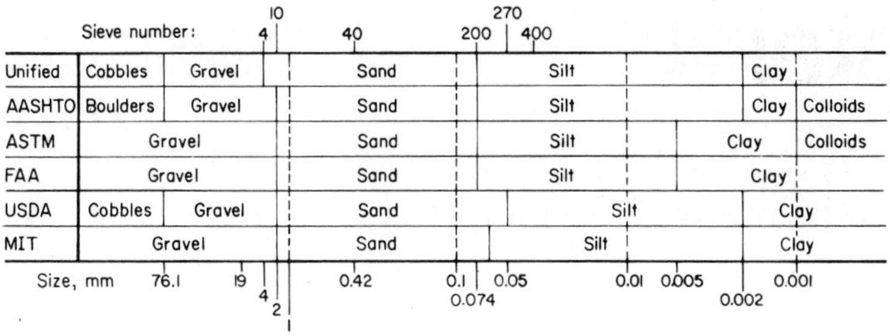

Sieve number: 10 4 | 40 | 270 200 | 400

Unified	Cobbles	Gravel	Sand	Silt	Clay
AASHTO	Boulders	Gravel	Sand	Silt	Clay	Colloids
ASTM	Gravel		Sand	Silt	Clay	Colloids
FAA	Gravel		Sand	Silt	Clay
USDA	Cobbles	Gravel	Sand	Silt		Clay
MIT	Gravel		Sand	Silt		Clay

Size, mm 76.1 19 4 2 0.42 0.1 0.05 0.074 0.01 0.005 0.002 0.001

Figure 4-1 Grain-size range in several soil classification systems.

modification this system is also common in Great Britain, Canada, Australia, India, Africa, South America, and Europe.

The AASHTO classification system started with the (then) U.S. Bureau of Public Roads in the years 1927–1929. The system was revised in 1945 to include additional subgroups in the A-2 group (see Table 4-2), to incorporate group indexing within subgroups, and to reduce the number of soil tests from five to three. As the name implies, this classification system is based upon the observed field behavior of soil subgrades beneath highway pavements.

The Federal Aviation Administration (FAA) of the U.S. Department of Transportation uses a separate soil classification system for establishing the thickness requirements for airport pavements.

The Unified, AASHTO, and FAA soil classification systems will be presented in this chapter, since these are the most widely used systems. Once the user has acquired some familiarity with soil classification, it is relatively easy to use any system and to classify a soil from one system into another.

4-2 NEED FOR SOIL CLASSIFICATION SYSTEMS

A soil classification system enables one to use the engineering experience of others. It also facilitates communication between widely separated groups of engineers using the same method of soil classification. In other words, it is a language for communication. The use of a classification system does not eliminate the need for detailed studies of soils or for testing for engineering properties. For example, the unit weight of a soil, compaction characteristics, performance under saturated conditions, susceptibility to frost action, etc., are not directly included in the classification process, yet they may be critical factors in a design involving a particular type of soil. Since the classification systems have indirectly incorporated engineering experiences with the group names, there is a general correlation of engineering properties with the classification grouping. For example, experience tells us that a sandy gravel is easily drained; thus, if we want an easily drained subgrade, we should try to use a sandy gravel. Classifying the soil immediately gives the engineer with some experience some idea of the probable soil behavior.

4-3 THE UNIFIED SOIL CLASSIFICATION SYSTEM

This system, originally developed for use in airfield construction, was reported by Casagrande in 1948. It had already been in use since about 1942, but was slightly modified in 1952 to make it apply to dams and other construction.

The principal soil groups of this classification system are given in Table 4-1. As shown in the table under the column heading "Group Symbols," the soils are designated by group symbols consisting of a prefix and a suffix. The prefixes indicate the main soil types and the suffixes indicate the subdivisions within groups as follows:

Soil type	Prefix	Subgroup	Suffix
Gravel	G	Well graded	W
Sand	S	Poorly graded	P
Silt	M	Silty	M
Clay	C	Clayey	C
Organic	O	$w_L < 50$ percent	L
Peat	Pt	$w_L > 50$ percent	H

A well-graded gravel is GW; a poorly graded sand is SP; a well-graded sand is SW; a silty sand is SM; a clay with a liquid limit > 50 percent is CH, etc. A verbal description should accompany the classification symbols, e.g., Brown, coarse, well-graded sand with trace of gravel, SW. Homework Problems 2-16 through 2-19 illustrate the use of visual/verbal soil descriptions.

A soil is well graded or *nonuniform* if there is a wide distribution of grain sizes present, i.e., if there are some grains of each possible size between the upper and lower gradation limits. This can be ascertained by plotting the grain-size curve as illustrated in Fig. 3-2 and either observing the shape and spread of sizes or computing the coefficient of uniformity C_U and the coefficient of concavity C_C, as given by Eqs. (3-1) and (3-2), respectively.

A soil is poorly graded, or *uniform*, if the sample is mostly of one grain size or is deficient in certain grain sizes. A beach sand is an example of a uniformly graded soil.

Table 4-1 illustrates that only the sieve analysis and the Atterberg limits are necessary to classify the soil. A sieve analysis is performed and a plot of the grain-size distribution curve is made. When *less than 12 percent passes the No. 200 sieve*, it is necessary to obtain C_C and C_U to establish whether the soil is well or poorly graded. When more than 12 percent of the material passes the No. 200 sieve, the uniformity coefficient C_U and the coefficient of curvature C_C have no significance and only the Atterberg limits are used to classify the soil.

The Unified Soil Classification System defines a soil as:

1. *Coarse-grained* if more than 50 percent is retained on the No. 200 sieve
2. *Fine-grained* if more than 50 percent passes the No. 200 sieve

The coarse-grained soil is either:

1. *Gravel* if more than half of the coarse fraction is retained on the No. 4 sieve
2. *Sand* if more than half of the coarse fraction is between the No. 4 and No. 200 sieve size

The coarse-grained soil is:

GW, GP or SW, SP	≤ 5 percent passes No. 200 sieve
GW-GM, GP-GM, GW-GC, GP-GC or SW-SM, SP-SM, SW-SC, SP-SC	5 < percent passing No. 200 sieve ≤ 12
GM, GC or SM, SC	> 12 percent passes No. 200 sieve

Classification of coarse-grained soils depends primarily on the grain-size analysis and particle size distribution. Note carefully that it is possible for a granular soil to have, say, 59 percent pass the No. 4 sieve and be called a "gravel" depending on the percent passing the No. 200 sieve. Add a little sand so that the percent that passes the No. 4 sieve increases to, say, 61 percent and the soil may be classified as a sand. The engineering properties would be about the same in both cases, but the classification is considerably different. A dual classification symbol could be used, such as GW-SW, but these were not included in the original classification to keep the system as simple as possible; also, others might not know what this means. A major classification change with a small increase or decrease in the percent passing the No. 4 or No. 200 sieve is another reason why it is necessary to include a verbal description along with the symbols, i.e., very sandy gravel, very gravelly sand, etc. Incidentally, *peat* is classified solely on the basis of visual appearance.

Classification of fine-grained soils also requires the use of the plasticity or A chart shown in Fig. 4-2. Each soil is grouped according to the coordinates of the plasticity index and liquid limit. On this chart an empirical line (the A line) separates the inorganic clays (C) from silts (M) and organic (O) soils. Although

Table 4-1 Unified soil classification chart

Including identification and description

Field identification procedures
(Excluding particles larger than 75 mm and basing fractions on estimated weights)

COARSE-GRAINED SOILS
More than half of material is larger than No. 200 sieve size
(The No. 200 sieve size is about the smallest particle visible to the naked eye)

GRAVELS — *More than half of coarse fraction is larger than No. 4 sieve size.* (For visual classifications, the 6 mm size may be used as equivalent to the No. 4 sieve size.)

- **CLEAN GRAVELS (Little or no fines)**
  - Wide range in grain size and substantial amounts of all intermediate particle sizes.
  - Predominantly one size or a range of sizes with some intermediate sizes missing.
- **GRAVELS WITH FINES (Appreciable amount of fines)**
  - Nonplastic fines (for identification procedures see ML below).
  - Plastic fines (for identification procedures see CL below).

SANDS — *More than half of coarse fraction is smaller than No. 4 sieve size.*

- **CLEAN SANDS (Little or no fines)**
  - Wide range in grain sizes and substantial amounts of all intermediate particle sizes.
  - Predominantly one size or a range of sizes with some intermediate sizes missing.
- **SANDS WITH FINES (Appreciable amount of fines)**
  - Nonplastic fines (for identification procedures see ML below).
  - Plastic fines (for identification procedures see CL below).

FINE-GRAINED SOILS
More than half of material is smaller than No. 200 sieve size

Identification procedures on fraction smaller than No. 40 sieve size

	Dry strength (Crushing characteristics)	Dilatancy (Reaction to shaking)	Toughness (Consistency near plastic limit)
SILTS AND CLAYS — Liquid limit less than 50	None to slight	Quick to slow	None
	Medium to high	None to very slow	Medium
	Slight to medium	Slow	Slight
SILTS AND CLAYS — Liquid limit greater than 50	Slight to medium	Slow to none	Slight to medium
	High to very high	None	High
	Medium to high	None to very slow	Slight to medium

HIGHLY ORGANIC SOILS	Readily identified by color, odor, spongy feel, and frequently by fibrous texture.

Table 4-1 (*Cont.*)

Including identification and description

Group symbols	Typical names	Information required for describing soils
GW	Well-graded gravels, gravel-sand mixtures; little or no fines.	Give typical name; indicate approximate percentages of sand and gravel, max. size, angularity, surface condition, and hardness of the coarse grains; local or geologic name and other pertinent descriptive information; and symbol in parentheses.
GP	Poorly graded gravels, gravel-sand mixtures; little or no fines.	
GM	Silty gravels, poorly graded gravel-sand-silt mixtures.	
GC	Clayey gravels, poorly graded gravel-sand-clay mixtures.	For undisturbed soils add information on stratification, degree of compactness, cementation, moisture conditions, and drainage characteristics.
SW	Well-graded sands, gravelly sands; little or no fines.	
SP	Poorly graded sands, gravelly sands; little or no fines.	EXAMPLE: Silty sand, gravelly; about 20 percent hard, angular gravel particles 12 mm maximum size; rounded and subangular sand grains coarse to fine; about 15 percent nonplastic fines with low dry strength; well compacted and moist in place; alluvial sand; (SM).
SM	Silty sands, poorly graded sand-silt mixtures.	
SC	Clayey sands, poorly graded sand-clay mixtures.	
ML	Inorganic silts and very fine sands, rock flour, silty or clayey fine sands with slight plasticity.	Give typical name; indicate degree and character of plasticity, amount and maximum size of coarse grains; color in wet condition, odor if any, local or geologic name, and other pertinent descriptive information; and symbol in parentheses.
CL	Inorganic clays of low to medium plasticity, gravelly clays, sandy clays, silty clays, lean clays	
OL	Organic silts and organic silt-clays of low plasticity.	
MH	Inorganic silts, micaceous or diatomaceous fine sandy or silty soils, elastic silts.	For undisturbed soils add information on structure, stratification, consistency in undisturbed and remolded states, moisture and drainage conditions.
CH	Inorganic clays of high plasticity, fat clays.	
OH	Organic clays of medium to high plasticity.	EXAMPLE: Clayey silt, brown; slightly plastic; small percentage of fine sand; numerous vertical root holes; firm and dry in place; loess; (ML).
Pt	Peat and other highly organic soils.	

Table 4-1 (*Cont.*)

Including identification and description

Use grain-size curve in identifying the fractions as given under field identification	Determine percentages of gravel and sand from grain-size curve. Depending on percentage of fines (fraction smaller than No. 200 sieve size), coarse-grained soils are classified as follows:	GW, GP, SW, SP, GM, GC, SM, SC, Borderline cases requiring use of dual symbols.	Laboratory classification criteria		
		Less than 5 percent More than 12 percent 5 to 12 percent	$C_U = \dfrac{D_{60}}{D_{10}}$ Greater than 4		$C_C = \dfrac{(D_{30})^2}{D_{10} \times D_{60}}$ Between 1 and 3
			Not meeting all gradation requirements for GW		
			Atterberg limits below "A" line, or I_P less than 4	Above "A" line with I_P between 4 and 7 are borderline cases requiring use of dual symbols.	
			Atterberg limits above "A" line with I_P greater than 7		
			$C_U = \dfrac{D_{60}}{D_{10}}$ Greater than 6		$C_C = \dfrac{(D_{30})^2}{D_{10} \times D_{60}}$ Between 1 and 3
			Not meeting all gradation requirements for SW		
			Atterberg limits below "A" line or I_P less than 4	Above "A" line with I_P between 4 and 7 are borderline cases requiring use of dual symbols.	
			Atterberg limits above "A" line with I_P greater than 7		

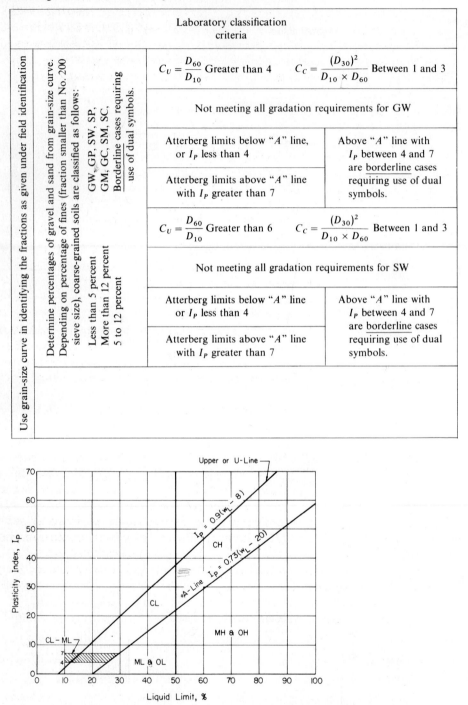

Figure 4-2 Plasticity or *A* chart for use in the Unified Soil Classification system.

the silty and organic soils overlap areas, they are easily differentiated by visual examination (dark color, presence of organic materials) and odor.

Most inorganic clays lie fairly close to the A line. Kaolin clays (described in Chap. 6) tend to plot below the A line as inorganic silts (ML or MH) because of the similarity of engineering properties. Some active clays such as bentonite may lie above the A line and/or very close to the U line even though the liquid limit may be several hundred percent. In general, however, soils that have the same classifications tend to have the same engineering behavior.

The upper limit line (U line) shown on Fig. 4-2 represents approximately the upper range of plasticity index and liquid limit coordinates found thus far for any soils. Any soil plotting to the left of the U line should be suspect and the limits rechecked as the first step in the classification sequence.

Example 4-1 Given the classification data for the following three soils, classify the soils using the Unified Soil Classification System.

	Soil		
Percent passing	A	B	C
No. 4 sieve	40	69	95
10	30	54	90
40	22	46	83
100	20	41	71
200	15	36	55
w_L, percent	35	39	55
w_p, percent	22	27	24
Visual observation	Dark tan, very gravelly	Greyish brown, some odor	Blue-grey, traces of gravel

SOLUTION Since more than 12 percent passes the No. 200 sieve, we can immediately eliminate GW, GP, SW, and SP as possible classifications for all three soils.

(a) *For soil A:*
(1) Less than 50 percent passes the No. 4 sieve; therefore, the soil must be predominantly gravel, thus G.
(2) Considering the location of the $w_L = 35$ percent and $I_p = 13$ (computed) on the A chart, we find a CL.
(3) From the preceding two observations and the visual description of this soil, soil A is: Dark tan, clayey gravel, GC.

(b) *For soil B:*
(1) Less than 50 percent passes the No. 200 sieve; therefore the soil is coarse-grained (either sand or gravel).

(2) Compute the percent passing the No. 4 sieve and retained on the No. 200 sieve as

$$69 - 36 = 33 \text{ percent sand}$$

$$100 - 69 = 31 \text{ percent gravel}$$

Therefore, of the coarse fraction more than half is sand.

(3) More than 12 percent passes the No. 200 sieve, and from the Atterberg limits, the soil plots below the A line ($w_L = 39$ and $I_P = 12$); thus, the $(-)$ No. 40 fraction is an ML. Noting that the percentage of sand and gravel are nearly equal, soil B is: Greyish-brown, very gravelly, silty sand with trace of organic material, SM.

(c) *For soil C:*

(1) With 55 percent passing the No. 200 sieve, the soil is fine-grained.

(2) Using $w_L = 55$ percent and $I_P = 31$, the soil plots above the A line and also above the line of $w_L > 50$; therefore soil C is: Blue-grey, sandy clay with a trace of gravel, CH.

It was originally suggested by Casagrande (1948) that the fine-grained soils might better be classified as:

Soil	$w_L =$	0		35		50	
Clay			CL		CI		CH
Organic			OL		OI		OH
Silt			ML		MI		MH

This intermediate plasticity classification is not used at present and is not recommended.

With the coarse-grained soils, an additional subgrouping of GU and SU is sometimes used, for example in Great Britain. The " U " indicates a uniform gravel or sand, such as certain gravel deposits and some dune and beach sands which consist of primarily one or two sizes.

4-4 THE AASHTO SOIL CLASSIFICATION SYSTEM

The original BPR classification system of the late 1920s has been revised several times. It classifies soils into eight groups, A-1 through A-8, and originally required the following data:

1. Grain-size analysis
2. Liquid and plastic limits and the calculated I_P
3. Shrinkage limit
4. Field moisture equivalent—the maximum moisture content at which a drop of water placed on a small surface will not be immediately adsorbed

5. Centrifuge moisture equivalent—a test to measure capacity of soil to hold water (dry soil is soaked in water for 12 h, then centrifuged for 1 h. The final water content thus obtained is the CME)

The revised (Proc. 25th Annual Meeting of Highway Research Board, 1945) system retained the eight basic soil groups but added two subgroups in A-1, four subgroups in A-2, and two subgroups in A-7. Soil tests (4) and (5) were deleted, so that the only soil tests required are the *grain-size* analysis and the *liquid* and *plastic limits*. This revised classification system was adopted by AASHTO as standard M-145. Table 4-2 illustrates the current AASHTO soil classification system. Soil group A-8 is not shown, but is peat or muck based on a visual classification. Shown are groups A-1 through A-7 with two subgroups in A-1, four subgroups in A-2, and two subgroups in A-7, for a total of 12 soil subgroups in this classification system (exclusive of peat and/or muck). Figure 4-3 can be used to obtain the graphic ranges of w_L and I_P for the A-4 to A-7 groups and for the subgroup classification in the A-2 subgroup. A group index can be computed using Eq. (4-3) or Fig. 4-4 to further compare soils within a subgroup.

In general, this classification system rates a soil as:

1. Poorer for use in road construction as one moves from left to right in Table 4-2, i.e., A-6 soil is less satisfactory than A-5 soil
2. Poorer for road construction as the group index increases for a particular subgroup, i.e., an A-6(3) is less satisfactory than an A-6(1)

4-5 GENERAL DESCRIPTION OF AASHTO SOIL CLASSIFICATION SUBGROUPS

The A-1 through A-3 soils are *granular* with not more than 35 percent of the material passing the No. 200 sieve.

A typical material in group A-1 is a well-graded mixture of gravel, coarse sand, fine sand, and a binder [$(-)$ No. 200] material which has little to no plasticity. Subgroup A-1a, which may contain appreciable gravel, is a coarser-graded material than A-1b, which is predominately coarse sand. The binder material in this group may have a small amount of plasticity ($I_P \leq 6$).

A-3 soil is fine, relatively uniform sand, typically a fine beach sand or desert blown sand. The A-3 subgroup may also include stream-deposited mixtures of poorly graded fine sands with traces of coarse sand and gravel. The silt or rock flour fractions, if any, passing the No. 200 sieve are *nonplastic* (NP).

Group A-2 is also granular but with appreciable (but not more than 35 percent) material passing the No. 200 sieve. These materials are on the borderline between the materials falling in groups A-1 and A-3 and the silt-clay materials of groups A-4 through A-7. Subgroups A-2-4 and A-2-5 include various materials in which not more than 35 percent is finer than the No. 200 sieve and which have the

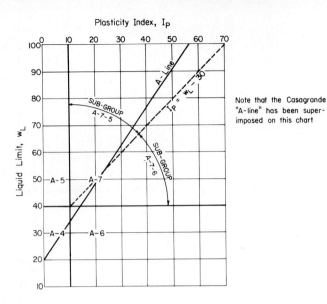

Note that the Casagrande "A-line" has been superimposed on this chart

Figure 4-3 Liquid limit and plasticity index ranges for silt-clay soils (*A*-4 through *A*-7).

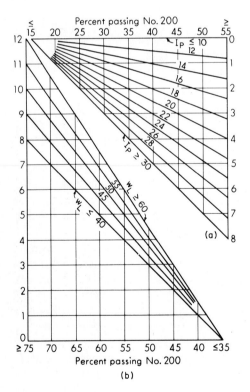

Figure 4-4 Charts to obtain group index of a soil. Group index equals the sum of readings from vertical of (*a*) and (*b*).

Table 4-2 AASHTO soil classification system

Note that *A*-8, peat or muck, is by visual classification and is not shown in the table.

General classification	Granular materials (35% or less passing No. 200)							Silt-clay materials (More than 35% passing No. 200)			
	A-1		A-3	A-2				A-4	A-5	A-6	A-7
Group classification	*A-1a*	*A-1b*	*A-3*	*A-2-4*	*A-2-5*	*A-2-6*	*A-2-7*	*A-4*	*A-5*	*A-6*	*A-7-5; A-7-6*
Sieve analysis: Percent passing:											
No. 10	50 max.										
No. 40	30 max.	50 max.	51 min.								
No. 200	15 max.	25 max.	10 max.	35 max.	35 max.	35 max.	35 max.	36 min.	36 min.	36 min.	36 min.
Characteristics of fraction passing No. 40:											
Liquid limit				40 max.	41 min.	40 max.	41 min.	40 max.	41 min.	40 max.	41 min.
Plasticity index	6 max.		N.P.	10 max.	10 max.	11 min.	11 min.	10 max.	10 max.	11 max.	11 min.
Group index	0		0	0		4 max.		8 max.	12 max.	16 max.	20 max.
Usual types of significant constituent materials	Stone fragments, gravel, and sand		Fine sand	Silty or clayey gravel and sand				Silty soils		Clayey soils	
General rating as subgrade	Excellent to good							Fair to poor			

plasticity characteristics of the A-4 and A-5 groups. Subgroups A-2-6 and A-2-7 are similar to A-2-4 and A-2-5 except that the plasticity characteristics of the $(-)$ No. 40 sieve fraction are those of the A-6 and A-7 groups. For example, if the soil has

$$w_L \leq 40 \text{ percent} \qquad I_P \geq 11 \qquad \text{Group index } GI \leq 16$$

and not more than 35 percent passes the No. 200 sieve, the soil is an A-2-6 since the plasticity characteristics are those of an A-6 soil as shown in Table 4-2. If more than 35 percent of the material had passed the No. 200 sieve, the soil would contain "appreciable" fines and be classified as an A-6.

Groups A-4 through A-7 are considered to be fine-grained soils, and all have more than 35 percent of the material passing the No. 200 sieve.

Soil group A-7 is further subdivided to

$$A\text{-}7\text{-}5 \text{ if } I_P < (w_L - 30)$$

$$A\text{-}7\text{-}6 \text{ if } I_P > (w_L - 30)$$

Figure 4-3 can be used to quickly classify the A-7 subgroups.

Soil group A-8 is peat (very organic) or muck (thin, very watery, and with considerable organic material) and is identified by inspection of the deposit.

4-6 THE AASHTO GROUP INDEX

To establish the relative ranking of a soil within a subgroup, the group index GI was developed. The group index is a function of the percent of soil passing the No. 200 sieve and the Atterberg limits. The group index can be obtained as the sum of the values from Fig. 4-4a and b, a graphical presentation of the following equation:

$$GI = 0.2a + 0.005ac + 0.01bd \tag{4-3}$$

where a = that part of the percent passing the No. 200 sieve greater than 35 and not exceeding 75, expressed as a whole number (range = 1 to 40)

b = that part of the percent passing the No. 200 sieve greater than 15 and not exceeding 55, expressed as a whole number (range = 1 to 40)

c = that part of the liquid limit greater than 40 and not greater than 60, expressed as a whole number (range = 1 to 20)

d = that part of the plastic index greater than 10 and not exceeding 30, expressed as a whole number (range = 1 to 20)

The group index should be rounded to the nearest whole number and placed in parentheses, as

$$A\text{-}2\text{-}6(3)$$

In general, the larger the group index value, the less desirable the soil for highway construction use within that subgroup.

Example 4-2

GIVEN Same soil classification data as Example 4-1.

REQUIRED Classify the three soils using the AASHTO classification system.

SOLUTION
    (*a*) *Classifying soil A:*
        (1) Proceeding from left to right in Table 4-2, the soil will be either an *A*-1, *A*-3, or *A*-2, since only 15 percent passes the No. 200 sieve.
        (2) Based on $I_P = 13$ (computed), we eliminate *A*-1 and *A*-3.
        (3) With $w_L = 35$ percent and $I_P = 13$, the soil fits the *A*-2-6 classification.
        (4) The group index can be computed as

$$GI = 0.2(0) + 0.005(0)(0) + 0.01(0)(3) = 0.0$$

The group index is more conveniently obtained as the sum of the values from Fig. 4-4*a* and *b*:

$$\text{Fig. 4-4}a \cong 0$$
$$\text{Fig. 4-4}b \cong 0$$
$$\overline{}$$
$$GI = 0$$

        (5) From inspection of the sieve analysis data and the classification data, soil *A* is dark tan, *silty* or *clayey sandy gravel*, *A*-2-6(0).
    (*b*) *Soil B:*
        (1) Proceeding from left to right in Table 4-2, the soil can only be an *A*-4, *A*-5, *A*-6, or *A*-7, since 36 percent passes the No. 200 sieve.
        (2) Based on $I_P = 12$, the soil can only be an *A*-6 or *A*-7.
        (3) With $w_L = 39$ percent, the soil is an *A*-6.
        (4) The group index is

$$\text{Fig. 4-3}a = 0.5$$
$$\text{Fig. 4-3}b = 0.4$$
$$\overline{}$$
$$GI = 0.9 \rightarrow 1.0$$

        (5) From inspection of the sieve analysis data (31 percent gravel, 33 percent sand) and data just obtained, soil *B* is a greyish-brown, *very gravelly sandy silt* or *clay* with trace of organic material, *A*-6(1).
    (*c*) *Soil C:*
        (1) With 55 percent passing the No. 200 sieve, the soil is an *A*-4, *A*-5, *A*-6 or *A*-7.
        (2) With $w_L = 55$ percent and $I_P = 31$, the soil is an *A*-7-6 since $I_P > w_L - 30$ (also from Fig. 4-3).

(3) The group index is

$$\text{Fig. 4-4}a = 8$$
$$\text{Fig. 4-4}b = \underline{\quad 5.8 \quad}$$
$$GI = 13.8 \qquad \text{say 14}$$

(4) Soil C is a blue-grey *sandy clay with trace of gravel*, A-7-6(14).

4-7 THE FEDERAL AVIATION ADMINISTRATION (FAA) SOIL CLASSIFICATION SYSTEM

The FAA has a separate method of soil classification, primarily designed for airport pavement design. The FAA originated as the Civil Aeronautics Administration (CAA), who proposed this soil classification system in 1944. The original FAA soil classification system used the Nos. 10, 60, and 270 sieves together with the liquid and plastic limit tests. The latest revision substitutes the No. 40 and No. 200 sieves for the No. 60 and No. 270 sieve sizes, respectively. No other changes were made in the system. Table 3-2 shows the corresponding sieve sizes as

$$\text{No. 40} = 0.420 \text{ mm} \qquad \text{No. 60} = 0.250 \text{ mm}$$
$$\text{No. 200} = 0.074 \text{ mm} \qquad \text{No. 270} = 0.053 \text{ mm}$$

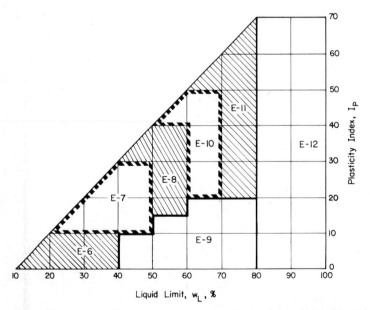

Figure 4-5 FAA Soil Classification System chart for use in classifying fine-grained soils with more than 45 percent passing the No. 200 sieve.

Table 4-3 Classification of soils in the Federal Aviation Administration system†

| Soil group | Mechanical (sieve) analysis | | | | Plasticity | | Subgrade class | | | |
| | Gravel‡ % > No. 10 | Material finer than No. 10 sieve | | | Liquid limit w_L, % | Plastic index I_p | Good drainage | | Poor drainage | |
		Coarse sand No. 10 > % > No. 40	Fine sand No. 40 > % > No. 200	Silt and clay % < No. 200			No frost	Severe frost	No frost	Severe frost
E-1	0–45	40⁺	60⁻	15⁻	25⁻	6⁻	Fa Ra	Fa Ra	Fa Ra	Fa Ra
E-2	0–45	15⁺	85⁻	25⁻	25⁻	6⁻	Fa Ra	Fa Ra	F1 Ra	F2 Ra
E-3	0–45	—	—	25⁻	25⁻	6⁻	F1 Ra	F1 Ra	F2 Ra	F2 Ra
E-4	0–45	—	—	35⁻	35⁻	10⁻	F1 Ra	F1 Ra	F2 Rb	F3 Rb
E-5	0–55	—	—	45⁻	40⁻	15⁻	F1 Ra	F2 Rb	F3 Rb	F4 Rb
E-6	0–55	—	—	45⁺	40⁻	10⁻	F2 Rb	F3 Rb	F4 Rb	F5 Rc
E-7	0–55	—	—	45⁺	50⁻	10–30	F3 Rb	F4 Rb	F5 Rb	F6 Rc
E-8	0–55	—	—	45⁺	60⁻	15–40	F4 Rb	F5 Rc	F6 Rc	F7 Rd
E-9	0–55	—	—	45⁺	40⁺	30⁻	F5 Rc	F6 Rc	F7 Rc	F8 Rd
E-10	0–55	—	—	45⁺	70⁻	20–50	F5 Rc	F6 Rc	F7 Rc	F8 Rd
E-11	0–55	—	—	45⁺	80⁻	30⁺	F6 Rd	F7 Rd	F7 Rc	F8 Rd
E-12	0–55	—	—	45⁺	80⁺	—	F7 Rd	F8 Re	F9 Re	F10 Re
E-13	Peat and/or muck based on field examination						Not suitable for subgrade use			

Decreasing order of Performance →

† From Tables 1 and 2 of *Airport Paving*, Department of Transportation, FAA, dated May 9, 1967, and revised April 1, 1970.

‡ When the sample contains material coarser than No. 10 in amounts equal to or greater than the maximum limit shown in the table, a lower group classification may be allowed provided the coarse material is reasonably sound and well graded.

and it can be seen that the sieve substitutions were relatively minor and make for much easier comparison of FAA-classified soils with those classified by the other, more widely used soil classification systems.

The FAA (Table 4-3) system classifies soils into groups E-1 through E-13. Groups E-1 through E-4 are granular soils of low plasticity with not more than 35 percent passing the No. 200 sieve and as much as 45 percent retained on the No. 10 sieve.

The E-5 through E-12 soils are primarily fine-grained, with more than 35 percent passing the No. 200 sieve, and of progressively increasing plasticity characteristics. Soil E-13 is peat or muck and corresponds exactly to an A-8 soil in the AASHTO classification system. Figure 4-5 is a convenient graphical solution for the fine-grained soils, E-6 through E-12. Soil E-5 is somewhat of a borderline soil with between 35 and 45 percent finer than the No. 200 sieve.

Depending on the soil group classification and on whether the pavement is rigid (portland cement concrete) or flexible (asphalt concrete), there are corresponding subgrade classes. The subgrade classes (also shown in Table 4-3) are based on the performance of a particular soil under the type of pavement under different conditions of drainage and frost potential.

In Table 4-3 F and R denote the subgrade classifications for "flexible" and "rigid" pavements, respectively. In airport pavement design the pavement structure includes the wearing course(s) of asphalt or portland cement concrete as well as any necessary base and subbase courses required to develop adequate resistance to the aircraft wheel loads. The Fa and Ra subgrade classification shown in Table 4-3 for the E-1 and E-2 soils does not require a separate base course; rather, the pavement surfacing may be laid directly on the prepared ground profile or subgrade. In general, as the subgrade number (as from F1 to F2 or Rb to Rd) increases, the depth of the pavement structure also increases.

The term *drainage* in Table 4-3 implies that the soil underlying the pavement surface does not retain water (it is granular and free-draining). The designation *no frost* means a climatic and geographical condition such that average frost penetration depth is not greater than the pavement wearing course.

Example 4-3 Reclassify the soils given in Example 4-1 in the FAA soil classification system. Also obtain the subgrade soil classification for an asphalt (flexible) pavement, assuming that the depth of frost penetration is 40 cm for all three soils. Also assume that any soil that has more than 20 percent of $(-)$ No. 200 material will be "poor drainage."

SOLUTION
    (*a*) *For Soil A:*
        (1) With 30 percent retained on the No. 10 sieve, the soil cannot be readily identified.
        (2) With 15 percent passing the No. 200 sieve, the soil can only be E-1 through E-5.
        (3) With $I_P = 13$ and $w_L = 35$ percent, soil must be an E-5.

(4) For a frost depth of 40 cm and good drainage [less than 20 percent is $(-)$ No. 200], the subgrade class is F2.

Soil A is: Dark tan, silty or clayey sandy gravel, E-5(F2).

(b) *Soil B:*

(1) Percent retained on the No. 0 sieve is approximately 70 percent, but this tells little about classification.

(2) Percent passing the No. 10 sieve and retained on the No. 40 is $54 - 46 = 8$ percent.

(3) With 36 percent passing the No. 200 sieve, soil can only be an E-5.

(4) With w_L of 39 percent and I_P of 12, soil is an E-5.

Soil B: Greyish brown, very gravelly sandy silt or clay with trace of organic material, E-5(F4).

(c) *Soil C:*

(1) With 90 percent $(-)$ No. 10, the soil will be something between E-6 and E-12.

(2) With $w_L = 55$ percent and $I_P = 31$, classify soil as an E-8 (refer to Fig. 4-5).

(3) With 55 percent passing the No. 200 sieve, the drainage will be poor, and with 40 cm frost depth the subgrade class is F7.

Soil C: Blue-grey, sandy clay with trace of gravel, E-8(F7).

4-8 FIELD IDENTIFICATION TESTS

Coarse-grained soils such as gravel, sand, sandy gravel, gravelly sand, etc., can be readily identified by inspection. Traces of silt and/or clay are somewhat more difficult to identify when mixed with these materials but may not be of much importance unless the quantity is over 5 to 10 percent. The sedimentation test described for fine-grained soils may be used to determine if significant quantities of silt, very fine sand, or clay are present.

Fine-grained soils can be identified using some, or all, of the following tests performed on approximately the $(-)$ No. 40 sieve sizes (remove the larger particles by hand rather than actually sieving).

1. *Dilatancy* (or pore water mobility reaction to shaking). Prepare a pat of moist soil with a volume of 1 to 3 cm^3, using enough water to make the soil soft but not sticky. Place the pat in the open palm of one hand and jar the hand vigorously with the other. If the soil is fine sand, silt, or silty fine sand, the inertia forces due to jarring will force the water to the surface of the soil pat, and it will appear wet or glossy. When the sample is manipulated, this surface water will disappear. In soils with substantial clay this test produces no reaction.

2. *Dry strength* (resistance of dry lumps to crushing). Mold a pat of soil to about the consistency of putty by adding water as necessary. Allow the pat to completely dry and then test the crushing strength by breaking or crumbling be-

tween the fingers. The dry strength increases with increasing plasticity. High dry strength is characteristic for clays of the CH group, lesser dry strength for CL and MH soils, and very low to nonexistent strength for OL and ML soils. Fine sand, silt, and sand-silt mixtures possess almost no dry strength. Note that one may do this test approximately on naturally dried in situ soil.

3. *Toughness* (consistency near the plastic limit). Take a specimen of about 1 cm^3 and mold it to the consistency of putty. Proceed to roll the soil in the palm (or on a smooth surface) into a thread about 3 mm in diameter. When the pat of soil crumbles and loses its plasticity, the plastic limit has been reached. The higher the resistance of the 3-mm thread to pulling apart, the higher the position of the soil on the plasticity chart with respect to the *A* line. A weak thread which is easily crumbled indicates silts or inorganic clays of low plasticity. Highly organic clays are also very weak but may feel spongy at the plastic limit.

4. *Sedimentation.* Place about 50 g (more for gravelly soils) in a glass jar, such as a beaker, test tube, glass graduate, or other jar on the order of 150 mm deep, with water to fill the jar. Vigorously shake for several minutes and allow to stand. Gravel and coarse sand will settle almost instantly. Medium to very fine sand will take not more than 1 to 3 min; silt will take not more than 15 min. Clay will take only slightly longer unless a deflocculating agent (see Sec. 6-7) is added. The relative thickness of the sediments is an indication of percentages of various grain sizes.

5. *Color.* In general, dark colors such as black, grey, and dark brown indicate organic soils.

6. *Odor.* Organic soils usually have a distinctive smell of decaying materials. This test should be applied to fresh samples which are still wet. Roots, pieces of weeds, wood, plants, etc., may be visually present as further aid.

7. *Feel.* Sands and silts dry rapidly and can be dusted from the hands easily. Clay tends to leave considerable discoloration after drying and the hands may have to be washed to remove all traces. Clay tends to be smooth to the touch or to leave a smooth streak when a spatula blade is moved across a wet mass. Silts and sands are rough and gritty and leave grain marks when a spatula blade is moved across a wet lump.

4-9 SUMMARY

This chapter has presented three systems of soil classification and field identification tests widely used in foundation engineering. The Unified system is used both in the United States and, with only minor modification (if any), abroad. The AASHTO system is used by most of the state highway departments in the United States and some abroad. The FAA system is used for civil airport pavement design in the United States.

These three classification systems all require:

1. A sieve analysis using at least the Nos. 10, 40, and 200 sieves. If the Unified

Table 4-4 Comparison of the AASHTO, Unified, and FAA soil classification groups

AASHTO	Unified	FAA
A-1a	GW, GP, SW, GM	E-1
A-1b	SW, SP, SM, GC	E-1
A-3	SP	E-1, E-2
A-2-4	CL, ML	E-1, E-2, E-4
A-2-5	CL, ML, CH, MH	
A-2-6	CL, ML	
A-2-7	CL, ML, CH, MH	
A-4	CL, ML	E-5, E-6
A-5	CL, ML, CH, MH	E-9
A-6	CL, ML	E-5, E-7, E-8, E-10, E-11, E-12
A-7	CL, ML, CH, MH	E-7, E-8, E-9, E-10, E-11, E-12
A-8	Peat and muck or organic	E-13 (also peat and/or muck)

system is used and less than 12 percent passes the No. 200 sieve, enough additional sieves must be used to plot a reasonable grain-size curve.

2. Determination of the liquid and plastic limits.
3. Applying the process of elimination to classify the soil using the grain size and plasticity data.

Table 4-4 compares the soil classifications of the three systems used in this chapter, and it is readily seen that the systems overlap considerably.

Note that there is not complete agreement on the particle size division between gravel, sand, silt, and clay. The following division, however, seems to be the most widely accepted and is satisfactory for geotechnical work.

Particle size	4.76 mm		0.74 mm		0.002 mm	
Sieve No.	4		200		—	
	Gravel		Sand		Silt	Clay

These arbitrary definitions are based on particle size, and it is possible for clay mineral particles, or platelets, to be slightly larger than 0.002 mm. Likewise, it is possible for particle sizes of 0.002 mm or less to be rock flour or colloids instead of clay minerals, as further discussed in Chap. 6.

Another problem in soil classification occurs between countries (refer to Table 3-2). The following stacks of sieves might be used for grain-size analysis in the United States and Great Britain:

United States			Great Britain		
No. 4	4.76 mm	← sand and gravel division →	No. 7	2.411 mm	
10	2.00		14	1.204	
20	0.841		25	0.599	
40	0.420	← for Atterberg limits →	36	0.422	
60	0.250		72	0.211	
100	0.149		100	0.152	
200	0.074		200	0.076	
Pan			Pan		

These typical sieve stacks also illustrate the small differences between the two countries for the sand and gravel size division and the sieve numbers used to obtain the soil fraction for determining the Atterberg limits.

HOMEWORK PROBLEMS

Use the following data for Probs. 4-1 through 4-4.

In all classification problems, in addition to the group classification symbols, give a description of the soil (sandy clay, sandy clay with trace of gravel, etc.) as appropriate for the sieve analysis data.

	Percent passing				
	Soil				
Sieve	A	B	C	D	E
4	48	72	51	97	—
10	37	68	43	93	98
40	25	41	32	85	94
100	19	37	22	76	75
200	13	25	4	53	61
w_L, percent	32.6	41.3	NP	53.4	48.3
w_P, percent	21.5	22.3	—	31.6	23.1
Visual:	Dark brown	Dark tan	Light brown	Dark grey with woody odor	Reddish brown

4-1 Classify soils A, B, D, and E from the above group using the Unified Soil Classification system. Do not plot a grain-size curve unless necessary.

4-2 Classify the assigned soils from the above group using the AASHTO classification system.

4-3 Classify soils A, C, and E using the FAA soil classification system. Assume a frost depth of 0.5 m. If more than 20 percent passes the No. 200 sieve, assume "poor" drainage characteristics.

4-4 Plot a grain-size distribution curve for soil C, compute C_U and C_C, and classify in the Unified Soil Classification system.

4-5 Classify soils A and B of Problem 3-2 using both the Unified and AASHTO systems.

4-6 Classify soils A and B of Problem 3-6 using both the Unified and AASHTO systems. Take $w_L = 32.6$ and 46.5 percent, respectively, and $w_P = 16.3$ and 26.5 percent for soils A and B, respectively.

4-7 Redo Prob. 4-6 above for the FAA system but omit the subgrade classification.

Chapter 5

Geological Properties, Formations of Natural Soil Deposits, and Ground Water

5-1 INTRODUCTION

Soil may be defined for engineering purposes as "the unconsolidated material above solid rock." Within this, we may particularly distinguish *topsoil*, the top 0.01 to 0.5 m of unconsolidated material, which contains organic materials and plant nutrients and which supports plant life. Topsoil is of particular interest to the agricultural engineer.

Soil is the deposited byproducts from weathering of the rock crust and/or rocks suspended, or exposed, in the soil matrix. Since the unconsolidated soil material constitutes such a large part of the earth's surface, both on the continents and beneath the oceans, lakes, and other water-covered areas, few engineering projects except perhaps rock tunneling operations can be conducted without encountering some type of soil. As it is generally impractical to carry building foundations to rock below the soil mantle, the founding of foundation structures in or on soil is one of the most important aspects of geotechnical engineering.

This chapter will be concerned with the geological aspects of soil formation in terms of landforms, soil formation, and groundwater. The reader is encouraged to supplement the following material by obtaining a textbook of geology, preferably one oriented to engineering geology, and carefully studying it. In any case the reader should make a habit of observing exposed soil and rock formations when traveling highways, when hiking, when on surveying projects, or whenever there are other visual opportunities.

This chapter will present a brief introduction to the geology of rocks, including some additional detail of the several rock groups. This will be followed by a discussion of soil formation via rock weathering. The geological factors in the formation of the several types of soil deposits will be considered, and finally, a discussion of groundwater will be presented.

5-2 THE EARTH

According to generally accepted theories, the earth was formed about 4.5 billion years ago from a huge molten ball of cosmic gases and debris. The cooling of this mass formed the *atmosphere*, *hydrosphere*, and *lithosphere*. The atmosphere is the gaseous envelope surrounding the hydrosphere, or zone of water (as in the ocean basins and lakes), and the lithosphere, or earth's crust and interior mass.

The earth's crust consists of both rock and weathered rock (as soil) and is generally believed to extend downward 10 to 15 kilometers (km) or more, as shown in Fig. 5-1. Figure 5-1 also illustrates several other facts generally accepted by geologists.

The principal elements which make up the earth's *outer crust* are approximately as follows:

Element	Symbol	Percent by weight	Percent by volume
Oxygen	O	46.6	93.8
Silicon	Si	27.7	0.9
Aluminum	Al	8.1	0.5
Iron	Fe	5.0	0.4
Magnesium	Mg	2.1	0.3
Calcium	Ca	3.6	1.0
Sodium	Na	2.8	1.3
Potassium	K	2.6	1.8

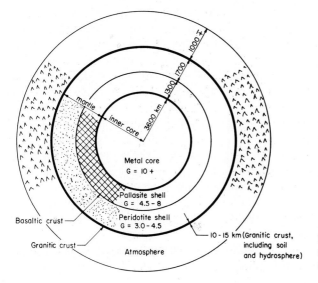

Figure 5-1 The earth, including atmosphere, with approximate dimensions.

These elements seldom exist alone but rather occur in combination, as *minerals*. The principal minerals which tend to be exposed to weathering to produce soil at or near the earth's surface are:

Mineral	Approximate percentage

Feldspar
  Orthoclase [K(Al)Si$_3$O$_8$]—pink, white, and grey-to-green
  Plagioclase [Na(Al)Si$_3$O$_8$]—white, grey, green, and red; and
    may contain Ca instead of Na
	30

Quartz (SiO$_2$, or silicon dioxide)	28

Clay minerals (see also Chap. 6) and micas
  Muscovite [K(Al$_2$)Si$_3$Al(O$_{10}$)(OH)$_2$]—light-colored mineral
  Biotite [K$_2$(Mg, Fe)$_6$(SiAl)$_8$O$_{20}$(OH)$_4$]—black, brown, or green color
	18

Calcite (as CaCO$_3$) or dolomite [as CaMg(CO$_3$)$_2$]	9

Iron oxides
  Hematite (Fe$_2$O$_3$)—red shades
  Limonite (2Fe$_2$O$_3 \cdot$ 3H$_2$O)—varying shades of yellow
	4

Pyroxene and amphibole
  Pyroxene—Calcium, magnesium, iron, and aluminum silicate
  Amphibole (hornblende)—sodium, calcium, magnesium, iron, and
    aluminum silicate
	1

Others including
  Kaolinite (clay)—hydrous aluminum silicate [Al$_2$Si$_2$O$_5$(OH)$_4$] as a
    principal weathering byproduct of feldspar
  Olivine (greenish color)—magnesium, iron silicate [(MgFe)$_2$SiO$_4$]
	10

5-3 PHYSICAL PROPERTIES OF MINERALS

The physical properties especially useful in mineral identification are:

Hardness—what materials a mineral will scratch and what materials will in turn scratch it

Color—green, white, colorless, grey, etc.

Streak—the color of the line of mineral powder formed when the surface is scratched with a hard object

Luster—the appearance of a freshly broken surface as seen in reflected light (bright, greasy, shiny, metallic, dull, etc.)

Specific gravity—related to the weight of a quantity of mineral

Cleavage—breaking along defined planes

Fracture—breaking along irregular fracture lines

The Mohs' hardness scale is used as a basis for evaluating the hardness of minerals as follows, in order of increasing hardness:

1. Talc (softest)
2. Gypsum
3. Calcite
4. Fluorite
5. Apatite

6. Feldspar
7. Quartz
8. Topaz
9. Corundum
10. Diamond (hardest)

Any mineral in the hardness scale will scratch the minerals below it, i.e., diamond will scratch all nine minerals below it. Hardness kits are available containing small specimens of the ten minerals in the Mohs' hardness scale. In lieu of the kit the following may be used:

	Hardness
Fingernail	$2\frac{1}{2}$ (will scratch talc and gypsum)
Copper penny	3
Glass	$5-5\frac{1}{2}$ (scratch apatite to talc)
Knife blade	$5\frac{1}{2}-6$ (may scratch feldspar)
Steel file	$6\frac{1}{2}-7$ (scratch feldspar to talc)

Figure 5-2 illustrates several of the more common minerals.

(a)

(b)

(c)

Figure 5-2 Several of the more commonly occurring minerals. Paper clip scale is 2.3 cm. (*a*) Orthoclase (pink) feldspar; (*b*) plagioclase (white) feldspar; (*c*) quartz—the white piece on the left is from a granite intrusion with the width shown; the piece on the right is pink, found in a stream bed.

5-4 THE ROCK AND SOIL CYCLE

Geologists classify all rocks into three basic groups: *igneous, sedimentary,* and *metamorphic.* Rocks are mixtures of several minerals or compounds, and vary greatly in composition. Limestone, for example, is primarily calcite, whereas granite contains feldspars, quartz, and varying amounts of ferromagnesians.

The approximately 1 billion years of documented geologic history (Table 5-1) indicates that the earth is continually changing. Weathering processes aided by crustal deformities (hills, valleys, etc.) reduce the solid rock(s) to fragments, creating *residual* soils, or in-place products of rock weathering. Initially the weathering process was applied to igneous rocks and/or deposits of mineral precipitates formed during the cooling of the molten rock. Gravity through sliding and creep, moving water as surface runoff, or wind and ice action may transport these weathered rock byproducts to new locations, producing sediments, or *transported* soil deposits.

These sedimented deposits through geologic time became indurated by consolidation due to the weight of the overlying sediments and/or cementation into *sedimentary* rocks. Much of the sedimentation took place in marine environments, so that considerable calcium, sodium, and magnesium salts (as carbonates, sulfates, chlorides, etc.) were present both as solution precipitates and from shell life which provided both sediments and cementing agents. Uplifts and other crustal movements either allowed additional sedimentation and induration pressures or exposed the sediments, along with underlying igneous and sedimentary rocks, to new weathering.

Where crustal movements caused increased overburden pressures and heat via energy dissipation and via cracks in the crust which allowed molten magma to flow close, some of the sedimentary (and some igneous) rocks metamorphosed into *metamorphic* rocks. Later crustal movements have exposed some of these rocks to renewed weathering and in some cases, at sufficient depth and geologic conditions, have returned them to molten magma to start the cycle anew. Figure 5-3 illustrates the rock-soil cycle just described.

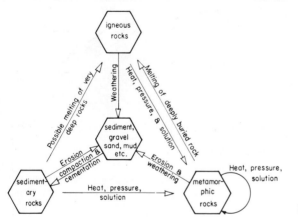

Figure 5-3 The rock-soil cycle.

Table 5-1 Geological time scale

Relative geologic time

Era	Period		Epoch	Time B.P.*	Type of life
Cenozoic	Quaternary — Neogene		Holocene		Man
			Pleistocene	2–3	
			Pliocene	12	
	Tertiary — Paleogene		Miocene	26	
			Oligocene	37–38	
			Eocene	53–54	
			Paleocene	65	
Mesozoic	Cretaceous		Late / Early	136	Mastodons
	Jurassic		Late / Middle / Early	190–195	Mammals
	Triassic		Late / Middle / Early	225	Dinosaurs
Palezoic	Permian		Late / Early	280	
	Carbon-iferous	Pennsylvanian	Late / Middle / Early		Coal-forming Swamps
		Mississippian	Late / Early	345	
	Devonian		Late / Middle / Early	395	
	Silurian		Late / Middle / Early	430–440	
	Ordovician		Late / Middle / Early	500	Fish
	Cambrian		Late / Middle / Early	570	
Precambrian				3600+	

* Estimated time *before* the *present* (B.P.), millions of years.

The earth's crust consists of approximately 95 percent igneous rock and only about 5 percent sedimentary and metamorphic rocks. However, of the rocks exposed to weathering at the surface, 75 percent are sedimentary rocks, and of these some 22 percent consist of limestones and dolomites. In order of area covered, the most important rocks (ranked in order of potential geotechnical problems) and their percentages are as follows:

1. Shales	52 percent	4. Granites	15 percent
2. Limestones and dolomites	7 percent	5. Basalt	3 percent
3. Sandstones	15 percent	6. All other rocks	8 percent

Available geologic evidence indicates that the sedimentary record is on the order of 5000 to 6000 m in depth. That is, sufficient weathering has taken place to place a depth of sediments of this thickness over much of the earth's surface. Had uplift and other crustal movements not taken place, this depth would have reduced the earth's surface to such an extent that a sheet of water would cover the entire surface. Much of the early sediments has long ago reformed into sedimentary rocks, so that the unconsolidated material is of much lesser thickness, generally well under 600 m.

5-5 IGNEOUS ROCKS

Igneous rocks are those rocks formed by the cooling of molten magma. Most of the magma now existing is at a considerable depth below the earth's crust, as qualitatively illustrated in Fig. 5-1, except in the active volcanic areas such as Yellowstone National Park, Wyoming; Hawaii; and Japan. As periodic stress adjustments produce cracks and faults in the rock crust, magma may find a path either part way (producing hot springs and geysers under certain conditions) or in some cases all the way to the surface (producing volcanos). Part way flows into the crust form intrusive or plutonic rocks as illustrated in Fig. 5-4; this figure also

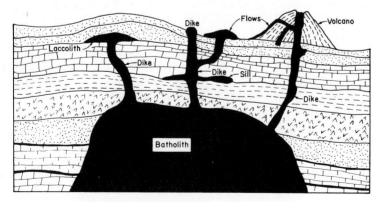

Figure 5-4 Igneous rock intrusions and extrusions.

gives the geological terms associated with these types of rock intrusions. Figure 5-5 illustrates two cases of later erosion exposing these rock intrusions.

Igneous rocks are classified according to texture, composition, color, and source. Several igneous rocks are:

Coarse-grained	Fine-grained	Lava rocks
Granite—light colored Diorite—intermediate color Gabbro—dark colored	Rhyolite—light colored Basalt—dark colored	Obsidian—black and glassy Pumice—light, frothy, and glassy Scoria—reddish to black with large voids

Figure 5-6 illustrates several of the more commonly occurring igneous rocks. Granite tends to be the predominant igneous rock; it is best known for its use as a building and monument stone. It ranges from greyish white to medium red, is rich in quartz, and tends to be intermixed with darker grains of mica and hornblende. A large range of grain sizes exists from relatively fine to very coarse-grained visible mineral crystals in the rock matrix. If the rock is very coarse-grained, it may also be called a porphyry, and if the crystals are abnormally large, the rock is termed a pegmatite. Rhyolite is essentially a granite with a fine-grained structure. The rate of magma cooling determines the size of the crystal structure—coarse is due to slow cooling and fine to rapid cooling.

Extrusive igneous rocks are formed when the molten rock hardens after reaching the surface. The most common extrusives are formed during volcanic eruptions, which, in addition to liquid lava, eject solid particles of volcanic ash and rock fragments termed *bombs*. The crystal structure of extrusive rocks tends to be fine-grained due to the rapid cooling. Some volcanic rocks may be quite porous (pumice and scoria) because they solidified while steam and other gases were still bubbling out. Figure 5-6 also illustrates samples of several extrusive igneous rocks.

Basalt, a fine-grained, dark-colored mineral aggregation often called trap rock, is one of the most abundant extrusive rocks. Basalts are rich in ferromagnesium minerals and are typically dark grey, dark green, brown, or black. It is a very hard, fine-grained rock, and if crushed is excellent for road construction.

Obsidian is a lustrous, glassy black to reddish brown, extremely fine-grained (there is actually no visible grain structure) rock formed by rapid cooling of molten lava. Pumice is a porous, light-colored rock with such a low mass that it may float on water. It is primarily glass and is formed as frothy lava is thrown into the air during a volcanic eruption.

5-6 SEDIMENTARY ROCKS

Rocks exposed through the earth's soil mantle are especially vulnerable to the agents of weathering. Weathering reduces the rock mass to fragmented particles which can be more easily transported by wind, water, and ice. When dropped by

(a)

(b)

Figure 5-5 Igneous rock intrusions which have been exposed by erosion (*a*) Dike on the order of 15 m wide by about 10 km in length. (*b*) Stone Mountain, Ga.; large granite laccolith (light colored).

Figure 5-6 Several of the more common igneous rocks. Paper clip scale is 2.3 cm. (*a*) Light grey granite from Stone Mountain, Ga.; (*b*) black fine-grained basalt; (*c*) granites (porphyrys)—upper is red, left is pink, lower right is mottled; (*d*) rock with seams of obsidian; (*e*) lava rock. Note coarse-grained, porous structure.

the agents of transportation, they are termed *sediments*. Sediments are typically deposited in layers or beds termed *strata*, and when compacted and cemented together (a process called *lithification*), they form sedimentary rocks. These rocks, of which the most common are shale, sandstone, and limestone, make up about 75 percent of the rocks exposed at the earth's surface.

Sedimentary rocks are generally classified as *clastic* or *chemical*. Clastic rocks are formed from rock grains of varying size. Typical clastic rocks include:

Shale. This is the most abundant of the sedimentary rocks. It is formed from silts and clays which have hardened into rock, with the principal induration agent being pressure. Shale may be *arenaceous*, with large amounts of sand; *argillaceous*, with large amounts of clay; *carbonaceous*, with large amounts of organic matter; or *calcareous*, with large amounts of lime as from shell life. Calcareous shale is used in the manufacture of portland cement, and carbonaceous shale may yield petroleum or coal. Shale may also be called claystone or siltstone based on the primary constituents.

Sandstone. This rock is composed essentially of pressure-cemented grains of sand (quartz). Sandstone may also contain grains of calcite, gypsum, feldspar, or iron compounds. Sandstone is used as an abrasive, as a building stone, and, when composed mostly of quartz, for glass making. A widely used sandstone is the St. Peter sandstone found from Minnesota through Wisconsin and underlying most of the state of Illinois. This sandstone varies from a few meters to more than 200 m thick but is commonly 30 to 60 m thick. It is widely found at subsurface depths of 50 to 100 m. It is believed to be a marine sand deposited during the Ordovician period some 400 million years B.P., when much of the central U.S. was, or was being, covered with a sea. Outcrops occur at Ottawa, Illinois, along the Illinois River, and at several other locations in the state. This sandstone is a major groundwater aquifer. It has considerable commercial value, being almost pure quartz, and the outcrops are extensively mined. It is very porous and loosely cemented, and often can be crushed by hand. It is quarried by blasting and using hydraulic washing to break the grains apart and to remove the few impurities coating the grains, such as iron oxide. It is the widely used "Ottawa sand" standard for civil engineering testing laboratories. It is also widely used for glass making and as molds for metal castings. Figure 5-7 shows an outcrop of St. Peter sandstone from near Ottawa, Illinois. Figure 2-8 shows grains of Ottawa sand compared with several other sands, and it can be seen that the sand is particularly well rounded; the light color indicates nearly pure quartz.

Conglomerate. This is a rock composed of cemented pebbles intermixed with sand. If the grains are angular, the rock is termed *breccia*; if it is formed from glacial deposits, it may be called *tillite*.

Chemical sedimentary rocks include:

Figure 5-7 Outcrop of St. Peter (Ottawa sand) sandstone near Ottawa, Ill. Note the extreme weathering that has taken place. The near white color is due to the sand being nearly pure quartz (silica).

Limestone. This is a chemical sediment consisting primarily of calcite (calcium carbonate, $CaCO_3$). There are several varieties of limestone depending on the makeup and physical appearance, as containing shells, fossils, sand, etc. Limestone quickly reacts with dilute (say 0.1 N) hydrochloric acid. The acid reaction may be used as a principal identification test for limestone. Limestone may contain silica precipitates or nodules of flint (dark color) or chert (light color).

Dolomite. This is limestone in which some of the calcite has been replaced with magnesium $[CaMg(CO_3)_2]$. Dolomite is very similar to limestone, and due to this similarity the only reliable determination, for any but the more experienced geologist, is the acid reaction test, since for dolomite the reaction is slow to nonexistent with dilute hydrochloric acid. Both limestone and dolomite tend to have the same grain structure and color; colors range from white to very dark grey, including greens, yellows, etc., depending on mineral impurities.

Evaporites. These are sedimentary rocks produced by minerals precipitated from sea water. They include gypsum ($CaSO_4 \cdot 2H_2O$), anhydrite ($CaSO_4$), and rock salt (NaCl and $CaCl_2$). Travertine limestone is a porous calcite precipitate from fresh water.

Biochemical or organic sedimentary rocks include:

Coquina. This is limestone containing shells and shell fragments (also called fossil limestone)
Reef limestone. This is limestone containing coral fragments
Chalk. This is limestone consisting of calcareous shells of microorganisms
Coral. This is marine limestone formed from the skeletons of marine invertebrate animals
Coal. The carbonized plant remains; various stages include:

1. Peat—decaying and semicompact organic matter
2. Lignite—second stage, more compact, may be called brown coal
3. Bituminous (soft coal)
4. Anthracite (hard coal and final stage)

Sedimentary rocks are generally characterized by stratification as in Fig. 5-8. The typical texture of several rock samples is shown in Fig. 5-9.

Ripple marks, indicating sedimentation under water or from wind deposits, are often found in sedimentary rocks. Mud cracks, which occur when a mud dries and shrinks, may be found also. If the cracks fill with sediments during a sudden rain before the mud adsorbs water and swells the crack closed, the sediments in the cracks will later form identifying inclusions in the rock. Many sedimentary rocks are brilliantly colored, as in the Grand Canyon, the Painted Desert, and Yellowstone Park. Color is produced by the weathering of hematite, limonite, and manganese compounds contained in the rocks. Some sedimentary rocks contain fossils which are used by the geologist in dating the rock, i.e., since certain types of shell life existed at certain times, a rock containing a shell could only have sedimented during the period of existence of that particular shell.

5-7 METAMORPHIC ROCKS

Metamorphism through high temperatures and pressures acting on either sedimentary or—less commonly—igneous rocks that have been buried deep in the earth produces metamorphic rocks. During the process of metamorphism the original rock undergoes both chemical and physical alterations which change the texture, as well as the mineral and chemical composition.

The rearrangement of minerals during metamorphism results in two basic rock textures: *foliated* and *nonfoliated.* Foliation results in the rock minerals becoming flattened or platy and arranged in parallel bands or layers. Typical foliated rocks include:

Slate. This is metamorphosed shale, characterized by a very fine texture, splitting into thin slabs; typically colors are grey, black, reds, and greens. Slate is widely used for roofing, blackboards, sidewalks, and pool tables.

(a)

Coal seam

Coal seam

(b)

(c)

Figure 5-8 Representative stratification of sedimentary rocks. (Refer also to Figs. 6-15 and 6-16.) (*a*) Stratified limestone. Note that the upper layers are dark grey and the lower layers light grey. Stains are from wet weather springs leaching iron oxides. (*b*) Stratified shale, limestone, and coal. Geologist's hammer rests against a coal seam approximately 40 cm thick. Ridges about 1.5 m above the lower coal seam and above the upper seam are limestone layers. Other material is shale. (*c*) Stratified limestone which is well weathered after less than 10 years.

Figure 5-9 Several of the more common sedimentary rocks. Paper clip size scale is 2.3 cm. (*a*) Banded reddish brown sandstone; (*b*) sample of St. Peter sandstone with iron oxide staining; (*c*) conglomerate from near Peoria, Ill.; (*d*) fossilized limestone (contains fossil shells); (*e*) shales—left is clayey, right is sandy; (*f*) limestone conglomerate; (*g*) limestones; the one on the right is very dark grey, the one in the middle is light grey and the one on the left is travertine limestone—note the extreme porosity.

Schist. Schist is a medium- to coarse-grained rock containing considerable mica. Although they are commonly formed from shale, schists may also be formed from igneous rock. *Mica schists* are rock with mica as the predominating mineral; chlorite schist has the mineral chlorite, etc. Figures 5-10*d* and 5-16*d* illustrate mica schist.

Gneiss. Gneiss is a highly metamorphosed (generally from granite) coarse-grained and banded rock. The rock is characterized by alternating bands of darker minerals, such as chlorite, biotite, mica, and graphite. The bands are typically folded and contorted and may resemble schists, but cleavage is very difficult, whereas with schist, slab separations may sometimes be effected with a knife blade.

Nonfoliated metamorphic rocks include:

Quartzite. This is metamorphosed quartz sandstone. It is one of the most resistant of all rocks. When formed of pure quartz, the rock is white; impurities may give red, yellow, or brown tints.

Marble. Marble is metamorphosed limestone or dolomite. Marble may be white when pure, but impurities cause a wide range of colors and tints. Marble is commonly used for building stone and monuments.

Anthracite. This is metamorphosed bituminous or soft coal.

Figure 5-10 illustrates the typical physical features of several of the more common metamorphic rocks.

Figure 5-10 Several common metamorphic rocks. Paper clip scale is 2.3 cm. (*a*) Pink quartzite with surface worn somewhat smooth; (*b*) yellow quartzite freshly fractured; (*c*) gneiss; (*d*) dark green mica schist; (*e*) Slate; note two thin pieces on the edge in the upper left; (*f*) white marble; note the fine-grained texture.

5-8 CRUSTAL MOVEMENTS

The crust of the earth has undergone considerable structural change during past periods of earth history. Geologic evidence indicates that large land areas of all the continents have been covered periodically by shallow seas. This evidence has been obtained from study of the fossils found in both sediments and exposed rocks. Figure 5-11 illustrates the approximate outline of the present land area of the North American continent which was covered by the sea at some time between the Cambrian to about the Pliocene periods (570 to 12 million years B.P.).

Geological evidence indicates that the Appalachian Mountains were formed and reformed during the Paleozoic and approximately up to the Cenozoic era (225 to about 63 million years B.P.). Considerable erosion has taken place, and subsequent uplift has formed the Piedmont plateau and moved the sediments overlying the coastal plain outward onto the continental shelf eastward into and under the Atlantic Ocean. The Rocky Mountains are of somewhat later origin (believed to be from the late Mesozoic or early Cenozoic era). Large mountain masses are found on all the continents. Figure 5-12 is an approximate surface outline of the United States which illustrates the locations of the several crustal movements locating the present mountains.

Crustal movements produce structural deformities termed folds, faults, and joints, as identified in Figs. 5-13 and 5-14. A *syncline* bends the rock layers into a concave shape upward; the *anticline* is convex upward. A *geosyncline* is an areal depression, often adjacent to a mountain, which fills with sediments and volcanic debris and later may be uplifted as a mountain. A *monocline* is a single fold, and it should be noted that both anticlines and synclines may be adjacent depending on

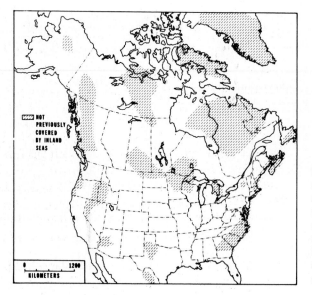

Figure 5-11 Map of North America showing approximate zones not covered by the sea at some time in the geological past. Zones once covered by the seas and now uplifted are characterized by sedimentary rocks, particularly limestones, sandstones, and large quantities of shale.

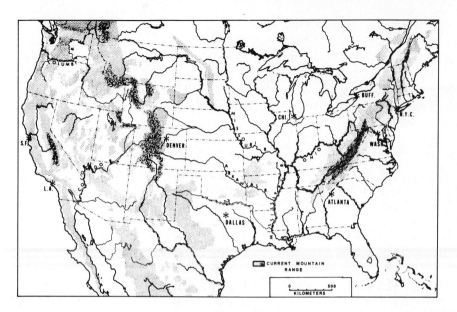

Figure 5-12 Outline of current mountain ranges in the United States. In mountainous areas outcrops of igneous rocks are likely to be found. Also, these areas are characterized by rock faults and fractures.

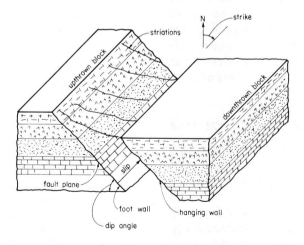

Upthrown block appears to have moved upward.
Downthrown block appears to have moved downward.
If hanging wall is on the upthrown side we have a reverse fault.
If hanging wall is as above we have a normal fault.

Figure 5-13 Fault elements. Only a fresh fault might appear as above. Old faults will have erosion across the zone so that only a gradual difference in elevation might appear (if any).

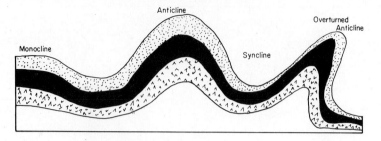

Figure 5-14 Crustal folding and some of the terms used to describe the configurations.

the amount of folding. Figure 5-15 illustrates field observations of both small and large folding of strata.

Terms used to orient the geometry of inclined and folded rock beds include *strike*, the angle of the bed axis from compass north, and *dip*, the angle of the bed axis from the horizontal plane measured at right angles to the direction of the strike. Both of these terms are illustrated in Fig. 5-13.

When the stresses within the rock crust exceed the ultimate strength of the rock, fracturing occurs. If very little movement occurs along the fracture zone, the fracture is termed a *joint*. *Normal faults* occur when movements have taken place along the fracture in the vertical direction, as indicated in Fig. 5-13, with the hanging wall over the fracture and on the downthrown block. *Strike-slips* occur when the fault movements are lateral. Many faults are slip faults, where both vertical and lateral movements occur. In east Africa there is a slip fault with a fracture zone some 6000 km in length. The Beartooth Mountains of Montana and Wyoming are partially due to a fault block 64 by 128 km raised vertically some 1000 m. The term *scarp* is used to describe the edge of the "cliff" formed by the abrupt difference in elevation of this type of fault. The San Andreas fault in California, beginning near the Salton Sea near the Mexican border and running along the coast northwesterly some 960 km to Point Arena in northern California, where it appears to enter the Pacific Ocean, is a strike-slip fault since the relative movement is primarily parallel to the fracture zone (and ground surface). The cracked and fractured rock zone extends several kilometers on each side of this fault. Varying opinions place relative displacements of the San Andreas fault as from 10 to 540 km laterally. A maximum relative lateral movement, on the order of 7 m, appears to have occurred during the 1906 earthquake, with some earth movements all along approximately 400 km of the fault line. Faults of this type are worldwide, including the Great Glenn fault in Scotland, the Alpine fault in New Zealand, and the Dead Sea Rift. A fault movement in the Cayman Trough south of Cuba in the Caribbean Sea running roughly east-west on the order of 2000 km to Guatemala is believed to be the cause of the great earthquake in Guatemala, Central America, on February 4, 1976, which killed more than 20,000 persons, injured some 77,000, and left more than 1 million homeless.

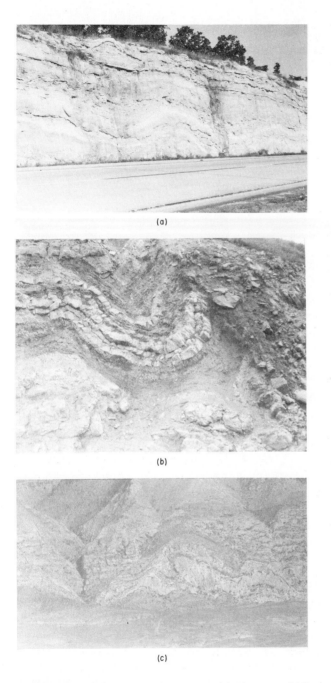

(a)

(b)

(c)

Figure 5-15 Examples of stratum folding. (*a*) Limestone folding near the Ozarks in Missouri; (*b*) limestone and shale folding in Montana; (*c*) folding on a large scale in the Big Horn Mountains.

Figure 5-16 Earth discontinuities. (a) and (b) Large faults in Montana and Wyoming. Note particularly in (a) the pronounced scarp line. (c) A joint in a limestone formation; (d) Jointing through a micaceous schist (dark band) underlying a granite formation in the Black Hills of South Dakota.

Lesser-known faults may be the most trouble for geotechnical engineers, since the great faults are more commonly known' and usually there is more surface evidence of their existence. Lesser faulting is extremely widespread; for example, both northern and southern Illinois, southeastern Missouri, Tennessee, Pennsylvania, northern Georgia, and other locations in the United States contain small to medium-sized fault zones. If the relative movement is small and/or took place so far in the past that erosion has removed any possible surface evidence, the only way to detect a fault may be a careful study of borings *taken into the bedrock*. In any case, a fault is an existing structural weakness that increases the probability of crustal movements in these zones as crust stresses build up over a period of time. Earthquakes are produced when the stresses become too large and relative rock mass movements suddenly occur. Figure 5-16 illustrates two large local faults in the Beartooth mountains and two cases of localized rock jointing.

From a geological viewpoint fault zones are a mixed blessing, since many valuable ore deposits are found as precipitates along the fault line.

5-9 ROCK WEATHERING AND SOIL FORMATION

Weathering of rocks is one of the most important of all geologic processes. It provides the material from which sedimentary rocks are formed and produces soil, without which both plant and animal life on earth would be impossible. Rock fragments produced by weathering are removed by *erosion*. Weathering may be either mechanical (or physical) or chemical.

A Mechanical Weathering

Mechanical weathering takes place when rock is reduced to smaller fragments without any chemical change taking place. Rock weathering is very much type-of-rock and time dependent. It may be caused by any or all of the following factors acting for significant periods of time.

Climate effects (including both temperature and rainfall) These are probably the principal factors involved in rock disintegration. Daily temperature fluctuations may not be too important, but freeze-thaw cycles over a long period of time cause rock fatigue even in milder climates. Severe temperatures producing local freezing of short duration may be significant, since water in rock pores will increase in volume approximately 9 percent at 0°C and will exert tremendous pressures. As the freezing pressure will tend to extrude ice from the pores and reduce the expansion pressures, local effects will be greater when the temperature drops considerably below 0°. Differential temperatures, not necessarily below freezing, coupled with the different thermal coefficients of the constituent rock minerals can have a fatiguing effect and produce rock fragments. In fact, some believe that temperature effects are one of the most significant mechanical agents in the weathering process.

Exfoliation Exfoliation is the spalling off of the exterior surface of exposed rocks. Rocks underlying thick soil strata are under large compressive forces. Surface stress adjustments accompanying regional uplift, coupled with erosion from surface water runoff reducing the overburden stresses, cause the outer rock shell to separate (or spall) from the main rock. Again the different stress responses of the constituent rock minerals may accelerate the process. Exfoliation may also be caused by relatively sudden temperature changes, especially of igneous rocks.

Erosion by wind and rain This is a very important topography-dependent factor and a continuing event. Flowing water carrying tiny particles of rock in suspension can erode or abrade the most solid of rock over geological time periods. This is especially significant in areas of rugged topography where high velocities may be obtained, as in mountainous areas. This is evidenced by the fact that stones found in stream beds tend to be subangular to highly rounded. Extreme cases of erosion are the Grand Canyon of the Colorado River in Utah, Arizona, Nevada, and New Mexico and the Cheddar Gorge of the River Avon in the south of

(a)

(b)

Figure 5-17 Sedimentation and erosion. (*a*) Severe erosion of sedimented deposits laid some 25 to 35 million years ago on the "Badlands" of South Dakota. Deposits consist of sands, gravels, fossils, volcanic dust, and loess materials as well as silts and clays. (*b*) Closeup of one of the erosion faces of (*a*), showing typical sedimentations. The darker band near the top is volcanic dust. A thin black band is believed to be carbon contamination from a prairie fire. The deposits have been laid so long that many of the lower deposits have lithified to shale. The thin white band is limestone. This deposit extends about 150 m above the floor.

England. Lesser erosion models include Niagara Falls, where the Niagara River flows over a bed of Niagara limestone which is relatively hard but is underlain by shale and soft Clinton limestone, which has eroded away to form the falls lying between the United States and Canada. Large canyons or gorges are found, even with small streams, in the western United States, Canada, Australia, Africa, and elsewhere which display the eroding effects of water acting over geological time periods.

Figure 5-17a and b illustrates that erosion is not limited to rocks and that large areas can be involved, and that with sufficient erosion an area can be rendered uninhabitable.

Abrasion Strictly, abrasion is the wear caused when two hard materials undergo relative movement while in contact. This can be caused by one of the materials being suspended in water, as sand, for example, but in the context of this text the term will be used to describe the pushing of large quantities of soil or ice under pressure across the underlying rock by glaciers, grinding or abrading both materials to smaller sizes.

Organic activity Cracking forces exerted by growing plants and roots in voids and crevasses of rock can force fragments apart. Animals, such as insects and worms, burrowing into the ground may bring rock fragments to the surface or otherwise expose the fragments to additional weathering.

B Chemical Weathering

Chemical weathering involves alteration of the rock minerals into new compounds. It may include the following processes.

Oxidation. A chemical reaction may take place when rocks are in contact with rainwater. It is readily noticeable in rocks containing iron as the brown to red staining of the weathered surface. Oxidation has produced the stains on the rock surfaces of Figs. 5-8a and 5-15a and the bright colors (bandings) shown in Fig. 5-17. Reactions may yield hydrated iron oxides, carbonates, and sulfates. If these reactions result in an increased volume, there will be a subsequent disintegration of the rock.

Solution. Certain rocks, notably limestones, are partially to completely dissolved in rainwater, especially if the rainwater contains appreciable carbon dioxide in the form of weak carbonic acid or has a $pH < 7$. Even a very weak acid solution acting over geologic time periods can decompose many rocks. In the case of limestones, the reader may readily observe that over time periods of only 5 to 10 years there can be considerable weathering, as along highway cuts. Figures 5-8c, 5-15b, and 5-16c illustrate limestone deterioration after periods of less than 10 years.

Figure 5-18 Limestone sinkholes and karst topography. (*a*) Small sinkholes in pasture; (*b*) large sinkhole which has just formed with an ideal round shape (seldom obtained); (*c*) typical sinkhole area where erosion has somewhat smoothed out the initial steep sides of (*a* and *b*).

Caves are widely formed, as are limestone sinkholes (karst formations) in areas with many limestone formations and considerable rainfall. Figure 5-18 illustrates typical karst topography as found in parts of north central Kentucky and south central Indiana. Land sinking and subsequent erosion tend to produce the rolling topography shown in Fig. 5-18c rendering the land unsuitable for anything but grazing.

Leaching. Water reacting with the cementing material of sedimentary rocks may cause the particles to loosen, with the smaller particles and the cementing agents carried away either to deeper strata or as surface runoff. Cementing agents carried to deeper strata by percolating rainwater may be a factor in future formation of new sedimentary rocks. In areas of little rainfall, water vapor may carry the cementing agents such as sulfates, carbonates, etc., to the ground surface, creating a salt crust which may make the soil unfit to support plant life.

Hydrolysis (formation of H^+ ions). Chemical weathering agents may be acting simultaneously. Consider, for example, the formation of clay from the weathering of orthoclase (usually pink in color) feldspar in the presence of ordinary water and carbonic acid formed by water mixing with carbon dioxide:

$$
\underset{2 \text{ parts}}{2(K)AlSi_3O_8} + \underset{1 \text{ part}}{H_2CO_3} + \underset{1 \text{ part}}{H_2O} \rightarrow
$$

$$
\underset{\substack{1 \text{ part} \\ \text{Clay mineral}}}{Al_2Si_2O_5(OH)_4} + \underset{\substack{1 \text{ part} \\ \text{Potassium} \\ \text{carbonate}}}{K_2CO_3} + \underset{\substack{4 \text{ parts} \\ \text{Quartz}}}{4SiO_2} \qquad (5\text{-}1)
$$

In this case the H^+ ion from the water forces the K^+ ion out of the feldspar. The H^+ ion then combines with the aluminum silicate to form the clay mineral. A plant root in the soil may attract local soil-water and become surrounded by an excess of H^+ ions, which initiates the hydrolysis process. Any fragment of orthoclase feldspar in close proximity can be broken down to form the clay mineral, according to Eq. (5-1). The potassium carbonate may be further broken down and leached away, it may become plant food, or the clay mineral may attract the potassium ions to form *kaolinite* clay.

5-10 GENERAL CONSIDERATIONS IN ROCK WEATHERING

The rate of weathering depends on the particle size. Small particles weather, in general, at a faster rate than large ones due to their larger surface area. Type of material, climate, moisture, exposure conditions, and plant and animal/insect activity are important factors affecting the rate of weathering.

Most weathering takes place near the ground surface; however, exfoliation due to loss of overburden pressure may be taking place at a depth of many meters. Downward-percolating rainwater or underground water may be producing chemical weathering far below the ground surface without the effects ever being exposed.

Generally both the rate and amount of weathering increase with time due to both the reduction of rock size and more material being exposed to the process. The very important effects of rainfall and temperature are summarized here.

A Rainfall

Low rainfall areas. Water only penetrates the soil to a limited depth; weathering takes place, but the byproducts (carbonates, sulfates, etc.) are not removed from the soil profile and the resulting pH tends to be alkaline. Water tends to be removed by evaporation, which tends to concentrate calcium, sodium, and potassium salts in the surface zone of the soil profile.

High rainfall areas. Water percolates through the soil, and weathering material is removed by leaching. Soluble substances are removed, and clay tends to be dispersed in the lower soil profile. The pH of the soil tends to be acidic.

B Temperature Effects

High average temperatures in moist areas increase vegetation and chemical weathering. High average temperatures in arid areas decrease vegetation and chemical weathering, and mechanical weathering tends to dominate.

Low average temperatures in moist areas cause soil freezing and permafrost and slow weathering.

The geotechnical engineer will be particularly concerned with the high rate of weathering of exposed shale, and to a lesser extent sandstone. Limestone weathering takes place with considerable rapidity in the presence of water whether it is exposed or not.

5-11 SOIL FORMATIONS PRODUCED BY WEATHERING

Soil may be classified according to the method of formation of the deposit as residual soil or transported soil. A *residual* soil is one which was formed in its present location through weathering of the parent (or bed) rock. These soils are rather widespread in tropical areas, where they may be termed *laterites*, and in other less tropical areas where glaciers have not been present, as in the southeastern and southwestern parts of the United States, most of Australia, India, Africa, and southern Europe. Residual soil deposits vary from a few centimeters to 100 or more meters in depth depending on the geological age and weathering conditions. These soils are formed by weathering and leaching from the top downward of the water-soluble materials. As leaching action naturally diminishes with depth, the residual soil will be less and less altered until the parent rock is reached.

If a vertical cut is made in a residual soil, a horizontal arrangement of layers can sometimes be seen, especially in a fresh cut. The vertical section is a soil profile, and the individual layers are *soil horizons*. Figure 5-19 illustrates a simplified arrangement of the soil horizons for geotechnical engineering use. In general, the horizons are:

Horizon	Comments
A	Top zone consisting of topsoil and organic matter, and in humid areas, highly leached materials; in arid areas it may be rich in various water-soluble salts remaining as water vapor from the lower depths evaporates. It generally is highly weathered, dark-colored material including various shades of blacks and browns of a few centimeters to 1 or 2 m thick and grading into the *B* horizon.
B	Zone underlying the *A* horizon and containing considerable leached materials (water-soluble salts such as carbonates, sulfates, and chlorides) and clay minerals. This zone may be on the order of 0.5 to several meters thick and grades into the *C* horizon.
C	Transitional zone of freshly weathered parent material (rock); it may consist of considerable rock fragments. This zone may be absent or of very shallow depth and grades into the *D* horizon.
D	Parent (or bed) rock.

The soil from the *B* horizon is considered the best for borrow since it contains both granular material and binder. The *A* horizon contains too much organic material and too little binder to be of value as a construction material. The *C* horizon may be too open graded, or deficient in material passing the No. 100 sieve and clay sizes, for use as borrow, although it may be blended with *B* horizon material and made satisfactory.

Residual soils tend to be characterized by

1. Presence of minerals that have weathered from the parent rock
2. Particles tending to be angular to subangular as illustrated in Fig. 5-19*b* and *c* as compared with the rounded particles in the transported deposits of *d* and *e*
3. Large angular fragments of rock tending to be found dispersed throughout the mass, as in Fig. 5-19*c*

An important residual soil found in many mountainous areas is termed a *saprolite*. A saprolite is a condition of chemical rock weathering such that the rock tends to crumble but still retains the original structure and texture. A climatic condition of moderately heavy to heavy rainfall, such as in the southern Appalachian mountains, the Australian Alps and areas of western Australia, India, South America, and Hawaii, produces this type of soil. Saprolitic material has been reported to depths of 100 m (Carroll, 1970).

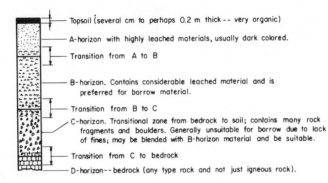

— Topsoil (several cm to perhaps 0.2 m thick -- very organic)

— A-horizon with highly leached materials, usually dark colored.

— Transition from A to B

— B-horizon. Contains considerable leached material and is preferred for borrow material.

— Transition from B to C

— C-horizon. Transitional zone from bedrock to soil; contains many rock fragments and boulders. Generally unsuitable for borrow due to lack of fines; may be blended with B-horizon material and be suitable.

— Transition from C to bedrock

— D-horizon -- bedrock (any type rock and not just igneous rock).

(a)

(b)

(c)

(d)

(e)

Figure 5-19 Soil horizons and several natural soil deposits. (*a*) Hypothetical soil horizon profile with several subdivisions; (*b*) residual soil in West Virginia (scale: ball point pen is 14 cm), (*c*) residual soil in Montana with 30-cm scale shown, (*d*) deposit in dry part of stream bed with 30-cm scale shown. (*e*) moraine deposit near Peoria, Ill., with 30-cm scale shown.

Transported soils were formed from rock weathering at one site and are now found at another site. The transporting agent may be

1. Water (principal transporting agent)
2. Glaciers
3. Wind
4. Gravity

Water, wind, and glacier deposits are very widespread. Often the deposits are given names indicative of the mode of transportation causing the deposit. Deposits formed by these several modes of transportation will be considered in more detail in the following sections.

A question naturally arises: What is the classification of the deposits in areas covered by marine deposits from several million years ago? While in the strictest sense these are transported deposits, the deposition took place so long ago that some to considerable lithification has since taken place. In present conditions the indurated soil is being weathered anew, producing a material that is more residual than transported. With these soils, however, the horizon concept may have little meaning due to the previous stratification, which may include sands, limestones, clay and silt layers or shales, etc. In these soils the value of a layer for construction purposes will depend on the properties of that stratum, and may depend very little on leaching from above or on the properties of the adjacent strata. Figure 5-17 illustrates this particularly well.

5-12 RUNNING WATER AND ALLUVIAL DEPOSITS

Once water has fallen on the land as precipitation, it follows one of the many paths making up the hydrological cycle (see Sec. 5-16). That portion following the path of *runoff* is of interest in this section as an agent causing *erosion* and *transportation*. Erosion and transportation depend on the velocity of the moving water, which in turn depends on the gradient, amount of water passing a point, and nature of the stream banks. In general, the gradient decreases from the headwaters (upper end) to the mouth. The mouth may terminate in another stream, a lake, or the ocean. Terminal velocity in lakes or the ocean will approach zero a short distance from shore based on considerations of continuity and the flow equation

$$Q = Av \tag{5-2}$$

where Q = discharge quantity
$\quad A$ = area of flowing water
$\quad v$ = velocity

Erosion is caused by friction of the flowing water, including the effects of any suspended material, on the flow channel. Since the eroded materials contain flaws and have varying degrees of resistance to erosion, no stream will be straight as seen from above, except for very short distances—generally less than 10 times the

(a)

(b)

Figure 5-20 Stream configurations. (*a*) Well-developed stream meanders, with oxbows shown. Also shown is a location (at the extreme right) where a cutoff will soon develop and a new oxbow form. This stream is about 7 m wide and is essentially a model but shows in a small, easily seen area many of the typical "older" stream features. (*b*) Well-developed stream valley. This stream displays a slough and stream braiding (the several branches of the stream in the center background).

width of the effective channel. The center of the channel, or principal flow, tends to swing back and forth, or *meander*, from side to side. Over geological time, this usually results in wide valleys cut between rock bluffs or banks, with the valley floor consisting of transported soils or sediments. Sedimentation occurs when the lowered velocity of the water on the inside of a meander will no longer support the suspended material. The outside of the meander, being of higher velocity, erodes into the bank, thus increasing the meander as illustrated in Fig. 5-20a. Flood stages may cause formation of *natural levees* parallel to the banks as the stream rises and overflows, and the velocity falls as the channel area suddenly enlarges. The reduced velocity causes transported materials to precipitate along the banks. Trees and shrub growth along the bank may considerably aid the formation of natural levees. The formation of natural levees may raise the river and the levees 3 to 5 m or more above the valley floor until a very large flood overtops the levee, with the resulting erosion cutting a new channel through the levee wall and into the valley floor. River terraces (Fig. 5-21b) may be formed when the stream cuts into a previously deposited sediment or as the stream bed is lowered over geological periods due to normal erosion or to crustal deformation.

Erosion of the neck of a meander may result in cutting the meander, leaving a curved and isolated channel or *oxbow* as in Fig. 5-20a. The oxbow may be a *slough* if one end remains open to the stream so that the backwater stands as in Fig. 5-20b. An *oxbow lake* is formed if the oxbow fills with water.

These several depressions may later fill with fine-grained sediments, muds, and organic material during and between subsequent valley flood stages, and the result is a particularly poor deposit of highly plastic (w_L often 60 to 100 or more) and/or organic silts, silty clays, clays, or peat. Soil exploration for foundation sites should proceed with caution to locate and identify these deposits.

The continuous slow change in channel position results in the entire valley floor consisting of *alluvium* or *sediments*. This continual reworking is gradually moving all the material downstream and reducing the stream gradient. The material downstream is, of course, progressively finer due to several factors, including more abrasion and the lower gradient with its resulting reduced velocity which can only support the finer weathered material. Since the valley floor is nearly flat and near the high water level of the stream, at flood stage the valley is essentially a flood plain and susceptible to widespread shallow flooding. These areas are poor building sites due to the periodic flooding unless the stream channel is confined by supplemental man-made levees. Buildings in a flood plain cannot obtain flood insurance in many areas; thus, sites near streams should always be checked for possible flooding.

Lake deposits are also called *lacustrine* deposits. *Varves* are a particular type of lake deposit formed during glacial periods from seasonal ice melting which temporarily increased the runoff velocity so that precipitated sand layers alternate with silt or silt-clay layers of precipitates made at low velocities.

A *marine* deposit is obtained when the sediment precipitates through salt water.

Deltas are sediments (Fig. 5-22) precipitated at the mouths of streams into

(a)

(b)

Figure 5-21 River bank formations. (*a*) Natural levee along the Wabash River; (*b*) well-developed stream terrace on Yellowstone River in Montana.

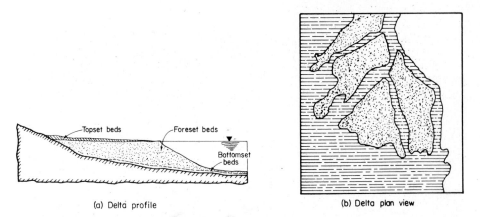

Figure 5-22 River or lake delta formation. Note that an alluvial fan will be similar in plan to the delta but will be a deposit at the exit from uplands onto a plain, valley, or desert flatland. (*a*) Delta profile; (*b*) delta plan view.

bays, oceans, or lakes. *Fans* are a similar type of deposit but found in arid areas where mountain stream runoff flows into wide valleys or onto the plain at the stream mouth.

Lake, marine, and delta deposits tend to be relatively fine-grained, with silt and clay sizes predominating. Most of these deposits will be loose and highly compressible. Organic material is sometimes present, as are seams of fine to medium-coarse sand. Some of these deposits are 75 to 150 m in thickness.

5-13 GLACIAL DEPOSITS

Glacial deposits form a very large group of transported soils. At various times a large part of the North American continent, as illustrated in Fig. 5-23, has been covered by glacial ice, as has northern Europe, including much of Germany, Poland, Northern Russia to the Ural Mountains, all of the Scandinavian countries, the British Isles, and Greenland. The approximate outline of the European glaciation is shown in Fig. 5-24. Greenland is still very nearly covered with glacial ice, as are parts of northern Canada and Alaska, most of Antarctica, the higher mountains of Scandinavia, the Swiss Alps, the Himalayas, and some of the Andes of South America; some small glaciers exist on the highest mountains in the United States. A survey by Flint (1970) showed that some 10 percent of the presently existing land surface was covered at one time by glaciers, amounting to some 15×10^6 km^2.

The eroding action of the glacial ice both scraped up soil at the interface of the ice and soil or rock and pulverized, crushed, and abraded the parent bedrock into silt, sand, and gravel-sized material. This could be accomplished due to the great depths, and resulting enormous pressure, of glacier ice. Thicknesses were on the order of:

Location	Approximate thickness, m
North America and Canada	300–2000
British Isles	350–900
Europe	500–3000

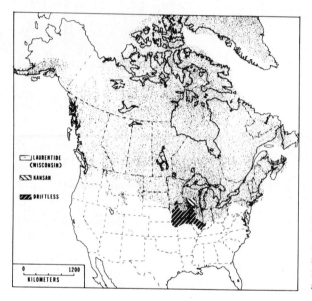

Figure 5-23 Approximate outline of glaciation on the North American continent. The most recent is the Laurentide (or Wisconsian) glaciation, which disappeared some 10 000 to 13 000 years ago. The Kansan glaciation covered about the same area but extended somewhat further south locally into Kansas, Missouri, and Illinois as shown.

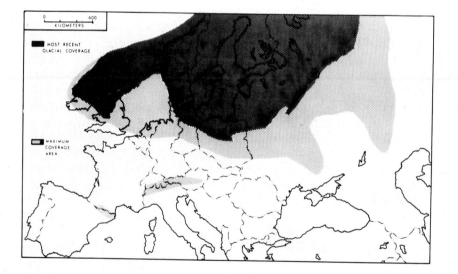

Figure 5-24 General location of European glaciation. The outline is only approximate but gives an indication of the amount of glaciation and about where to expect glacial deposits.

Figure 5-25 Photographs of several glacial features as noted. (*a*) Small mountain glacier; also shown are small valley moraines and glacially formed lakes. (*b*) Larger terminal moraine; note ridges rather than a single ridge. This moraine is about 0.5 km wide by about 5 km long. (*c*) Glacial topography confined to a small area; shown are the terminal moraine, a drumlin, and several erratics.

An ice thickness of 1000 m would create an ice pressure of some 8800 kPa on the underlying soil.

Soil deposits pushed into ridges around the periphery of the glacier are called *moraines. Terminal moraines* (Fig. 5-25*b*) are ridges of material scraped or bulldozed to the front of the glacier; *lateral moraines* develop along the sides. These formations are generally 0.5 to several kilometers wide, may be 25 to 100 or more meters high, and may be 60 to 100 km long. The moraine may not be a single nicely rolled ridge, but rather a highly serrated, above ground level, earth mass. There may be discontinuities in the ridge where glacial melt forms channels carrying outwash, and small lakes may temporarily form in depressions behind the ridge, producing lacustrine sediments. *Ground moraine* (also termed glacial till or simply till) was the deposit of ice-suspended material through the bottom of the glacier; it ranges from a few centimeters to 150 m or more in thickness. *Eskers* (Fig. 5-26) are ridges formed when water-suspended material flowing in ice tunnels precipitated; they vary from about 10 to 30 m high and are from about 0.5 to several kilometers in length. *Drumlins* are isolated mounds of glacial debris varying from about 10 to 70 m high and 200 to 800 m long. Most drumlins are on the order of 30 m or less in height and 300 m or less in length. They often occur in drumlin fields (several), as in Fig. 5-27. *Erratics* are large boulders picked up by glaciers, transported to a new location, and dropped, as in Fig. 5-25*c*.

Eskers, drumlins, and glacial outwash tend to have commercial value as sand or gravel sources, since the material often contains very little ($-$) No. 200 sieve material. To determine suitability as a sand or gravel source requires some soil exploration, since not all these formations are suitable. Sometimes localized areas of a lateral or terminal moraine may contain suitable borrow, as shown in Fig. 5-28, where a borrow location is established in the Shelbyville Moraine near Peoria, Illinois. Note the characteristic ridge profile and the grading which may be obtained in these deposits.

Melting glacial ice formed streams flowing away from the glacier which carried fine sand, silt, and clay material to lakes to form varves or downstream as fluvial sediments as the ice melting increased or decreased with the seasons and the stream velocity fluctuated. As the glaciers melted, material suspended in the ice precipitated onto the underlying soil or rock to form glacial till. Till deposits are characterized by containing all sizes of particles with no obvious arrangement.

Figure 5-26 Two eskers. The esker on the left is some 6 km in length and in places 27 m high. It is North Dakota and is one of the largest ones ever reported. The esker on the right is about 2 km long, but higher points are on the order of 30 m.

Figure 5-27 Drumlins. The field on the left is near Sauk Centre, Minn.; the field on the right is from Wyoming. One can distinguish a drumlin from an eroded rock outcrop because the interior rocks and gravel will be rounded due to glacial abrasion.

One analysis stated that the till around Boston, Massachusetts, consists of 25 percent gravel, 20 percent sand, 40 to 45 percent fine sands, and less than 12 percent clay (Leggett, 1962). The glacial deposit is called *stratified drift* if the profile is sorted according to size.

Glacial till or drift has a highly variable thickness which depends on the location, such as in buried valleys (old stream erosion traces) or in end moraines. Some typical thickness values are as follows:

Location	Estimated thickness, m
Great Lakes region, U.S.A.	12
Illinois	0–180 and averaging 35
Central Ohio	29
But in buried valleys	60–230
Ontario, Canada	0–75+
New Hampshire	10
Southeastern Wisconsin	14
Central Quebec	2–3
Denmark	2–40
Sweden	0–200
Finland	2–3

Glacial deposits range from excellent to poor foundation materials. In many locations, even though the deposits are unsorted, the material is dense and contains considerable sand and gravel. Many of these deposits are permanently above the water table. Submerged valleys, lenses of saturated silt and/or clay, and the presence of suspended boulders cause the most problems. The boulders cause difficulty in both soil exploration and pile driving. The presence of small gravel creates problems in obtaining undisturbed soil samples for laboratory testing. *Boulder clay* is a term used to describe deposits containing considerable cohesive material with randomly suspended boulders.

(a)

(b)

(c)

Figure 5-28 A borrow pit in the Shelbyville morains near Peoria, Ill. (*a*) End view of one of the ridges, which is about 20 m high. (*b*) Closeup showing material distribution of the left side of (*a*). (*c*) Closeup showing material distribution of the right side of (*a*). Darker material is clay which is slightly wet from a rain two days before. The sand-gravel material quickly dried and is lighter in color.

5-14 WIND DEPOSITS

Wind, or aeolin, deposits are primarily *loess* and *dune* sands. Loess covers large areas of the central United States, Russia, Europe, and Asia. These deposits are believed to have been at least partly caused by changes in air density in the vicinity of melting glaciers and flowing outwash streams causing windborne particles to precipitate. Loess deposits are characterized by being of buff color, of low density (often less than 14 kN/m^3), of low wet strength, and with the ability to stand on vertical cuts. Loess deposits range in thickness from a few centimeters to more than 30 m. Commonly, along the Illinois River the loess depth is 5 to 8 m. The thickness tends to be greater near the east side of the streams, and the deposits thin rapidly with distance eastward from the stream, reinforcing the theory of how they were formed.

Loess is a quartzose, somewhat feldspathic, clastic sediment composed of a uniformly sorted mixture of silt, fine sand, and clay particles. Typically the particles range from 0.002 to 1 mm, with the largest percentage between 0.005 and 0.150 mm (No. 100 sieve). It tends to deposit in a loose arrangement which becomes rather stable due to cementation from clay particles, organic activity, and calcium carbonation. This structure is particularly susceptible to saturation and may collapse when saturated. On vertical cuts, large blocks tend to slough when wet as illustrated along the base of the cut in Fig. 5-29. Water percolating vertically through root and worm holes may cause considerable vertical erosion.

Figure 5-29 A loess deposit near Vicksburg, Miss. Deposits here are considerably more than 30 m thick. This vertical cut, which has been standing more than 10 years, is about 8 m high. Note the typical weathering pattern of vertical chimney formations (grooves) and spalling.

Dune sands are sand deposits formed by wind action rolling the sand, which is too large for air transport, along the ground until an obstruction is met, whereupon a dune (or mound) forms. Later winds may demolish the dune and redeposit it at a new location further downwind. Dune sands tend to be well rounded from abrasion. Dune deposits are found in desert areas such as areas of California, the Sahara Desert in northern Africa, large areas of the Mideast such as Saudi Arabia, and the Gobi Desert in Asia. A few local dunes are found along the southeastern shores of Lake Michigan.

5-15 GRAVITY DEPOSITS

Gravity deposits are primarily *talus*, found at the base of cliffs. They may also include landslide deposits if the slide has removed the soil sufficiently from the original site.

Talus is the weathered rock/soil deposit formed at the base of cliffs when rock weathering causes the face of the cliff to loosen and fall away, producing a pile of rock fragments at the cliff base. These fragments are likely to be rather loose and porous and may require removal where a dam is to abut against the cliff. Figure 5-30 illustrates both a talus deposit and a gravity deposit formation on the sides of a mountain.

5-16 SUBSURFACE WATER

Groundwater is one of the most important mineral resources extracted from beneath the earth's surface. Probably 30 percent of the daily water consumption worldwide is obtained from groundwater; the remainder is obtained from surface water in streams or lakes.

The geotechnical engineer is concerned with groundwater when solving problems of water supply, drainage, excavations, foundations, and control of earth movements. Because of the many engineering projects in which groundwater is a significant parameter, the engineer should have a good understanding of the modes of its occurrence and movement.

Subsurface water is derived from several sources, and impurities in it may be indicative of its origin and/or history. Some groundwater is a direct contribution from magmatic or volcanic activity during the process of rock cooling. This water may be termed *juvenile* (just beginning to freely circulate) water. Water trapped in the interstices of sediments which are later covered by more impermeable sediments may be retained until tapped by accident or intent. This water is called *connate* water and is often salty since most of the sediments were deposited beneath sea water.

The most important source of groundwater is that portion of the precipitation which sinks into the ground, called *meteoric* water. Water is drawn into the

(a)

(b)

Figure 5-30 Gravity deposits as found in regions of rugged topography. (*a*) Slope deposit gradually moving to bottom of mountain; (*b*) talus deposit.

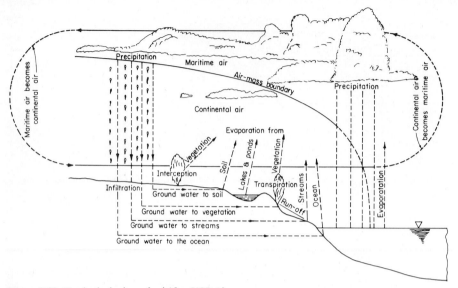

Figure 5-31 The hydrologic cycle. (*After USDA.*)

atmosphere by evaporation and widely distributed by wind currents. Condensation returns this water to the earth's surface as rain, snow, sleet, hail, frost, and dew. That part falling on land surfaces becomes subdivided as follows:

1. Part is reevaporated back to the atmosphere (probably 70 percent).
2. Part runs off into streams and thence to lakes or the ocean.
3. Part is used by plant and animal life.
4. Part sinks into the ground to become groundwater (probably less than 20 percent of the condensation falling on the surface).

Figure 5-31 illustrates the hydrologic cycle. The amount of subsurface water obtained depends on:

1. *Surface gradient.* Steep slopes encourage surface runoff both in quantity and rate.
2. *Vegetation.* Thick foliage may intercept large amounts of condensation before it even reaches the ground surface.
3. *Climatic conditions.* Amount of rainfall and daily temperature influence the evaporation rate.
4. *Porosity and permeability of the mantle.* This means the percentage of pore space and the facility with which the water can move through the earth mass.

Water entering the mantle may be partly held by surface tension forces in the upper soil layers (vadose zone), to later evaporate or be used by plant life. Below this zone is the saturation zone, extending to a considerable depth but depending

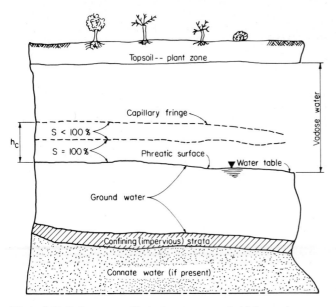

Figure 5-32 Soil-water profile in the upper mantle of the earth.

on the stratigraphy, in which the interstices and cracks are completely filled with water. The saturation zone includes (as in Fig. 5-32) a depth in which the water is held by surface tension, or capillary zone, and a lower zone where the water is free to move, or flow, under the influence of gravity. The *phreatic* line delineates between these two zones and defines the *water table*. The water table must be penetrated to provide a dependable well or permanent stream. The water table tends to follow the contours of the ground surface, rising under hills and descending beneath valleys. It tends to be close to the surface in moist climates and at greater depths in arid regions. If the water table is not replenished, usage lowers it. The slope of the water table is the *hydraulic gradient*. Figure 5-33 illustrates groundwater and conditions for streams to replenish (influent) or take from (effluent) the groundwater supply.

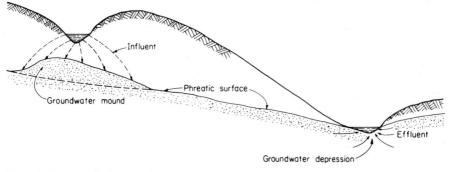

Figure 5-33 Groundwater and streams.

Table 5-2 Typical porosities of some rocks (after Leggett, 1962)

Type	n
Soil and loam	< 60
Chalk	< 50
Sand and gravel	25–35
Sandstone	10–15
Oolitic limestone	10
Limestone and marble	5
Slate and shale	4
Granite	1.5
Crystalline rocks, generally	< 0.5

A Aquifers

A permeable material through which the groundwater actually flows is called an *aquifer*. Sand or sand and gravel strata are particularly excellent aquifer materials due to their large porosity and permeability. Table 5-2 lists typical values of porosity (n values) of several types of rocks. Some porous sandstones are important aquifers, such as the St. Peter sandstone of Illinois, Wisconsin, and parts of Indiana with entrance in Wisconsin, and the Dakota sandstone underlying large areas of the Dakotas, Minnesota, Kansas, Nebraska, and parts of Colorado with entrance near the Black Hills of the Dakotas. It should be noted, however, that materials of high porosity may not be good aquifers. Mississippi River sediments often have porosities on the order of 80 to 90 percent, but the permeability is so

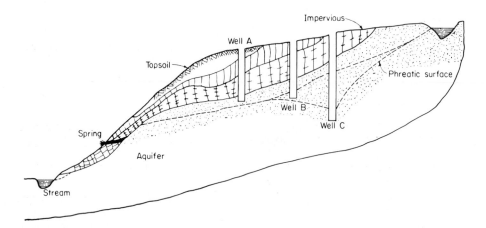

Figure 5-34 Conditions for wells, a spring and stream supplied by groundwater. The spring flows through a crack in the impervious upper layer and may be artesian if some pressure head remains after head loss through the crack. Well *A* is dry unless the water table rises. Well *B* becomes nonproductive when well *C* drawdown lowers the ground water table as shown.

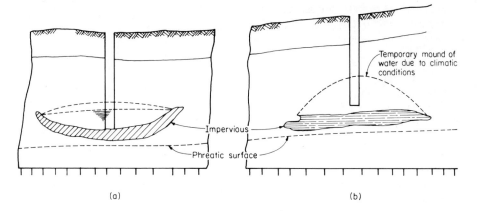

(a)   (b)

Figure 5-35 Two different conditions producing a perched water table. Note that in (b) the well may dry up through a combination of production and natural drainage.

low that little water would be obtained from a well. This is generally true of all silts, silt-clays, very fine silty and/or clayey sands, and loam soils.

Limestone which has weathered sufficiently to contain large solution cavities may be an excellent source of underground water. Chalk is also an excellent source; it was the source of early artesian water in France and supplies considerable water for domestic use in the southern part of Great Britain. Generally igneous, metamorphic, and other sedimentary rocks are poor aquifers unless they are badly cracked or fissured to provide both a water reservoir and flow channels. Figure 5-34 illustrates conditions for water supply via wells or springs. The "perched" water table of Fig. 5-35 is a common occurrence. In the situation of Fig. 5-35b, a well or spring may be intermittent. In the conditions of Fig. 5-35a, the supply may be permanent. Note, however, that boring through the impermeable layer containing the perched water table, carelessly or by design, may allow it to drain and be permanently lost.

B Artesian Water

Artesian water is obtained from an aquifer which is under a hydrostatic pressure. Conditions necessary to produce artesian water are as follows (see also Fig. 5-36):

1. The water must be contained in a permeable layer so inclined that one end can intake water at the ground surface.
2. The aquifer is capped by an impermeable layer of clay, shale, or other dense rock.
3. Water cannot escape from the aquifer either laterally or from the lower end.
4. There is sufficient pressure in the confined water to raise the free surface above the aquifer when it is tapped via a well (or any boring).

Thirty to one hundred years ago artesian wells where the water overflowed

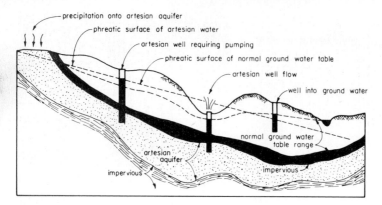

Figure 5-36 Conditions for artesian water.

the well at the ground surface were widely obtained. At present, due to indiscriminate use (or ignorance of the source) and/or allowing continuous flowage from the earlier artesian wells, most of the aquifers are no longer under pressure or the pressure has fallen to such values that pumps must be used. This represents an irrevocable loss of a natural resource, since replenishment is estimated to take on the order of 150 to over 1000 years.

C Groundwater Discharge

Large quantities of groundwater are used by man-made wells. Probably even larger quantities are lost through seepage directly into a lake, ocean, or stream, or from springs and through ordinary evaporation. Springs (Fig. 5-37) usually are found on hillsides or at the edges of valleys, but they may issue beneath the sea, lake, or stream. Streams may be mere trickles or torrents. For years the principal source of water for the city of Huntsville, Alabama, was a large spring. Big Spring in Carter County, Missouri, discharges some 11 m^3/s. Silver Springs in Florida, which is an artesian spring, discharges some 23 m^3/s, Thousand Springs along the Snake River in Idaho discharges some 140 m^3/s from the several springs making up the "thousand springs." Several rivers and numerous smaller streams originate as springs.

A *mineral* spring is any spring with considerable mineral content, which gives the water a distinctive taste (all groundwater contains some dissolved minerals). Hot springs are found in volcanic regions or, as in the case of Hot Springs, Arkansas, or Warm Springs, Georgia, are caused by the rising of artesian water from a great depth, where it may be presumed that a deep-seated igneous intrusion has heated the surrounding rock and water. These springs may be fissure springs due to the likelihood of artesian water flowing through a crack in the confining rock.

(a)   (b)

(c)   (d)

Figure 5-37 Several springs. (*a*) Spring from a limestone cave near Decatur, Ala. It is about 3.5 m wide × 1.5 m deep, and the flow rate is about 0.15 m/s. (*b*) Spring from a limestone hole near Roanoke, Va. The flow volume is about 0.25 cm/s. (*c*) Spring through a talus in Montana. (*d*) Hot spring. Steam bubbles can be seen. The spring is surrounded by travertine limestone, with flow in the small channel in the foreground.

Geysers are hot springs which erupt intermittently; they are found in areas of dying vulcanism where a substantial heat source is available. Notable geyser regions include Yellowstone National Park in Wyoming, North Island of New Zealand, and Iceland.

Wet-weather or intermittent springs flow during periods of wet weather when the recent rainfall builds the water table to a sufficient height. As the water table recedes, the spring ceases to flow. These springs are nuisances for road construction (see Figs. 5-8*a* and 5-15*a*) and are frequently found in subdivisions where landscaping has altered the topography such that springs form in the newly exposed areas and in the streets or at the basement level of houses. The former condition results in early pavement breakup and visibly wet spots in the street after heavy rains or in the spring due to groundwater recharge from melting snow. The latter springs result in wet basements until the stream flow is intercepted; the condition may be particularly expensive to correct if the basement was constructed during dry weather so that a spring was not evident.

In arid regions groundwater, generally as water vapor, is drawn upward to the surface, where it evaporates, leaving behind a coating of any dissolved salts. These deposits are termed *alkalies* (any bitter-tasting salts), also *caliche* (crusty deposits of calcium carbonate). These deposits will ruin agricultural land and tend to form where land is irrigated and the quantity of irrigation water is not sufficient to leach these materials into the *B* horizon.

D Groundwater Erosion

Groundwater is an effective erosion agent because it is charged with carbonic acid (contained in rainwater), which dissolves carbonate rocks such as limestone, dolomite, marble, rock salt, and gypsum. Limestone, being particularly widespread, tends to concentrate the effects of subsurface erosion.

As water percolates into the earth through cracks in the rock, the chemical and physical action of the water enlarges the cracks. This tends to isolate the rock into blocks and form cavities around and in them into which the water drains. Large limestone caves are formed by this process. Where the cavities are more local in extent, the surface may collapse into the underground cavity, forming *sinks* or *sinkholes*. When the outlets are adequately plugged, small lakes may form and later fill with sediment. An area with many sinkholes is termed a *karst* region after a region of Italy and Yugoslavia with this characteristic topography. Karst areas in the United States include the Shenandoah Valley of Virginia, southern Indiana, and north central Kentucky; portions of Missouri in the Ozark Mountain area; and the central portions of Florida.

Karst areas should be treated cautiously, since a foundation site on a potential sinkhole, or on a sinkhole later filled with sediments, may produce a disaster. These sinkholes, often less than 10 000 m^2 in plan (on the order of 3 to 6 m in initial depth) may be missed in a soil exploration program unless the engineer recognizes that the area contains these filled-in sinkholes.

Groundwater precipitates may fill cracks or fissures to form mineral veins. Calcite and quartz veins are common and may carry concentrations of metallic minerals such as copper, silver, or gold.

E Recharge of Groundwater

Most groundwater recharge is obtained from precipitation. For the normal water table in moist areas, the recharge via this mode keeps pace with groundwater removal. In wet years, the water table may even increase, with some loss during dry periods.

Where ground intake is particularly heavy, as in areas of large population or high industrial densities, the water table may suffer irreparable harm unless artificial recharge methods are used, such as putting the used water back into the aquifer via wells or recharge ponds whose bottoms exit onto the aquifer, or pumping river water into the aquifer.

5-17 SOURCES OF GEOLOGICAL INFORMATION

From a study of the geological concepts and photographs illustrating these concepts, it is evident that many landform features are areal rather than local. Identification of large-sized features often requires study of the plan, or at least of an oblique view from above. In many cases a perspective from a hill or mountain top is not possible, and often in the growing season plant growth will obscure the geology.

Because of these several problems it is often necessary to obtain aerial photographs of the area or topographic quadrangle sheets. Aerial photographs may be obtained from the Agricultural Conservation Service of the U.S. Department of Agriculture. Many, if not all, of these photographs may be obtained for stereoscopic viewing (in adjacent pairs). Topographic maps, commonly called quadrangle sheets, can be obtained from the U.S. Geological Survey. Annual publications give the areas where quadrangle sheets have been completed.

State geological survey publications are particularly useful for identifying state geology, mining operations, etc. Often the state geological publications which have been made for the separate counties are particularly well detailed.

5-18 SUMMARY

This chapter has presented a great deal of geological data and terms. Terms have been italicized for ease of location, but it is not expected that the reader will be able to learn all these terms nor is it generally necessary.

The reader should have an awareness of:

1. The principal earth-forming minerals
2. The range of specific gravities making up the earth's mass
3. The three rock groups (igneous, sedimentary, and metamorphic) and the mode of formation of each
   a. Recognize granite, gabbro, basalt, and feldspar
   b. Recognize shale, sandstone, some of the limestones, and coal
   c. Recognize marble, quartzite, and slate
4. The processes of rock weathering to produce *residual* and *transported* soil deposits
5. The role of glaciers and typical glacial deposits
6. The role of wind and typical wind-formed deposits
7. The role of water and typical fluvial water deposits
8. Groundwater, groundwater development, the concept of the water table, water well development, sources of springs, and development of artesian water.
9. The role of subsurface water in erosion.

The reader should supplement this chapter material with a good textbook on geology. The local library usually contains state geological survey publications which go into considerable detail for that state's geology. In addition to the usual and numerous textbooks on general geology, the reader should be aware that separate geology textbooks are available which describe the geology of the several continents as well as separately considering glaciation and regional or areal geomorphology (landforms).

The reader should make it a habit, whenever the opportunity arises, to observe landforms and rock formations using exposed profiles in highway cuts as well as natural topography and stream problems. Observe rocks exposed at the ground surface and in stream beds and/or attempt to identify them.

HOMEWORK PROBLEMS

5-1 Why are some igneous rocks fine-grained whereas others are coarse?

5-2 Discuss the typical profile likely to be obtained in an end or terminal moraine.

5-3 What type of profile may be obtained in an esker? Why?

5-4 Why is the B horizon soil preferable as a fill rather than the C horizon material?

5-5 Why is the soil-horizon concept not applied to transported soils?

5-6 Sketch the necessary conditions for a boring through a perched water table to drain it.

5-7 Sketch the necessary conditions under which a boring through an excavation would have to be plugged so that the excavation could be made.

5-8 Sketch the conditions showing how in a soil exploration program a filled-in slough or oxbow could be missed.

5-9 Sketch the conditions showing how a soil exploration program may miss a sedimented-in sinkhole. Explain how you might be able to discern that you are in an area of filled-in sinkholes.

5-10 Explain how you could increase the yield from a well into a chalk formation which is 1.5 m in diameter. Hint: What is an adit?

5-11 Explain how you might distinguish between a drumlin and a weathered mound of earth.

5-12 Explain how you might distinguish moraine (ridged) topography from ordinary weathering of residual soil.

5-13 How would you proceed to identify a rock fault area?

5-14 What are several problems which may develop in jointed rock?

5-15 Obtain the state geological survey bulletin from the library (if available) and write a rough description of the geology of the assigned area of the state.

Chapter 6

Soil Structure
and Clay Minerals

6-1 SOILS AND SOIL FORMATION

Chapter 4 considered soil in terms of whether it was cohesionless or cohesive. These classifications used a visual grain description together with consistency index properties to describe the soil as gravel, sand, silt, clay, or a mixture, as sandy clay, silty sand, etc.

Chapter 5 considered soil formation as a cyclic geological process involving the weathering of rocks. The soil was then considered to be either a residual or a transported soil, and the agents of transportation and resulting deposits were considered in some detail.

This chapter will consider soil in terms of its composition, structure, and, in the case of clay, the clay minerals and their very important influence on soil behavior. Soil at this point may be defined as follows:

1. The unconsolidated material of the earth's crust used to build upon or used as a construction material.
2. The loose, or unconsolidated, materials overlying bedrock produced by rock weathering.

In addition to soil, bedrock at shallow depths is often of interest, particularly in terms of quality, to the geotechnical engineer.

6-2 SOIL STRUCTURE AND FABRIC

Soil structure is both the geometric arrangement of the particles, or mineral grains, and the interparticle forces which may act upon them. Soil structure includes gradation, arrangement of particles, void ratio, bonding agents, and associated electrical forces. The structure is the property which produces a response to external changes in the environment, such as loads, water, temperature, and other factors.

Soil fabric is a more recently introduced (mid-1960s) term to describe the "structure" of clays. Fabric denotes the geometric arrangement of the mineral particles in a clay mass as observed by optical or electron microscopes. The geometric arrangement includes particle spacing and pore size distributions.

6-3 GRANULAR SOIL STRUCTURE

The arrangement of the individual soil particles in a granular soil may be termed *packing*. The packing of grains of soil, or any other particulate medium, is strongly influenced by both particle size distribution and particle shape.

Figure 6-1*a* illustrates ideal packing of spheres in a volume which is one sphere thick. In Fig. 6-1*b* the same number of spheres have been rearranged into a more dense configuration termed rhombic packing. According to theoretical considerations—which are simple for the simple cubic packing of Fig. 6-1*a*—if we fit balls inside a cube of side $= 2R$, we obtain

	Volume	Unit weight	Void ratio e
Simple cubic	$8R^3$	$\dfrac{\gamma\pi}{6}$	0.91
Rhombic	$4\sqrt{3}\,R^3$	$\dfrac{\gamma\pi}{3}\sqrt{3}$	0.65
Pyramidal	$4\sqrt{2}\,R^3$	$\dfrac{\gamma\pi}{4}\sqrt{2}$	0.35

This range represents the theoretical maximum and minimum void ratios of any particulate mass consisting of equal spheres of radius R.

Figure 6-2*a* illustrates the ideal particle size distribution for optimum packing. Approximations to this condition are desirable in many geotechnical engineering problems where stability is a concern. Typically these problems include highway and railroad fills, levees, and dams where optimum packing (translate: maximum density) tends to develop maximum shear resistance and minimum subsidence.

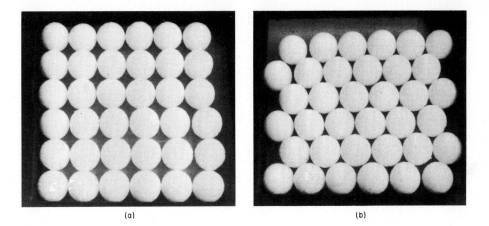

Figure 6-1 Photograph illustrating ideal packing of spheres. (*a*) Simple cubical packing; (*b*) rhombic packing with the same number of spheres occupying a smaller volume.

Ideal particle size distributions seldom exist in real soils; however, the packing of Fig. 6-2*a* establishes the upper limit, while the packing of Fig. 6-2*b* establishes the lower limit as obtained by a "one" size particle distribution. It should be noted that with equal spheres the variation of unit weight is independent of sphere radius; with real soil the variation of γ will depend on both the "one" size and the numerical value of that size. This is due both to "size" not being uniform and to "size" being a range of particle sizes which have passed one sieve opening and been retained on a smaller sieve size.

Geotechnical considerations, as stated previously, require optimum particle packing. In optimum packing, shear strength is increased because there is more particle contact, providing additional lateral support. Compression and/or subsidence is reduced since there are fewer available soil voids to allow volume change. Further, with packing there is less tendency for particles to readjust to new equilibrium positions under stress. Where a site is founded on a loose granular deposit, it will be necessary to increase the packing via compaction. Where the deposit is shallow, a vibratory compactor may be used. For deep deposits

(a) Ideal packing with particles ranging from large to small (well graded).

(b) Same large grain structures as (a) but small sizes removed (poorly graded).

Figure 6-2 Particle packing.

vibroflotation (a vibratory device) may be used or piles may be driven and/or withdrawn. In either case vibration energy is used to displace the soil particles into a more dense configuration.

6-4 OTHER CONSIDERATIONS OF GRANULAR SOIL STRUCTURE; RELATIVE DENSITY

Cohesionless soils tend to form a *single-grained* structure, as illustrated in Fig. 6-3, which may be either loose (Fig. 6-3*a*) or dense (or packed). Single-grained soil structures are formed when the soil grains independently settle out of a soil-water suspension, as opposed to "floc" settling (see Sec. 6-7). Generally grains larger than about 0.01 mm will form single-grained structures. This size is large enough that the interparticle forces and ionic forces in the water are not sufficient to overcome gravity forces acting on the soil grains. Piles of pure silt, sand, or gravel or silty sand, sandy gravel, etc., mixtures are single-grained structures. Very small soil particles on the order of 0.001 mm and smaller are *colloids*. Colloids are sufficiently small to be more affected by both interparticle forces and ions in soil-water suspensions than by gravity forces. Clay minerals are particles smaller than about 0.002 mm, but their behavior is such that they will be separately considered.

True cohesionless soils can only be found in transported soil deposits where the wind or water has removed the colloidal and/or clay mineral contaminants. Typical cohesionless deposits include sand and gravel bars in streams, select glacial outwash deposits, some drumlins and eskers (discussed in Chap. 5), sand dunes and beach sands, and similar deposits. Other deposits can sometimes be washed to remove the cohesive materials and produce sands and gravels.

(a) (b)

Figure 6-3 Single-grained soil structure of real soil. Note, however, that in the field geologically aged deposits often contain significant interparticle growths at the contact points (cementation) from precipitated calcium, iron, magnesium, aluminium, etc., salts or oxides. (*a*) Loose structure; (*b*) soil of (*a*) rearranged into a more dense configuration.

Certain conditions of deposition can produce a very loose (metastable) soil structure. This type of structure may be able to support a substantial static load but may collapse under relatively small dynamic, or vibratory, loads. This type of soil structure is most likely to have been formed geologically by an upward flow of water in the deposit which has since gradually lessened or disappeared altogether. The diminished flow of water allowed the soil grains to settle out of the flow into a very loose structure. Soil exploration may not detect this state, nor does a laboratory sample built to the same void ratio necessarily produce the same behavior. A careful examination of geologic evidence may give an indication of the potential problem. Careful observation of the penetration test may also give an indication, since the first blows of the test would cause a soil collapse and relatively large penetration per blow where the later penetration per blow rate would be less.

Where the soil grains are from about 5 to 0.05 mm (coarse to fine sand), the presence of a small amount of water can alter the engineering behavior considerably. The surface tension of water at conditions of $S \ll 100$ percent is sufficient to restrict particle movement in this size range, producing an "apparent cohesion" which disappears when the soil dries out. Practically, apparent cohesion allows near vertical sand cuts or greater mobility of rubber-tired vehicles over damp sand. Apparent cohesion inhibits soil packing or produces what is commonly termed *bulking*. Figure 6-4 qualitatively illustrates the effect of bulking on unit weight. At a water content of zero the unit weight is nearly the maximum possible for some energy input. At intermediate water contents, bulking, or surface tension accumulations, restricts particle movement and the unit weight decreases. As more water is added, some of the pores become saturated, with loss of surface tension and increased packing. At $S = 100$ percent all the surface tension effect is lost, and together with some particle lubrication effect, the maximum unit weight is obtained. In field densification operations, where adding water does not affect the surrounding soil detrimentally, flooding a sand (to ensure $S = 100$ percent) will assist considerably in increasing the unit weight.

Optimum packing of a granular soil results in the greatest unit weight and minimum void ratio e_{min}. Conversely, minimum packing results in the loosest state, minimum unit weight, and maximum void ratio e_{max}. The loosest state is approximately obtained by pouring dry sand (Bowles, 1978) into a calibrated mold. Sometimes the sand is allowed to fall through water to produce a known weight and volume. With G_s, the volume of the mold, and the weight of soil in the

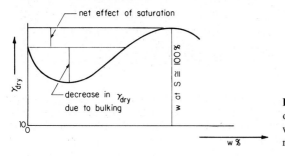

Figure 6-4 Qualitative effect of bulking on resulting dry unit weight. By placing water content at $S = 100$ percent, the maximum dry unit weight is obtained.

Table 6-1 Typical values of e_{max}, e_{min}, and unit weight for several soils

	Loose			Dense		
	Dry unit weight	e_{max}†	ϕ‡	Dry unit weight	e_{min}†	ϕ‡
Gravel	16.0–18.0	0.62–0.44	32–36°	18.0–20.0	0.44–0.30	35–50°
Coarse sand	15.0–17.5	0.73–0.50	32–38	17.5–19.6	0.50–0.33	35–48
Clayey sand	14.0–16.5	0.86–0.58	28–32	16.5–18.5	0.58–0.40	35–40
Silty sand	12.6–15.5	1.05–0.68	28–32	15.5–17.5	0.68–0.49	32–38
Fine sand	14.0–18.5	0.86–0.40	27–33	15.5–18.0	0.68–0.44	33–39
Sandy gravel	15.0–18.0	0.73–0.44	30–38	18.0–22.0	0.44–0.18	36–45
Gravelly sand	15.0–18.0	0.73–0.44	30–38	18.0–22.5	0.44–0.16	36–50
Silt	14.0–15.5	0.86–0.68	20–30	15.5–17.5	0.68–0.49	25–32

† Depends on G_s.
‡ Use higher values for angular particles.

mold, the void ratio is easily computed. The densest state is obtained by vibrating a confined weight of sand and measuring the volume. Table 6-1 gives some ranges of void ratios and other data for several soils as an indication of the values one might expect to obtain. Tabulated values such as these are acceptable for preliminary design but should never be used for any final design.

The *relative density* is a measure of the in situ void ratio e_n related to the laboratory values of the maximum and minimum void ratios as

$$D_r = \frac{e_{max} - e_n}{e_{max} - e_{min}} \tag{6-1}$$

Relative density can also be expressed in terms of the maximum (γ_{max}), minimum (γ_{min}), and in situ (γ_n) dry unit weights as

$$D_r = \frac{\gamma_{max}}{\gamma_n} \frac{\gamma_n - \gamma_{min}}{\gamma_{max} - \gamma_{min}} \tag{6-2}$$

This equation is preferable to Eq. (6-1) due to the greater ease of determining unit weights and because it does not require a determination of the specific gravity. Considerable importance was attributed to the relative density by early proponents, who attempted to relate various soil properties such as void ratio, angle of internal friction, and, thus, indirectly settlement and strength characteristics to this index property. Relative density is sometimes used at present in liquefaction studies (Sec. 14-7) as a field compaction specification requirement, and to assess the competence of in situ granular materials for foundations.

The major reason for using relative density is that undisturbed sampling of in situ cohesionless sands and gravels is nearly impossible and, as a consequence, penetrometer testing is widely used. A large data base presently exists—albeit with considerable scatter—relating penetration tests to relative density. Table 6-2 gives some simple field identification tests which may be used to estimate D_r.

Table 6-2 Terms and field identification in relative density

Soil state†	D_r	Field identification
Very loose	0–0.20	Easily indented with finger, thumb, or fist
Loose	0.20–0.40	Somewhat less easily indented with fist. Easily shoveled
Medium compact	0.40–0.70	Shoveled with difficulty
Compact	0.70–0.90	Requires pick to loosen for shoveling by hand
Very compact	0.90–1.00	Requires blasting or heavy equipment to loosen

† Not all authorities agree on either the terminology describing the soil state or the value of D_r to which it applies. These values are as good as any proposed and with the subjective nature of the index property may be used with confidence.

Considerable research, with the latest reported in ASTM (1973), indicates that D_r is not a very reliable soil index property. It is quite possible for two sands with identical values of in situ void ratios e_n and D_r to have significantly different engineering behavior due to grain shape, cementation, confinement, and stratification resulting from deposition and stress history.

Since D_r depends on laboratory determination of γ_{max} and γ_{min}, or the corresponding void ratios, a large error may result in not accurately determining both of these values. Generally, a statistical determination of γ_{max} will produce a rather consistent (average or standard deviation) value about 0.45 kN/m³ too small, and conversely for the minimum unit weight a value about 0.45 kN/m³ too large. This is illustrated in Ex. 6-1.

Example 6-1 A medium coarse, gravelly sand was tested in the author's laboratory by a group of 10 students. Each student did three tests each for the maximum and minimum unit weights and reported the extreme (not the average) values obtained. The sand was returned to the source container and well mixed for further use. The data were as follows (the Δ-values were later computed for obtaining the standard deviation $\bar{\sigma}$):

Test	1	2	3	4	5	6	7	8	9	10
γ_{min}	14.36	14.58	15.02	15.12	14.78	14.51	14.40	15.08	15.15	14.43
Δ	0.0	0.22	0.66	0.76	0.42	0.15	0.04	0.72	0.79	0.07
γ_{max}	18.57	18.70	18.86	18.39	18.50	18.32	18.45	18.42	18.49	18.75
Δ	0.29	0.16	0.0	0.47	0.36	0.54	0.41	0.44	0.37	0.11

REQUIRED Compute the standard deviation and assess the error in D_r if the in situ value of $\gamma_n = 16.0$ kN/m³.

SOLUTION The standard deviation is computed based on the maximum and minimum values of unit weight, not on the average, since the definition of D_r is based on the extreme values. With this concept, the Δ values are obtained as the difference between the smallest unit weight of 14.36 and the other nine values; thus

$$14.58 - 14.36 = 0.22$$

Similarly, with the maximum γ, obtain

$$18.86 - 18.57 = 0.29$$

The standard deviation is computed as

$$\bar{\sigma} = \sqrt{\frac{\sum (\Delta^2)}{9}} = 0.52 \qquad \text{for } \gamma_{min}$$

and 0.37 for γ_{max}. Thus, one would expect that on the average the minimum unit weight would vary as

$$\gamma_{min} = 14.36 + 0.52 \text{ kN/m}^3$$

(range from 14.36 to as much as 14.88.)

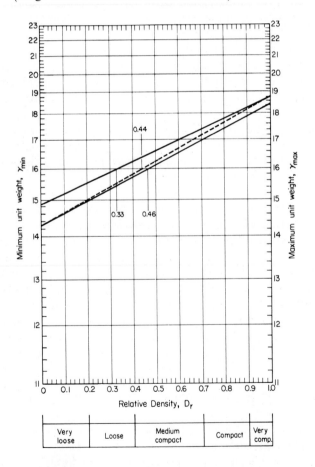

Figure E6-1

Since the range of values of γ_{min} and γ_{max} must be investigated, it will be useful to make a plot of D_r versus γ as shown. The scale is adjusted so that the curve is linear, since Eq. (6-2) directly plots a curve. The unit weight scale is adjusted, once and for all, by obtaining the range of γ and computing D_r for the scale increment; the spacing of this increment is made such that the slope is constant, using a constant increment for the D_r scale. For example, for the unit weight increment 11 to 12, the spacing may be 30 mm; then from 12 to 13, use 25 mm; from 13 to 14, use 22.9 mm; etc. This is shown on Fig. E6-1.

From Fig. E6-1 obtain the "true" value for D_r (assuming $\gamma_n = 16.00$ is exact) of 0.44 and extreme values of 0.46 and 0.33, respectively. The probable maximum error is

$$\text{Probable error} = \frac{0.44 - 0.33}{0.44} \times 100 = 25 \text{ percent}$$

6-5 STRUCTURE OF COHESIVE SOILS

A cohesive soil may be defined as an aggregation of mineral particles which has a plasticity index defined by the Atterberg limits and which forms into a coherent mass on drying such that force is necessary to separate the individual microscopic grains. The ingredients necessary to give a soil deposit cohesion are *clay minerals*, sometimes termed argillaceous materials. The amount of cohesion depends on the relative sizes and amounts of various soil grains and argillaceous materials present. Generally when over 50 percent of the deposit consists of particles 0.002 mm and smaller, the deposit is termed "clay." With this relative percentage the larger soil particles are suspended in a fine-grained soil matrix. Where 80 to 90 percent of the deposit material is smaller than the No. 200 sieve (0.075 mm), as little as 5 to 10 percent clay can give the soil a cohesive label.

Seldom does a pure clay deposit exist naturally; it is nearly always contaminated with silt and/or fine sand particles as well as colloidal (<0.001) sizes. Colloids, sometimes called rock flour, are the byproduct of rock abrasion and do not possess clay mineral properties (Sec. 6-6) even though the size range is similar.

A complete description of the structure of a fine-grained cohesive soil requires a knowledge of both the interparticle forces and the geometrical arrangement, or fabric, of the particles. It is nearly impossible to measure the interparticle force fields surrounding clay particles directly; therefore, the fabric is the principal focus in studies of cohesive soils. From the fabric studies, estimates are theorized or postulated of the interparticle forces. The interparticle forces appear to be developed from three different types of electric charges:

1. *Ionic bonds.* Bonds due to a deficiency in electrons in the outer shells of atoms making up the basic soil units
2. Van der Waals bonds. Bonding due to alternations in the number of electrons at any one time on one side of an atomic nucleus

3. *Others*. Includes hydrogen bonds and gravitational attraction between two bodies

Recent studies of clay soils with the scanning electron microscope (SEM) show the individual clay particles to be aggregated or flocculated together in submicroscopic fabric units which are called *domains* by numerous recent researchers (Yong and Sheeran, 1973; Collins and McGown, 1974). Domains in turn group together to form submicroscopic groups called *clusters*. These groupings are due to the interparticle forces acting on the small basic units. Clusters group together to form *peds* and groups of peds of macroscopic size. Nonscientific terms for the peds include soil "crumbs" and soil "aggregations." Peds and other macrostructural features such as joints and fissures constitute the macrofabric soil structure. A sketch of this system is shown in Fig. 6-5. Photographs of various clays using the SEM technique indicate a very complex structure. This may in part account for the complex engineering behavior of clay soils.

Macrostructure, including the stratigraphy, of fine-grained soil deposits has an important influence on soil behavior in engineering practice. Joints, fissures, root holes, varves, silt and sand seams and lenses, and other discontinuities often control the behavior of the entire soil mass. The strength of a soil mass is significantly less along a crack, or fissure, than through intact material, particularly in laboratory tests where the defect may extend completely through a small sample. If an in situ defect happens to be unfavorably situated with respect

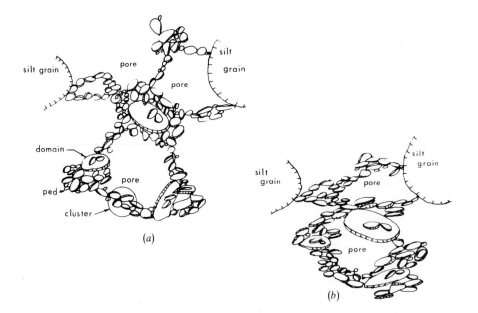

Figure 6-5 Structure of a clay soil. (*a*) A porous, flocculated sediment interspersed with silt grains. (*b*) The sediment after it has been subjected to overburden or other stresses which have resulted in a reorientation of the domains, clusters, and peds into a more parallel (dispersed) state.

to applied stresses, failure or instability may occur unless the surrounding material provides sufficient confinement. The drainage of a clay layer can be markedly affected by the presence of even a very thin layer(s) of silt or sand. Therefore, in any engineering problem involving stability or settlements, the geotechnical engineer must carefully investigate the clay macrostructure.

Microstructure is more important from a fundamental than from an engineering viewpoint, but it is useful as an aid in the general understanding of soil behavior. The microstructure of a clay is the complete geological history of that deposit, including both stress changes and environmental conditions during deposition. These geological imprints tend to affect the engineering response of the clay very considerably. Recent research on clay microstructure suggests that the greatest single factor influencing the final structure of a clay was the electrochemical environment existing at the time of deposition. Flocculated structures, or aggregations, of varying degrees of packing and interconnections result during sedimentation. The general degree of packing appears to be particularly sensitive to whether the deposition took place in a marine, brackish, or fresh water environment. The ion concentrations in these three waters would range from high in the case of marine to low in fresh water. The degree of packing also appears to be influenced to a large degree by the clay mineralogy as well as by the amount and angularity of the fine sand or silt grains present. Silt particles in a cohesive deposit have been observed to have thin skins of apparently well-oriented clay particles. Both silt and clay particles/aggregations often contain thin films of amorphous

Flocculated

Honeycomb

Cardhouse

Dispersed

Figure 6-6 Structure of clay soil using earlier terms of structure orientation. The flocculent structure might be obtained from sedimentation in water with a low salt content. The honeycomb structure could be obtained from sedimentation in a marine (high salt content) environment. The "card-house" description was widely used prior to SEM studies. The "dispersed" state is a convenient description for reorientation from compaction as in Chap. 7.

materials (organic, silica, or iron compounds) on their surfaces. Leaching into or out of these fine-grained deposits may change the soil characteristics considerably as the peds and clusters become coated or the amorphous material is leached away.

Early descriptions of cohesive soil included honeycomb, flocculent, and dispersed structures. These terms, illustrated in Fig. 6-6, are still widely used to describe the total cohesive soil structure. The honeycomb structure may very well be a situation where the clusters form particular groupings during sedimentation, and the flocculent structure may be a situation where either silt grains attract coatings of clay minerals or peds form and produce the porous and random flocculent structure.

Present evidence tends to the theory that the cluster arrangement between peds of the flocculent structure and the somewhat analogous cluster arrangement between honeycomb cells produce the initial distortions (settlements) under stresses. Some research evidence also indicates that water flow through these soils may dislodge domains and/or clusters, which increases the pore spaces and flow. Deposition downstream, however, may plug other pores in the same mass and result in a decreased flow rate. This concept is of particular importance in permeability studies and may produce some governing factors as considered in Sec. 8-5.

6-6 CLAY AND CLAY MINERALS

Clay minerals are predominately silicates of aluminum and/or iron and magnesium. Some of them also contain alkalies and/or alkaline earths as essential components. These minerals are predominately crystalline in that the atoms composing them are arranged in definite geometrical patterns. Most of the clay minerals have sheet or layered structures. A few have elongated tubular or fibrous structures. *Clusters* are books of sheetlike units or bundles of tube or fiber units. Soil masses generally contain a mixture of several clay minerals named for the predominating clay mineral with varying amounts of other nonclay minerals.

Clay minerals are very small (less than 2 μm) and very electrochemically active particles which can be seen using an electron microscope only with difficulty. In spite of their small size, however, the clay minerals have been studied extensively (Grimm, 1968; Mitchell, 1976) due to their economic importance, particularly in ceramics, metal molding, oil field usage, and engineering soil mechanics. The clay minerals exhibit characteristics of affinity for water and resulting plasticity not exhibited by other materials even though they may be of the clay size or smaller. For example, finely ground quartz does not exhibit plasticity when wetted. It should be particularly noted that any fine-grained "clay" deposit is likely to contain both clay minerals and a wide range of particle sizes of other materials which are essentially "filler."

There are two fundamental building blocks for the clay mineral structure. One is a silica unit (Fig. 6-7), in which four oxygens form the tips of a tetrahedron

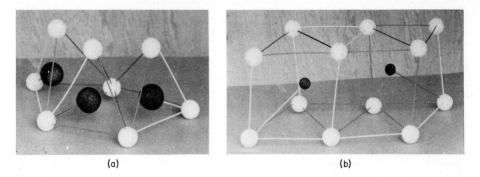

(a)   (b)

Figure 6-7 Plastic models of the two fundamental clay mineral building blocks. (*a*) Silica unit, with the four white balls representing oxygens and the black interior ball the silicon atom. (*b*) The octahedral unit, with the white balls representing hydroxyls and the small black balls the Mg or Al atom. Several octahedral units are interconnected, but not all bonds are shown.

and enclose a silicon atom, producing a unit approximately 4.6 Å high.† The other unit is one in which an aluminum or magnesium (and sometimes Fe, Ti, Ni, Cr, or Li) atom is enclosed by six hydroxyls having the configuration of an octahedron which is about 5.05 Å high. Figure 6-8 gives the block diagram equivalent of these two units. The octahedral unit is called *brucite* if the metallic atom is mainly

† Angstrom unit Å = 10^{-10} m.

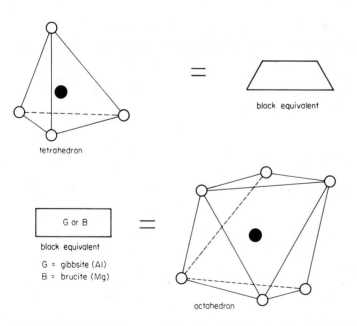

tetrahedron

== block equivalent

G or B

block equivalent

G = gibbsite (Al)
B = brucite (Mg)

octahedron

Figure 6-8 Simplified diagrams of the silica tetrahedron and octahedral units. Carefully note the orientation of the diagram for the tetrahedron to match the point.

magnesium and *gibbsite* if the atom is aluminum. All the possible combinations of these basic units to form clay minerals produce a net negative charge on the exterior of the clusters. A soil-water suspension will thus have an alkaline reaction (pH > 7) unless the soil is contaminated with an acidic substance.

The principal source of clay minerals is the chemical weathering of rocks which contain

Orthoclase feldspar
Plagioclase feldspar
Mica (muscovite)

all of which may be termed complex aluminum silicates (see Sec. 5-2 for chemical equations). However, according to Grimm (1968), clay minerals can be formed from almost any rock as long as there are sufficient alkalies and alkaline earths present to effect the necessary chemistry. The decomposition of orthoclase feldspar to give a clay mineral was given in Sec. 5-9.

Weathering action on rocks produces a very large number of clay minerals with the common property of affinity, but in widely differing amounts, for water. Some of the most common clay minerals are the following.

A Kaolinite

The name "kaolinite" is a modification of "Kauling," meaning high ridge of a hill near Jauchau Fu, China, where a white kaolinite clay was obtained several centuries past. The term kaolin actually describes several distinct clay minerals. Engineers use the term to describe a clay group characterized by low activity.

The kaolinite structural unit consists of alternating layers of silica tetrahedra with the tips embedded in an alumina (gibbsite) octahedral unit as in Fig. 6-9. This alternating of silica and gibbsite layers produces what is sometimes called a 1 : 1 basic unit. The resulting flat sheet unit is about 7 Å thick and extends

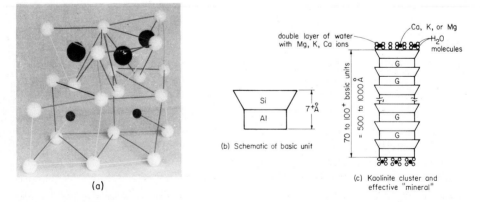

Figure 6-9 The kaolinite clay mineral.

infinitely (relative to 7 Å) in the other two dimensions. The kaolinite cluster is a stacking of 70 to 100 or more of these 7-Å sheets as a book with hydrogen bonds and van der Waals forces at the interface. The resulting formula is approximately

$$(OH)_8Al_4Si_4O_{10}$$

The bonding combination of hydrogen and van der Waals forces results in considerable strength and stability with little tendency for the interlayers to take on water and swell ("activity") and results in the mineral book being on the order of 500 to 1000 Å thick. Kaolinite is the least active of the clay minerals so far observed. Kaolinite can be produced by weathering of certain of the more active clay minerals as well as being directly formed as a byproduct of rock weathering.

Another 1 : 1 mineral of the kaolinite "family" is *halloysite*. It differs from kaolinite by being more randomly stacked, so that a single molecule of water may enter between the 7-Å units, giving the equation

$$(OH)_8Al_4Si_4O_{10} \cdot 4H_2O$$

Halloysite further differs from kaolinite in that the elemental sheets are rolled into tubes, as illustrated by the SEM of Fig. 6-10*b*. Dehydration by heat on the order of 60 to 70°C, and even air drying, will often permanently alter halloysite, either reducing it to $2H_2O$ or completely removing the water molecules and producing approximately kaolinite. The engineering properties of halloysite are considerably different from those of kaolinite, and since even air drying may affect the chemistry which is indirectly measured by the Atterberg limits, great care is necessary in obtaining realistic samples for the Atterberg limits and hydrometer analysis.

Kaolinite and halloysite clays are widely used for chinaware due to the absence of iron and subsequent iron discoloration on firing at high temperatures. Kaolin clay is widely used as an intestinal absorbent to combat intestinal infections, i.e., in antidiarrheal medicines and for digestive disorders.

Kaolinite tends to be found in regions of heavier rainfall, as in the southeastern United States, China, parts of Europe, South America, and other more local areas.

B Illite

Illite is a general term for a clay group discovered in Illinois. These clay minerals are of the general equation

$$(OH)_4K_y(Si_{8-y} \cdot Al_y)(Al_4 \cdot Mg_6 \cdot Fe_4 \cdot Fe_6)O_{20}$$

where y is between 1 and 1.5. Illite is derived principally from muscovite (mica) and biotites and is sometimes called mica clay.

The illite clay mineral (Fig. 6-11) consists of an octahedral layer of gibbsite sandwiched between two layers of silica tetrahedra. This produces a 1 : 2 mineral with the additional difference that some of the silica positions are filled with aluminum atoms and potassium ions are attached between layers to make up the

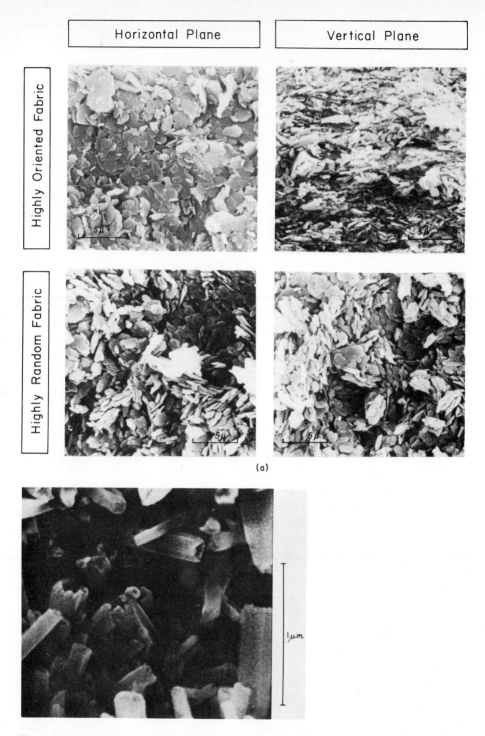

Figure 6-10 SEMs of clay minerals in the kaolinite clay group. (*a*) Kaolinite (*SEM courtesy of Dr. R. J. Kriezek, Northwestern University; read 5µ as 5 µm*); (*b*) Halloysite clay minerals (*from Tovey and Yan, 1973*).

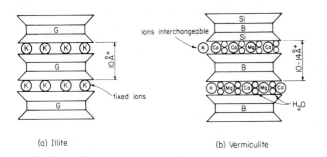

Figure 6-11 (*a*) Illite and (*b*) vermiculite clay minerals. Note, as shown, that seldom will only K ions, in the case of illite, or Mg and/or Ca ions, in vermiculite, be the sole ions found.

charge deficiency. This bonding results in a less stable condition than for kaolinite, and thus the activity of illite is greater.

Vermiculite is a clay mineral in the illite family which is similar except for a double molecular layer of water between sheets interspersed with calcium and/or magnesium ions and substitution of brucite for gibbsite in the octahedral layer.

Illite and vermiculite clays and clay shales are widely used in making light-weight aggregates (sometimes called expanded shale or "vermiculite"). The vermiculite in particular expands considerably on high heating because the water layers quickly turn to steam with resulting large expansions.

Illite clays tend to be found in areas of moderate rainfall, as in the central United States, England, and Europe.

C Montmorillonite

Montmorillonite was the name given (ca. 1847) to a clay mineral found at Montmorillon, France, with the general formula

$$(OH)_4Si_8Al_4O_{20} \cdot nH_2O$$

where nH_2O is the interlayer (*n* layers) of adsorbed water. The term *smectite* is also used for this clay mineral group.

The montmorillonite clay mineral is made of sheetlike units ordered, also as a 1 : 2 unit, as schematically illustrated in Fig. 6-12. The intersheet bonding is mainly due to the van der Waals forces and is, thus, very weak relative to hydrogen or other ion bonding. Various substitutions take place, including Al for Si in the tetrahedra and Mg, Fe, Li, or Zn, for Al in the octahedral layer. These exchanges result in a relatively large net unbalanced negative charge on the mineral, with resulting large cation exchange capacity and affinity for water with H^+ ions in the absence of metallic ions.

Bentonite is a montmorillonitic clay found in partially weathered volcanic deposits in Wyoming, Switzerland, and New Zealand. This clay mineral is particularly active in terms of swelling in the presence of water and has been widely used in drilling oil wells and in soil exploration as a drilling mud and as a clay

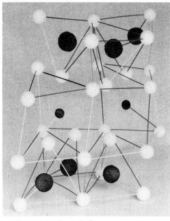

(a)

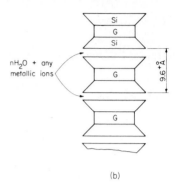

(b)

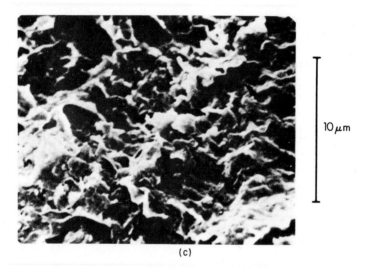

(c)

Figure 6-12 Montmorillonite and bentonite clay minerals. (*a*) Plastic model of montmorillonite; (*b*) schematic of montmorillonite; (*c*) SEM of bentonite (*from Matsuo and Kamon, 1973*).

grout. Bentonite is somewhat variable in its properties depending on the source and amount of weathering of the parent volcanic material. At present, bags of clays of high activity, commercially available for drilling and grouting, are loosely called "bentonite" although they are merely montmorillonite clays.

Weathering of montmorillonite clay minerals often produces kaolinite clay, and in areas where weathering has progressed both minerals are usually present.

Montmorillonite is found in the more arid regions of the world, as in the western United States, Australia, New Zealand, and southern Africa.

6-7 GENERAL CLAY MINERAL PROPERTIES

Several characteristics are similar for all the clay minerals.

A Hydration

Clay particles are almost always hydrated, i.e., surrounded by layers of water molecules called adsorbed water. This layer is often at least two molecules thick and is called the diffuse layer, the double diffuse layer, or simply the double layer. This water is firmly attracted and/or contains metallic ions. A diffusion of the adsorbed cations from the clay mineral extends outward from the surface of the clay into the adsorbed water layer. The effect of this is to produce a net $(^+)$ charge near the mineral particle and a $(^-)$ charge at a greater distance. This diffusion of cations is a phenomenon very similar to the diffused interface between a free water surface and the atmosphere where the diffused material is water molecules. This water is often so firmly attracted it behaves more as a solid than as a liquid, and some researchers report the density $\rho_w \to 1.4$ g/cm^3.

This layer of water may be lost at temperatures higher than 60 to 100°C and will reduce the natural plasticity (reduce w_L, say, 6 to 10 percent) of the soil. Some of this water may be lost by air drying. Generally if the double layer is dehydrated at low temperatures, the plasticity properties may be recovered by mixing with sufficient water and "curing" for 24 to 48 h. If dehydration occurs at higher temperatures, the plasticity properties are permanently lowered.

Clay minerals have sufficient attractive potential to H$^+$ ions that a layer of water up to about 400 Å can surround the particle as illustrated in Fig. 6-13. This qualitatively illustrates the difference between kaolinite and montmorillonite clays in terms of in situ water content and possible liquid-limit values.

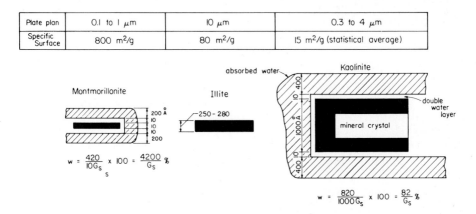

Plate plan	0.1 to 1 μm	10 μm	0.3 to 4 μm
Specific Surface	800 m^2/g	80 m^2/g	15 m^2/g (statistical average)

Figure 6-13 Relative sizes, absorption potential, relative range in water content, grain size, and specific surface for montmorillonite, illite, and kaolinite clay minerals.

B Activity

The edges of all the clay minerals have net negative charges. This results in attempts to balance the charges by cation attraction. The attraction will be in proportion to the net charge deficiency and may be related to the *activity* of the clay. The activity may be defined as

$$\text{Activity} = \frac{\text{plasticity index } I_p}{\text{percent clay}} \tag{6-3}$$

where the percent clay is taken as the soil fraction $< 2 \ \mu\text{m}$. Activity is also related to relative potential water contents as illustrated in Fig. 6-13. Typical activity values based on Eq. (6-3) are as follows:

Kaolinite	0.4–0.5
Illite	0.5–1.0
Montmorillonite	1.0–7.0

Although activity is numerically defined in Eq. (6-3), a better practical indicator of activity is the shrinkage limit. The shrinkage limit (defined in Chap. 3) is the threshold water content to initiate volume change. Activity in terms of volume change is a principal concern in evaluating the soil for use in earthworks and foundations.

The cation exchange capacity in terms of millequivalents per 100 g of clay is also used as an indication of activity. For example, 1 meq of $\text{Na}_2/100 \text{ g} = 0.031$ percent Na_2O. The exchange capacity of several clay minerals is as follows:

Clay	Exchange capacity, meq/100 g
Kaolinite	3–15
Halloysite (4H$_2$O)	10–40
Illite	10–40
Vermiculite	100–150
Montmorillónite	80–150

In practical terms, clays can have their activity in terms of the plasticity characteristics altered by substitution of metallic ions of higher order as in the following substitution scale:

$$\text{Li} < \text{Na} < \text{NH}_4 < \text{K} < \text{Mg} < \text{Rb} < \text{Ca} < \text{Co} < \text{Al}$$

According to this scale, Ca will replace Na or Mg more easily than either Mg or Na will replace Ca. Also, from a practical standpoint, the higher the exchange capacity, the more cations (in some form of admixture) will be required to effect a change in activity.

Many clays, particularly in the southwest part of the United States, tend to swell large amounts when they become saturated. This swelling can be considerably reduced by cation exchange where the usual cation is calcium, furnished by mixing lime with the clay. This process is termed *soil stabilization.* Other additives such as cement and fly ash (a byproduct of burning coal in steam power plants) will produce similar results due to the high concentration of Ca and Al ions in these materials.

A chemical analysis is not generally used to determine the chemical formula of a clay because of the presence of large numbers of different clay minerals. Instead a trial mix procedure is used where the soil is mixed with one or more additives in varying percentages to obtain the optimum percent of the particular additive for field use.

C Flocculation and Dispersing

Clay minerals almost always produce soil-water suspensions which are alkaline (pH > 7) due to the net negative charges on the mineral units. A few exceptions may occasionally occur when the minerals are contaminated with amorphous substances. Due to these charges, the H^+ ions in water, the van der Waals forces, and the small size of the particles, they tend to become attracted together on collision (or even in very close proximity) in a solution. Several particles thus attracted form a randomly oriented *floc* or structure of larger size which will settle out of suspension very rapidly to form a very loose sediment. In the laboratory, a 50- or 60-g clay sample will settle out of a 1000-mL suspension in about 30 min unless floc formation is controlled. To avoid flocculation a dispersed soil-water suspension may be neutralized by adding additional H^+ ions as furnished by acidic materials. The most common material for laboratory work is sodium hexametaphosphate (trademark Calgon, $NaPO_3$), which produces an acidic solution when the dry material is mixed with water. The solution can be checked for acidity by using blue litmus paper (which should turn pink). When the solution is neutralized, the clay particles do not form flocs on collision in the water. Addition of an alkaline material such as sodium hydroxide (NaOH) or alum $[KAl(SO_4)_2]$ will cause very rapid flocculation.

A freshly flocculated clay, in these circumstances, can be easily dispersed back into solution by shaking, indicating that the interparticle attractions are far less than the shaking forces. After the clay has been standing some time, however, dispersion is not easily accomplished, indicating a *thixotropic* (strength gain with aging) effect. Piles driven into saturated soft clays will remold the soil structure in a zone around the pile. Initial load capacity is often extremely low, whereas after 30 or more days of aging the design load may be developed by adhesion between the clay and the pile.

D Water Effects

The water phase of clay soils is not very likely to be chemically pure water. This water accounts for the plasticity properties of clay. In laboratory tests for the Atterberg limits it is specified by ASTM that distilled water be added as required.

The use of distilled water, which is relatively ion free, may produce results somewhat different from those obtained using a more contaminated water such as may enter the soil in situ.

A particular phenomenon of clay is that a clay mass which has dried from some initial water content forms a mass which has considerable strength. If these lumps are broken down to elemental particles, the material behaves as a cohesionless particulate medium. When water is again added, the material becomes plastic with some strength intermediate to the dry lump strength. If the wet clay is again dried, it forms hard, strong lumps. The role of water in this phenomenon is not fully understood, although in drying, surface tension certainly pulls the particles into maximum contact with the very minimum of interparticle spacing so that the interparticle forces are a maximum. It appears that the higher density resulting from packing and the close spacing resulting in the maximum effect of interparticle force attraction give this very high strength. We can readily observe that the strength of the clay varies from a very low value at $S \rightarrow 100$ percent to a very high value at $S = 0$. It is of interest to note that the use of water, which is a dipolar agent, will produce this effect, whereas a nonpolar agent such as carbon tetrachloride (CCl_4) does not. A dipolar agent is one which tends to develop a $+$ and $-$ charge on opposite sides of the molecule. The $+$ charge on one side of a dipole tends to attract the $-$ charge of any material present—including both clay particles and the negative side of other water molecules.

6-8 SUMMARY

This chapter has considered the definition of soil as commonly used for engineering purposes. We have looked at soil in terms of both structure and fabric. Fabric is an appropriate term for describing cohesive soils, but structure, more specifically single-grained structure, is preferable in describing the geometric arrangement of cohesionless soils.

We note that the smallest soil unit of cohesive soils is the *domain*, which in turn forms *clusters* and clusters form *peds* or visible soil crumbs.

The effect of water on granular packing to produce bulking of sands was noted. The concept of relative density D_r to describe packing of cohesionless soils was given, together with observation that large errors may be associated with this concept.

The clay mineral was considered in some detail. It is noted that there are three main groups of clay minerals of particular geotechnical engineering importance. These three groups are:

Kaolinite—least active
Illite—intermediate activity
Montmorillonite—most active

It was also given that the clay mineral develops the plasticity in cohesive soils and that water has a significant effect on clay minerals. It was noted that seldom is a single type of clay mineral found in a soil deposit and that because of this a chemical analysis is seldom directly used in soil stabilization studies. Rather soil stabilization using additives is based on obtaining the optimum amount of material by trial mixes.

Finally we have learned why sodium hexametaphosphate is used as a dispersing agent in the hydrometer test described in Sec. 3-8.

HOMEWORK PROBLEMS

6-1 Verify that the void-ratio for simple cubic packing is 0.91.

6-2 Derive Eq. (6-2) using Eq. (6-1).

6-3 Make a master graph sheet for a D_r versus unit weight graph as used in Fig. E6-1.

6-4 Referring to Ex. 6-1, redo and obtain the range of possible error in D_r using a standard deviation based on the *average* test results. Comment on this procedure vs. using the procedure based on the extreme values of the example.

6-5 Referring to Ex. 6-1, what is the percent chance (approximate) that a single test would give the correct value of D_r? Note that you also have to define what the correct value is! How many tests would be required to give a 95 percent reliability on D_r?

6-6 Given the following data:

$$\gamma_{max} = 21.3 \text{ kN/m}^3 \text{ but the average of 20 tests is 21.0}$$
$$\gamma_{min} = 15.7 \text{ kN/m}^3 \text{ but the average of 20 tests is 16.0}$$
$$\gamma_n = 18.8 \pm 0.3 \text{ kN/m}^3$$

what is D_r and what is the maximum error which might be obtained on a single test?

6-7 Given the following data obtained by two groups on the same soil:

	Group 1		Group 2	
Trial	γ_{max}	γ_{min}	γ_{max}	γ_{min}
1	18.08	15.94	18.19	15.83
2	18.36	15.83	17.78	16.04
3	17.45	15.36	17.97	15.75
4	18.15	16.00	17.92	15.83
5	18.17	15.50	18.09	15.93
6	17.97	15.79	17.97	16.13
7	18.09	15.83	18.23	16.05
8	18.14	15.77	18.17	16.00
9	17.90	15.61	18.19	15.75
10	18.11	16.08	18.03	16.00
11	18.19	15.83	17.98	15.68
12	18.22	15.82	17.98	15.90

REQUIRED:

(a) Compute the standard deviation of each group and of the total of 24 tests and compare.

(b) Compute the relative density of this soil based on $\gamma_n = 17.5 \pm 0.2 \text{ kN/m}^3$.

6-8 Explain how you would obtain the Atterberg limits on a soil where air drying sufficiently to sieve through the standard sieve would alter the limits too much to tolerate. What is the sieve number for this test?

6-9 A 3 percent by weight addition of lime will adequately alter the Atterberg limits and swell characteristics of a clay soil for an airfield. For a treatment to a depth of 1 m, how large an area will a 45-kg sack cover?

HINT: You will have to do some estimating!

Chapter 7
Compaction
and Soil Stabilization

7-1 GENERAL CONCEPT OF SOIL STABILIZATION

When the soils at a site are loose or highly compressible, or when they have unsuitable consistency indices, too high permeability, or any other undesirable property making them unsuitable for use in a construction project, they may have to be stabilized. *Stabilization* may consist of any of the following:

1. Increasing the soil density
2. Adding materials to effect a chemical and/or physical change in the soil material
3. Lowering the water table (soil drainage)
4. Removal and/or replacement of the poor soils

Any alteration of the physical or engineering properties of a soil mass will require investigation of economic alternatives such as relocation on the site or using an alternative site. At present most of the more desirable building sites near urban areas have been used, so that an alternative location may not be practical. Currently, sites such as abandoned sanitary landfills (garbage dumps), swamps, bays, marshes, hillsides, and other poor areas are being used for construction sites, with this trend expected to both continue and accelerate. When alternative sites are not available or environmental considerations, citizen opposition, and zoning regulations severely limit the options available, it becomes more and more necessary to modify or stabilize the site soil to obtain the needed properties. Economically feasible solutions may severely tax the ingenuity of the geotechnical engineer.

In cases such as earth dams, embankments, levees, or other fills, where select materials in sufficient quantities may not be available, selective use of the available materials, and understanding of both the function of the earth structure and the mechanics of the earth mass, can produce a satisfactory solution via use of zoned construction.

7-2 SOIL STABILIZATION

Stabilization is usually mechanical or chemical, but thermal and electrical means have been used on occasion. Mechanical stabilization includes compaction, various patented vibration techniques, and blasting. Compaction will be the major emphasis of this chapter, although other methods will be briefly considered. The reader should consult references such as Mitchell (1968) if a more in-depth study is necessary.

Chemical stabilization includes the mixing or injecting (termed grouting) of chemical substances into the soil. Typical chemical agents include:

Portland cement
Asphalt
Sodium chloride
Lime
Calcium chloride
Paper mill wastes

Grouting is a special stabilization technique involving injection of a thin slurry into the soil; it is used to

1. Decrease permeability (close soil voids) to affect water control
2. Increase shear strength
3. Reduce machinery vibrations by stiffening the soil

The grout slurry is commonly a mixture of

Portland cement and water (usually type III with a water-cement ratio between 0.5 and 5)
Portland cement and water with additional admixtures such as lime, fly ash, clay, sand, silt, etc.
Clay and water or lime and water

Portland cement, lime, and lime–fly ash mixtures are widely used in soil stabilization either to alter the plasticity of a soil to control volume change or to improve its strength.

Plasticity alteration involves a trial mix procedure of adding small amounts of lime or a lime–fly ash mixture in increasing amounts from, say, $\frac{1}{2}$ to perhaps 3 to 5 percent and performing the Atterberg limits to see which percent of the admixture is the optimum.

Strength alteration involves adding select percents, say, from about 4 to 8 percent, based on dry weights, of portland cement or cement–fly ash mixtures. It is usual to develop optimum moisture density curves using standard compaction methods which include the various percents of the admixture. Compression tests are run on a series of samples which are compacted at optimum moisture content (for maximum dry unit weight) after curing periods of 7, 14, and 28 days. The end product is a very low quality concrete with 28-day compressive strengths on the order of 1700 to 2000 kPa.

In both of these stabilization methods, fly ash should be used if economically feasible both to obtain the desired improvement and to provide an environmentally acceptable means of disposal of the enormous quantity of fly ash being produced in coal-fired power plants. A U.S. Department of Transportation study (DOT, 1976) cites numerous uses/methods for fly ash in subgrade stabilization.

Determination of optimum percent admixture is a trial mix procedure. The natural variability of soils is such that a chemical analysis cannot be economically justified. Additional study may be obtained from PCA (1958) and HRR (1968).

Other stabilization methods include drainage (or lowering the water table) and preloading. Preloading consists of applying a temporary surcharge to the soil for a period of 6 months to 1 or more years. The additional load causes settlement, with a decrease in void ratio and an increase in strength. The preload may be accomplished by placing a fill of the necessary depth (1 to 2 m) over the site. A disadvantage is that the fill may later have to be hauled away. Where the top stratum is impervious, an earth dike can be built around the site and water used for the surcharge. Protection to avoid drownings may be necessary if the water depth is much over 1 m.

7-3 SOIL AS A CONSTRUCTION MATERIAL

Soil is one of the most readily available of construction materials at a site, and, when it can be used, it is likely to be the most economical. Earth dams, river levees, and highway and railway embankments are all economical uses of earth as a construction material; however, as with any other construction material, it must be used with quality control. If soils are dumped or otherwise placed at random in a fill, the product will be a fill of low unit weight with a resulting low stability and high subsidence. Subsidence is used to describe the mechanism of vertical movements within a fill due to self-weight; settlement is the vertical movement of the underlying soil caused by the weight of the fill.

Early road fills were usually constructed by end-dumping fill from wagons or trucks with very little attempt to compact or densify the soil and no quality control via specifications. Failures of high embankments were common. Many earthworks, such as levees and earthen dams, are almost as old as mankind, but where the structures have survived it was a happy combination of inadvertent quality control and luck. For example, in ancient China and India, the structures were constructed by workers carrying small baskets of earth and dumping them in the embankment. Laborers bringing additional fill walked over the previously dumped loose earth, compacting it. In some cases herds of goats and sheep were driven across the loose earth for additional compaction with (presumably) the organic matter resulting being picked up by hand and removed. Even today, in some areas, elephants are used to compact earth embankments; however, Meehan (1967) reported that, although the elephant has sufficient mass, he refuses to distribute his walking pattern sufficiently to achieve uniform compaction.

7-4 SOIL COMPACTION

Compaction is the densification of soils by the application of mechanical energy. It may also involve a modification of the water content as well as the gradation of the soil. Cohesionless soils are compacted by some means of confining the soil coupled with vibration energy. In the field, hand-operated vibrating plates and motorized vibratory rollers of various sizes are very efficient in compacting sand and gravel soils. Large falling weights have been used recently in Europe to dynamically compact loose granular fills (Menard and Broise, 1975).

Fine-grained cohesive soils may be compacted in the laboratory by falling weights and hammers, by special "kneading" compactors, and even using static pressure as obtained in the common compression testing machine. In the field, common compaction equipment includes hand-operated tampers, sheepsfoot rollers, rubber-tired rollers, and other types of speciality equipment. Considerable compaction can also be obtained by proper routing of the hauling equipment over the loose soil.

The objective of compaction is the improvement of the engineering properties of the soil mass. There are several advantages which occur through compaction:

1. Reduction in settlements due to reduced void ratio
2. Increase in soil strength
3. Reduction in shrinkage

The principal disadvantages are that swell and frost heave potential are increased.

7-5 THEORY OF COMPACTION

A control specification for the compaction of cohesive soils was developed by R. R. Proctor while constructing dams for the Los Angeles Water District during the late 1920s. The original method was reported through a series of articles in the *Engineering News Record* (Proctor, 1933). For this reason the standard laboratory compaction test is commonly called the Proctor test.

Proctor defined the four variables of soil compaction as

1. Dry unit weight (he actually used the void ratio in the original work)
2. Water content
3. Compaction effort (or energy)
4. Soil type (gradation, presence of silt, clay, large particles, etc.)

Compactive effort is a measure of the mechanical energy applied to the soil mass. In SI units the compaction energy is in kilojoules per cubic meter (kJ/m^3), where $1 kJ = 1 kN \cdot m$. In fps units, $1 ft \cdot lb/ft^3 = 0.047\ 96\ kJ/m^3$. In the field, compactive effort is related to the number of passes of the particular piece of compaction equipment on a given volume of soil. Compactive effort is seldom a specification criterion in field work due to the difficulty of measuring it. Rather, the end product of dry unit weight of the soil, use of a particular type of compactor, or a specified number of passes with a given piece of equipment is used.

In the laboratory the compaction energy may be developed by impact (most common), kneading, or static means. During impact compaction a hammer is allowed to drop several times on a soil sample in a mold. The size of the hammer, height of the drop, number of drops, number of layers of soil, and volume of the mold are specified. The standard compaction (also Proctor, standard AASHTO or ASTM) test uses a 105-mm- (most common) or 152-mm-diameter mold and consists in:

	SI units		Fps units
Hammer	24.5 N		5.5 lb
Height of fall	0.305 m		12 in
Number of layers		3	
Blows/layer		25	
Mold volume (105-mm-diameter)	0.000 942 2 m³†		1/30 ft³
Soil	(−) No. 4‡ sieve		

† The author suggests using a 1000-cm³ (0.001-m³) mold with dimensions of 10.3 diameter by 12 cm with 3 layers at 26 blows for the standard SI version of this test. Comparison of tests using this new mold and the standard 1/30 ft³ with a soft conversion gives identical results. With any new mold it will be necessary to hold the L/d approximately constant; otherwise wall friction will affect the unit weight.

‡ Soil passing the 19-mm sieve is also used, but usually with the 152-mm-diameter mold, and ASTM should be consulted for procedures when the soil contains these larger particles.

The impact compactive effort for the standard compaction test is readily computed.

In SI units:

$$CE = \frac{24.5\ N \times 0.305\ m \times 3\ layers \times 25\ blows/layer}{0.000\ 942\ 2\ m^3 \times 1000} = 594.8\ kJ/m^3$$

In fps units:

$$CE = \frac{594.8}{0.047\ 96} = 12\ 400 \text{ ft} \cdot \text{lb/ft}^3$$

For the kneading and static compaction methods, the computation of compaction energy is extremely complex. In kneading compaction, the tamper kneads the soil by applying a given pressure for a fraction of time. The kneading action is supposed to simulate the compaction produced by a sheepsfoot roller in field compaction. In static compaction, the soil is simply pressed into a mold by a static pressure of some magnitude, and side friction on the mold becomes a significant factor since a lateral pressure is developed which is related to the vertical pressure according to Eq. (15-1). Impact and kneading compaction can only be applied to cohesive soils, static compaction to any soil.

In the standard compaction test several samples of the same soil, but at different water contents, are compacted with the same compactive effort into the standard mold of known volume. The wet unit weight and water content samples are taken. After the water content for each test is obtained, the dry unit weight is computed and a curve of dry unit weight versus water content is plotted as in Fig. 7-1. Generally five points are required to obtain a reliable curve, with the water content spread between points not over 3 percent. The calculations for the curve points are made as follows:

$$\gamma_{\text{wet}} = \frac{\text{weight of wet soil in compaction mold}}{\text{volume of compaction mold}} \qquad (7\text{-}1)$$

and after obtaining the water content,

$$\gamma_{\text{dry}} = \frac{\gamma_{\text{wet}}}{1 + w} \qquad (7\text{-}2)$$

The compaction curve is unique for a given soil type, method of compaction, and compaction effort. The peak point of the compaction curve corresponds to the maximum dry unit weight and optimum moisture content (OMC). Fill specifications are always based on the maximum dry unit weight and sometimes on both the weight and the OMC.

Typical values of maximum dry unit weights range from 16 to 20 kN/m^3, with the maximum values ranging from 13 to 24 kN/m^3 (see typical range in Fig. 7-2). Values of $\gamma_d > 23$ kN/m^3 are rare, since this value is about that of wet concrete. Typical optimum moisture contents range from 10 to 20 percent, with a maximum range of 5 to 30 percent. Shown on Fig. 7-1 are curves representing the degree of saturation of the soil at various water contents. The saturation curves depend on both S and G_s, those shown being for $G_s = 2.68$ as given on the figure for the soil. Note that at OMC for the soil shown on Fig. 7-1 the degree of saturation is approximately 75 percent. Observe also that the compaction curve, even at very high water contents, never reaches the zero air voids curve ($S = 100$ percent). This is true for both lower and higher compactive efforts (such as curve b) and may be

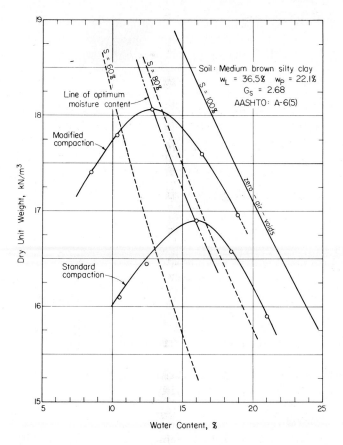

Figure 7-1 Standard and modified compaction test curves for a clayey glacial soil from near Peoria, Ill.

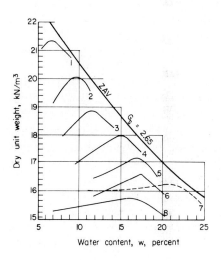

Figure 7-2 Typical standard compaction test curves for soils shown. (*After Johnson and Sallberg, 1960.*)

attributed to the fact that the grains are free to move and the water is not confined; thus, some air will always be present in some of the voids. Curve *b* of Fig. 7-1 is a compaction curve obtained using the *modified compaction* (also modified AASHTO or modified ASTM) *test*. This test uses the same compaction mold† and:

$$
\begin{aligned}
\text{Hammer} &= 44.5 \text{ N} \\
\text{Height of fall} &= 0.457 \text{ m} \\
\text{Layers and blows} &= 5 \text{ at } 25
\end{aligned}
$$

The reader should verify that the compaction energy for this test is 2698 kJ/m³. The modified compaction test was developed during World War II for airfield subgrade compaction to support the heavy military aircraft.

Curve *b* illustrates a consistent phenomenon found in soil compaction—that increasing the compaction energy results in increased dry unit weight and decreased OMC. It should be noted from the above compaction energy computation that the increase in compaction energy is not proportional to the increase in dry unit weight. The modified compaction test uses over four times the energy of the standard compaction test. For this additional energy input the dry unit weight increase is seldom over 10 percent.

A line drawn through the peak points of the family of compaction curves obtained from the different compactive efforts on the same soil will approximately parallel the zero-air-voids curve. This line is called the *line of optimums*. Typical compaction curves for different types of soils are illustrated in Fig. 7-2. One may note a basic similarity in the shape of all the curves for the several soils shown on this plot. The general trend shown for higher unit weights for soils with lesser amounts of clay is found worldwide.

These curves illustrate that at low water contents the soil is dry and not enough water is present to effect a breakdown or softening of the clay lumps. Compaction in this moisture content range depends primarily on whether the energy input is sufficient to crush the lumps. As the water content increases, the particles have sufficient water to slake the lumps and develop water films. The water films on the clay particles reduce the interparticle bonding by increasing the spacing so that the compaction energy tends to produce a more dispersed structure. A small contribution of water lubrication at the particle contact points may be developed. A water content is reached where this condition is optimized. Beyond this point the degree of saturation becomes such that the hammer blows cause large instantaneous pore pressures to develop. This results in local shear failure around the hammer, so that displacement of the soil mass occurs instead of particle packing. These shear displacements mean that no matter how much additional water or compaction effort is input, the soil can never obtain a state of $S = 100$ percent.

---

† A 152-mm-diameter instead of the 105-mm-diameter mold is sometimes used in certain ASTM versions of this and the standard compaction test.

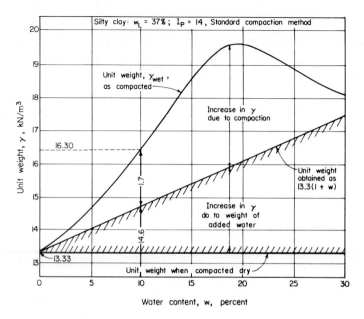

Figure 7-3 Moisture content vs. unit weight relationship showing the effect of addition of water as both increased unit weight and increased compaction unit weight. (*From Johnson and Sallberg, 1960.*)

The effect of water on compaction is also illustrated in Fig. 7-3 for a particular soil which has an initial dry unit weight of 13.30 kN/m³ at $w = 0$ percent. Water is added and a specimen is compacted. The water content and *wet* unit weight are computed as 10 percent and 16.3 kN/m³, respectively. This can be interpreted as

$$\gamma_{d(\text{initial})} + 0.10\gamma_d + \text{additional weight of wet compacted soil}$$

Substituting,

$$13.3 + 0.10\gamma_d + [16.3 - 13.3(1.1)] = 16.3 \text{ kN/m}^3$$

The dry unit weight corresponding to the water content of 10 percent is

$$\gamma_d = \frac{16.3}{1 + 0.1} = 14.8 \text{ kN/m}^3$$

If we had simply added 10 percent water to the dry soil with no compaction, the wet unit weight would have been

$$\gamma_{\text{wet}} = 13.3(1.1) = 14.63 \text{ kN/m}^3 \qquad \text{(not the 16.3 obtained by compaction)}$$

The actual water added is $16.3 - 14.8 = 1.5 \text{ kN/m}^3$ and not $13.3(0.1) = 1.33 \text{ kN/m}^3$. This added water ($1.5 \text{ kN/m}^3$) and the compaction process produced an additional increase in dry unit weight of

$$\Delta\gamma_d = 14.8 - 13.3 = 1.5 \text{ kN/m}^3$$

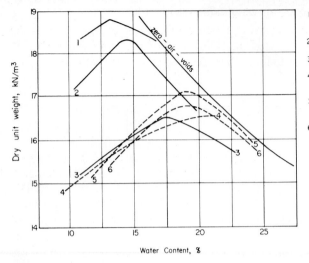

1 = laboratory static compaction at 13790 kPa.

2 = Modified AASHTO compaction.

3 = Standard compaction.

4 = Laboratory static compaction at 1379 kPa.

5 = Field compaction, rubber-tired load at 6 passes.

6 = Field compaction, sheeps-foot roller at 6 passes.

Figure 7-4 A comparison of field and laboratory compaction unit weights. (*Turnbull, 1950; see also Lambe and Whitman, 1969.*)

At the approximate OMC of 20 percent, the addition of 3.28 kN/m³ of water plus compaction energy produces an increase in dry unit weight of

$$\Delta\gamma_d = 16.4 - 13.3 = 3.1 \text{ kN/m}^3$$

At a water content of 25 percent, the change in dry unit weight is

$$\Delta\gamma_d = 15.0 - 13.3 = 1.7 \text{ kN/m}^3$$

thus the addition of 5 percent additional water results in a decreased dry unit weight from the value at OMC. The reader should extrapolate the wet unit weight curve and find the water content at which no increase in dry unit weight due to compaction is obtained (i.e., the alternative location of $\gamma_d = 13.3$ kN/m³). This curve clearly illustrates that water is needed to effect an increase in unit weight, but too much water is undesirable.

Compaction behavior of cohesive soils as described here is typical for both laboratory and field compaction procedures. The curves obtained from the two different procedures will vary somewhat with OMC and maximum dry density, but they will have similar shapes, as shown in Fig. 7-4.

The laboratory compaction tests were developed as a standard of quality control for field compaction. The correlation between field and laboratory compaction is not direct, since the laboratory uses impact energy and most field compaction methods use kneading, or a combination of kneading and static pressure. The Harvard Miniature Compaction test was developed by Wilson (1950) to attempt to overcome this deficiency, but this test has not proved very popular due to the small sample size. The miniature equipment is useful to compact small laboratory samples for student use in strength testing.

Numerous factors influence the compaction of soils, both in the laboratory and in the field. A complete discussion is beyond the scope of this text, and the interested reader should consult publications such as Johnson and Sallberg (1960, 1962), USBR (1968), or various laboratory texts such as Bowles (1978).

7-6 PROPERTIES AND STRUCTURE OF COMPACTED COHESIVE SOILS

The structure and, thus, the engineering properties of compacted cohesive soils will depend greatly on the method or type of compaction, the compactive effort, the soil type, and the water content. Usually the water content of compacted soil is referenced to the OMC for the given type of compaction. Depending on the relative position, this may be "dry of optimum," "near or at optimum," or "wet of optimum." Research on compacted clays has shown that when they are compacted dry of optimum, the structure of the soil is essentially independent of the type of compaction (Lambe, 1958; Seed and Chan, 1959). Wet of optimum, however, the type of compaction has a significant effect on the soil fabric and thus on the strength and compressibility of the soil.

The fabric of compacted clays is about as complex as the fabric of natural clays, discussed in Chap. 6. At the same compactive effort, with increasing water content the soil fabric becomes increasingly oriented (or dispersed). Dry of optimum, the soils tend to produce a flocculated (or cardhouse) fabric. This is qualitatively illustrated in Fig. 7-5, with the fabric at B more dispersed than at A. If the compactive effort is increased, the soil tends to become more dispersed even though the water content remains constant, as at point E. The sample fabric is considerably more oriented at C than at A for the same energy, since it is wet of

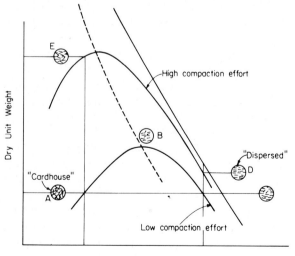

Figure 7-5 Qualitative effect of compaction on soil fabric and structure. (*After Lambe, 1958.*)

optimum. Also the fabric at D will be more oriented than at C for the same water content due to the increased compaction effort.

Flow of water through soil (permeability, discussed in Chap. 8) depends on the void ratio and fabric orientation and will decrease

1. At constant compactive effort and increasing water content, since tests have shown a dispersed fabric to be less permeable
2. When the compactive effort is increased, both causing a decrease in void ratio and producing a more dispersed fabric

Compressibility (Chap. 11) of compacted clays seems to be a function of the stress level imposed on the soil mass. At relatively low stress levels, clays compacted wet of optimum are more compressible. At high stress levels, clays compacted wet of optimum are less compressible. Langfelder and Nivargikar (1967) consider the several factors affecting the compressibility and strength of compacted clays.

Swelling of compacted clays is greater for those compacted dry of optimum. This may be due to the random structure orientation producing more sensitivity to water at the contact points. It may also be that at wet of optimum $S \rightarrow 100$ percent, which is the usual long-term condition causing swelling of in situ clays in subgrades and beneath floor slabs in buildings. In this case the soil starts at about the same water content as the final state, whereas an initial dry of optimum condition requires a state change to obtain $S \rightarrow 100$ percent.

The strength of cohesive samples compacted dry of optimum is larger and they tend to brittle failure, while those compacted wet of optimum tend to lower strengths and progressive failure (see Fig. 14-1b). The strength wet of optimum also depends somewhat on the type of compaction. Water content is a particularly significant parameter in soil strength due both to softening of the interparticle bonds and to swelling. Of particular interest is that the clay cores of dams might be compacted on the wet side so that a large differential settlement can be tolerated without cracking. The subgrade of a highway fill might be compacted on the dry side of optimum for strength considerations; if this is balanced against swell and/or strength loss if the fill becomes saturated, compaction on the wet side of optimum might be alternatively specified.

7-7 EXCAVATION AND COMPACTION EQUIPMENT

Soil to be used in a compacted fill is excavated from a borrow area. The borrow may be either on or off site. Power shovels, draglines, and self-propelled scrapers are used to excavate the borrow material. These devices may cut through layers of different materials, allowing the several soils to become mixed. The power shovel mixes the soil by cutting along a vertical surface, while the scraper mixes the soil by blading across a sloping surface where different layers may be exposed.

The borrow area may be near or several kilometers from the fill site. Various types of equipment are used to transport and spread the soil on the fill area. These include scrapers, conventional trucks, and trucks especially made for this purpose. It is preferable to spread the material when dumping to save spreading time. Often, however, the borrow requires processing prior to compaction, which may include running the soil through a processor which thoroughly mixes it, simply spreading the dumped earth and using a farm disc to pulverize it, adding water, or drying the soil. The layer of loose earth placed over a previously compacted layer is called a *lift*. Lifts range from 15 to 50 cm depending on the size and type of equipment and the type of soil to be compacted.

The compacting equipment used depends on the type of soil (and on the equipment available to the contractor). Equipment is available which applies pressure, vibration, vibration and pressure, and kneading energy. The equipment may be self-propelled or towed. Figure 7-6 illustrates several types of compaction equipment common to fill projects. Particular features of several types of compaction equipment include:

1. *Smooth-wheel rollers.* These rollers give 100 percent coverage beneath the wheel with contact pressures up to about 400 kPa. These may be used on all types of soil except when large boulders are present. These rollers are also called steel-wheel rollers. They are particularly suited for cohesionless soil with and without vibration devices attached. They are very efficient when using flooding for sand compaction. These rollers are also commonly used for finish rolling of subgrades and base courses and compacting asphalt pavements; rollers are self-propelled.

2. *Pneumatic or rubber-tired rollers.* These rollers give about 80 percent coverage; tire pressures go up to about 700 kPa. Several rows of four to six closely spaced tires with front and rear spacing alternated are used for the higher percent coverage cited. They may be towed but are generally self-propelled. These rollers may be used for either cohesive or cohesionless soil. The tires are sometimes misaligned vertically (wobbly wheel), to produce a kneading action on the soil.

3. *Segmented wheel compactors.* These produce about 60 percent coverage and generate contact pressures up to 1000 kPa. They are designed for cohesive soils, and the lift is generally restricted to 15 cm.

4. *Tamping foot rollers.* These rollers produce about 40 percent coverage and contact pressures from 1400 to about 8500 kPa depending on the diameter of the roller and whether the drum is filled for added weight. The tamping foot roller has small, rectangular feet similar to the sheepsfoot roller. Compaction begins in the soil below the foot projection (about 15 to 20 cm depth), and the depth of penetration is successively less on subsequent passes (called "walking out"). These rollers are only suitable for cohesive soils. They are commonly self-propelled, with either two rollers for front wheels or four rollers for both sets of "wheels."

5. *Sheepsfoot rollers.* These rollers produce about 8 to 12 percent coverage due to

Figure 7-6 Compaction equipment of several types. (*a*) Smooth wheel rollers; the roller on the left is equipped with a vibratory device; (*b*) grid roller; (*c*) sheepsfoot rollers towed by large agricultural type tractor; (*d*) steel wheel compactor in sanitary landfill; (*e*) self-propelled tamping foot roller; (*f*) using a disc to process fill soil prior to compacting.

the small "sheeps foot" projections of 3 to 8 cm^2 area. The drum can be filled with water or sand to increase the weight. Contact pressures range from about 1400 to 7000 kPa. Sheepsfoot rollers are commonly towed in parallel or two in parallel with a trailing roller to cover the gap between the front rollers. Six to eight passes (producing 40 to 60 percent coverage) will usually obtain the required density due to the spreading of the contact pressure from the sheepsfoot tips. This roller is only suitable for cohesive soil.

6. *Mesh or grid pattern rollers.* These produce about 50 percent coverage, and contact pressures range from 1400 to about 6000 kPa. This roller is useful for compacting rocky soil, gravels, and sands.

Table 7-1 Compaction equipment and soil compaction applications

Soil group	Soil type		Degree of compaction	Typical compaction equipment	Shallow fills and backfills					Deep foundations	
					No. of passes or coverages	Lift thickness, mm	Placement w, %	γdry, kN/m³	Field control	Comp. method	Field control
Pervious or free-draining	GW GP SW SP	Compacted	95 to 105 percent of standard compaction test or 70 to 85 percent of D_r	Steel-wheeled roller or vibratory compactor; Rubber-tired roller; Crawler-type tractor; Hand tampers (mass > 45 kg)	As required; 2–5; 2–5; As required	As required; 300 mm; 200; <150	Saturate by flooding	17–21	Field density tests at random locations	None available except for near surface as listed at left	Undisturbed samples from borings or test pits; SPT before and after compaction
	GW GP SW SP	Semi-compacted	90 to 95 percent of standard compaction test or 60 to 70 percent of D_r	Rubber-tired roller; Crawler tractor; Hand tampers; Controlled routing of construction equipment	2–5; 1–2; As required	350; 250; <200	Saturate by flooding	16–20	Field density tests at random locations	Vibroflotation, compaction piles, sand piles, explosives, and surface methods	
Semipervious to impervious	GM GC SM	Compacted	95 to 105 percent of standard compaction test	Rubber-tired roller; Sheepsfoot roller; Hand tampers	2–5; 4–8; As required	200; 150; <100	OMC based on lab compaction test	16–20	Field density tests at random locations to determine RC	Preload fills; Lower water table; Generally consolidation theory applies	
	SC ML CL OL MH CH OH	Semi-compacted	90 to 95 percent of standard compaction test	Rubber-tired roller; Sheepsfoot roller; Crawler tractor; Hand tampers; Controlled routing of construction equipment	2–4; 4–8; 2–4; As required; As required	250; 200; 150; <150; <200		14–19			

Notes: 1. Rubber-tired rollers with tire pressures of 550 to 700 kPa.

2. Sheepsfoot rollers with foot pressures on order of 1700 to 3500 kPa.

3. Crawler tractors weighing over 85 kN and track pressures greater than 45 kPa.

4. A *pass* is made with a sheepsfoot roller; *coverage* is made with rubber-tired rollers, steel-wheel rollers, and crawler tractors. Coverage includes 100 percent of surface, whereas sufficient compaction may be obtained with 3 to 5 passes which includes only 45 to 50 percent of surface area.

Vertical vibrators are attached to the steel-wheel rollers to densify granular soil more efficiently. The principle is that of confining the soil across most of the contact area, applying pressure from the weight of the device, and using vibration to break or dislodge the particle contact points so that particle slip can occur. Small hand-towed vibrating plates are available ranging from about 0.23 to 1.2 m square with masses from 450 to 5000 kg. These devices are for use in small areas such as culvert, pipe, or trench backfill; behind retaining walls, and adjacent to foundation walls for buildings. Effective compaction depth ranges from about 7 to 20 cm. Table 7-1 tabulates recommended compaction equipment for several soils based on the Unified Soil Classification system.

7-8 COMPACTION SPECIFICATIONS

The objective of compaction is to improve the engineering properties of the soil. The dry unit weight should be specified to accomplish this goal and not simply specify, "the soil shall be compacted to 95 percent of the unit weight obtained in the standard compaction test". Many standard soil specifications contain clauses such as this for ease of specification writing. Unfortunately, in many cases these standard clauses are used without any real concern for engineering properties. Assume that an engineering analysis has determined that the field compaction should be some percent of that obtained on the laboratory compacted sample after inspection of the moisture-dry unit weight curve to see what the OMC and maximum dry density value is. Presumably enough classification tests or strength tests have been performed so that judgment now applies, and it is decided that the relative compaction (percent of that obtained in the laboratory) is defined as

$$RC = \frac{\text{field dry unit weight}}{\text{maximum dry unit weight from laboratory test}} \times 100 \quad \text{percent}$$

(7-3)

which sets the value of the field dry unit weight to be obtained. The value of RC is typically from 105 to about 90 percent, usually based on γ, index properties, classification, and previous performance. Note that it is defined from the laboratory test and may be the standard, modified, or some other amount of compaction energy. Coupled with RC may be a compaction water content requirement, say, 3 percent wet of optimum, optimum ± 1.5 percent, 3 percent dry of optimum, etc., based on climatic factors, fill use considerations, and other factors such as those already considered.

The reader should note that there is a considerable difference between RC and relative density as defined by Eq. (6-1). Relative density (if used) applies only to cohesionless soils with little $(-)$ No. 200 fines; however, the author has successfully used a unit weight comparison instead of D_r on numerous occasions. The compaction test is generally used if the soil contains more than 12 percent fines, but there is no reason that some type of unit weight test cannot be used for all soils to establish compaction specifications.

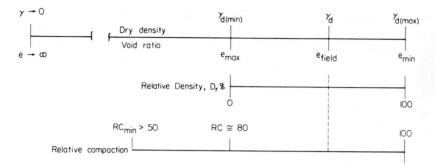

Figure 7-7 Relationship between relative density, density, and relative compaction. (*After Lee and Singh, 1971.*)

A relationship between relative density and relative compaction is shown in Fig. 7-7. A statistical study by Lee and Singh (1971) on 47 different granular soils indicated that the relative compaction corresponding to zero relative density is about 80 percent. It should be further noted that relative compaction can never be less than about 80 percent, since soil always has a unit weight and simply dumping it in a pile will produce unit weights on the order of 12 to 14 kN/m³.

Unit weight and/or water content (end product) specifications are used for most highways and buildings. As long as the contractor is able to obtain the specified relative compaction, it does not matter how it is obtained or what equipment is used. Project economics should ensure that the contractor will utilize the most efficient compaction procedures. The most economical compaction

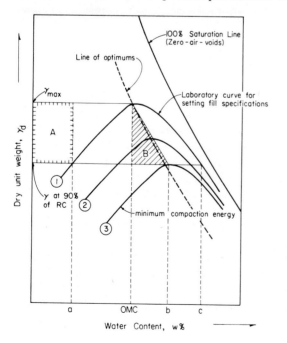

Figure 7-8 Dry unit weight vs. water content illustrates most economical field compaction conditions. Assuming that the placing of water content is not a significant parameter, any w in zone B will result in maximum compaction efficiency.

conditions are illustrated in Fig. 7-8, which shows three qualitative compaction curves for the same soil at different compactive efforts. A study of these curves, assuming that curve 1 represents the curve which can be obtained with existing equipment, indicates that to achieve, say, 90 percent relative compaction, the placement water content of the fill must be greater than water content *a* and less than water content *c*. These points are found where the 90 percent RC line intersects curve 1. If the fill water content is not in the range *a* to *c* as in zone *A* of Fig. 7-8, it will be difficult, if not impossible, to obtain the required RC without increasing the compactive effort by increasing the number of passes or using heavier equipment. This is why it is necessary to wet (by sprinkling) or dry (by discing) the soil prior to rolling.

Referring again to Fig. 7-8, from a purely economical standpoint, the most economical water content would be at *b*, where the minimum compactive effort is needed to attain 90 percent RC. However, consistently achieving the minimum RC for a project requires the use of a slightly higher compactive effort, as in curve 2. With this consideration (noting that there will be some inevitable fluctuation in water content due to environmental factors), the most economical fill water contents exist in the hatched zone *B* between the OMC and *b* and are "wet of optimum."

7-9 FIELD CONTROL OF COMPACTION

Field tests are required to determine if the specifications for relative compaction have been met. The test location(s) should be representative or typical of the compacted lift and borrow material. Typical job specifications may specify a field test for unit weight† for every 1000 to 3000 m³ of fill or when the borrow material changes significantly. It may be necessary to make the field test on the lower lift if for some reason inspection was temporarily suspended, especially if sheepsfoot rollers are used. Likely sites for tests may be randomly selected, determined by probing with a 12- to 15-mm bar (for soft spots), or based on judgment. On large projects a table of random numbers (Table 7-2) can be used to obtain test location coordinates to avoid technician bias in selecting the locations (Sherman et al., 1967).

Example 7-1

GIVEN A fill area 16 × 150 m and two lifts.

REQUIRED Use a table of random numbers to obtain coordinates for four field test locations in each of the two lifts.

---

† This is usually called a field density test. This terminology is likely to continue and will be one of the "problems" in using SI.

Table 7-2 Table of random numbers. Use entire number, first digit, second digit, first two digits, etc., of each entry as a random number

	1	2	3	4	5	6	7	8	9	10
1	16057	43688	44334	73470	36029	98628	65873	95981	16409	39478
2	90781	31691	58816	84699	42103	30877	64837	18638	74650	69832
3	16912	50831	52883	46964	28355	97607	96167	47999	21406	63691
4	83985	19760	93957	19489	91200	77460	93463	20248	29432	53652
5	23887	61081	78790	94252	92003	68007	95276	89500	74668	48696
6	58466	88644	27123	90111	31217	38330	30489	54761	83117	93031
7	29299	82138	53630	92397	59271	17913	25445	23067	26689	17551
8	49598	95030	77669	43385	74670	62512	78271	45552	12081	21582
9	33896	30936	83344	28733	14580	13850	40619	65704	32284	71527
10	56994	30339	10399	81934	15452	20487	39270	84252	84969	34925
11	36894	36092	56141	33005	61489	98271	70052	14176	19258	65804
12	84181	48602	27486	60619	29858	52449	66877	26789	50822	80501
13	18461	59131	59651	66667	25752	47027	70791	33144	31610	90766
14	35004	17366	10528	54429	85783	34960	22855	57764	32264	20082
15	43873	72014	26305	26562	82931	86516	44909	48288	90932	36015
16	48204	75791	63079	69512	25532	44530	94331	69846	64175	40864
17	94364	66167	41721	78013	68174	48069	65946	84485	32139	12772
18	32341	57576	82616	42470	11107	48341	46697	78290	41703	88563
19	44772	85906	20207	52697	52904	57420	86667	93784	21296	38184
20	77562	97250	77693	33952	31771	94514	99730	51652	52445	49397
21	62417	10349	84296	29440	43012	39920	18373	25506	23004	52726
22	76442	86840	65332	71065	72133	54460	12671	88171	66015	71785
23	96900	27897	80531	61432	45690	67476	10666	99470	89055	56170
24	93419	85823	24387	29005	24944	14714	49948	32762	40196	90955
25	88405	63859	21544	21142	38450	81002	36981	24422	39675	40194
26	89660	10750	97814	65802	62022	53586	72648	17047	76282	22567
27	42314	89765	64787	45291	91159	92491	26070	82484	61974	73654
28	46944	41836	69151	34955	91135	72669	61806	41477	17172	38816
29	83740	43131	21062	68972	26378	35988	47376	65070	97459	70006
30	82523	82256	20151	99233	38053	67565	28000	65743	30203	70695
31	87602	53569	31347	74647	90912	73002	41613	35895	40808	94110
32	82906	91833	67205	54785	29921	29927	97673	23682	52653	28407
33	90161	27905	58192	56878	99741	22540	11519	19039	79195	23449
34	79382	90238	88670	48204	71521	65893	90058	10930	12892	85028
35	66484	16313	82920	77322	31492	92498	36672	35206	94974	99123
36	27462	85739	51713	30093	45021	61949	95123	14020	58246	29479
37	66414	67612	35651	84658	87061	39212	27017	91553	94703	84771
38	51325	34019	48488	37111	17171	77727	76492	36703	44002	43310
39	10219	95606	70067	60742	19477	13451	15798	96236	11690	96916
40	99391	67465	84013	30452	62321	81155	63610	47089	74115	91706

SOLUTION Use Table 7-2 and interpret the first two digits of column 1 and the first two digits of column 2 of every fifth line to give the X and Y coordinates from the left end of Fig. E7-1. Obtain:

Lift	1		2	
Test	X	Y	X	Y m
1	16	4.3	77	9.7
2	23	6.1	88	6.3
3	56	3.0	82	8.2
4	43	7.2	6.6	16

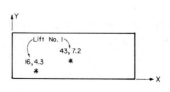

Obviously other combinations could be used, but these coordinates reasonably cover the fill area (at least statistically for four tests on each lift).

Field testing may be either *destructive* or *nondestructive*. Destructive testing involves excavation and removal of some of the fill material (leaving a hole to be filled and compacted by the contractor later), while nondestructive testing determines the unit weight and water content of the fill indirectly.

A Destructive Testing

The steps required for the common destructive field tests are as follows:

1. Locate the test site.
2. Excavate a hole in the fill at the desired depth. The size of the hole will depend on the maximum size of fill material and should be as large as the equipment limitations allow. Weigh the wet soil as W_t.
3. Take a water content sample (some organizations use the entire sample) to determine the fill water content.
4. Measure the volume of the excavated material by using the hole. Techniques commonly employed include the sand cone and balloon method (Fig. 7-9) and sometimes the pouring of oil or water or other fluid of known unit weight into the hole.

   In the sand cone method, dry, uniform sand [often $(-)$ No. 20 and $(+)$ No. 30 sieve size or "Ottawa" sand from commercial suppliers] of known unit weight is allowed to flow through a standard cone pouring device into the hole. The cone is used to approximately control the flow rate and height for a more reproducible unit weight. The "one size" sand is similar to the equal spheres concept of Chap. 6 and is also for controlling the unit weight. The hole volume is obtained by proportion using the unit weight of the sand and determining the weight used to fill the hole. The sand cone is considered to be the most reliable means of determining in situ unit weights and is the calibration standard for other methods.

   The volume of the hole is determined directly in the balloon method, where a rubber balloon is expanded into the hole from a calibrated cylinder using air pressure on the water. A "before and after" reading on the calibrated cylinder gives a direct reading of the hole volume. Care must be used in this method not to apply too large an expansion pressure on the balloon, since this will enlarge the hole and reduce the computed unit weight.

5. The wet unit weight is computed as $\gamma_{wet} = W_t$/volume of hole.
6. The dry unit weight is computed using Eq. (7-2).
7. A comparison of dry unit weight is made to $\gamma_{dry(max)}$ from the compaction curve, and RC is computed using Eq. (7-3).

(a)

(b)

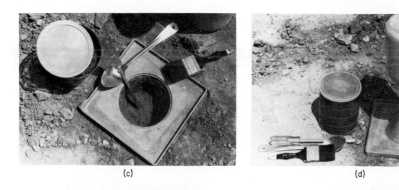

(c)

(d)

(e)

(f)

Figure 7-9 Field unit weight equipment. (*a*) Sand-cone equipment: jug, cone, template, digging tools, and can to recover wet soil. (*b*) The site is smoothed and the template positioned. (*c*) The hole is dug as large as practical; the can should be about 80 percent filled. (*d*) Cover the can and position the sand jug to fill hole with sand. (*e*) Balloon equipment; volumeasure, digging equipment, template, and can to recover wet soil. (*f*) The hole is dug and the volumeasure positioned on the template—note that the water level indicates the hole is nearly filled.

The principal problem associated with destructive testing is the time lag required to oven-dry water-content samples, which may be as long as 24 h. Recently devices have become available which allow moisture determinations almost instantly using chemicals. These determinations may require occasional comparison with oven-dried samples for "calibration." Other problems of somewhat lesser importance include:

1. Backfilling the test holes—this is particularly critical in dam construction
2. The setting up of a reasonable specification criterion for how many unsatisfactory field tests will produce a fill that is not acceptable, since it will be nearly impossible to find every test meeting RC and/or water-content specifications
3. No compaction test curve being available due to changes in borrow or in lack of blending, or in blending, to produce a fill different than that used for the laboratory compaction curve

B Nondestructive Field Testing

Nondestructive unit weight tests and water content determinations using radioactive isotopes have increased in popularity in recent years as the equipment has gotten safer, easier to use, and cheaper. Nuclear methods have several advantages over traditional tests, including:

Making more tests
Obtaining unit weight and water contents as direct readouts

Making more tests gives a much better statistical control of the fill and covers a much larger area. Principal disadvantages include:

High initial cost of the equipment
Potential danger from radioactivity
Necessity for careful and repeated calibration checks, usually obtained by constructing a block of soil about $40 \times 40 \times 20$ cm with known density and water content. Alternatively, calibrate using the sand cone or balloon test
Small effective depth of density determination (on the order of 2 to 5 cm depending on the strength of the radioactive isotope used)
Need for careful preparation of the fill surface, primarily because of the small effective depth of the test

Generally nuclear devices can be justified only for large projects where a large number of tests will be required. These devices operate essentially by emitting radiation into the soil for 1 to 3 min. The pickup is activated to detect radiation not absorbed by the soil and water. The amount of gamma rays recorded is related to the unit weight and/or the amount of neutron loss is related to the amount of hydrogen atoms present, which in turn is related to the water content.

Example 7-2

GIVEN Field unit weight test with the following data obtained:

Weight of wet soil removed from test hole = 1942 g
Weight of sand used to fill hole and cone = 2744 g
Density of sand = 1.60 g/cm^3
Weight of sand to fill cone = 1289.7 g (by calibration)

REQUIRED Find the in situ wet and dry unit weights if the entire sample is oven-dried to 1708.7 g.

SOLUTION By proportion the volume of the hole is

$$V_{hole} = \frac{2744 - 1289.7}{1.60} = 908.9 \text{ cm}^3$$

The wet unit weight is immediately computed as

$$\gamma_{wet} = \frac{1942}{908.9} (9.807) = 20.95 \text{ kN/m}^3$$

The dry weight is also directly obtained, since the entire sample was dried, to give

$$\gamma_{dry} = \frac{1708.7}{908.9} (9.807) = 18.44 \text{ kN/m}^3$$

The in situ water content is

$$w = \frac{1942 - 1708.7}{1708.7} (100) = 13.7 \text{ percent}$$

7-10 STATISTICAL FIELD UNIT WEIGHT CONTROL

Seldom do 100 percent of the field tests for unit weight meet specifications, just as steel or concrete quality control tests also do not. Since it would be unfair to the contractor to require all the test locations not meeting specifications to be excavated and replaced, it is necessary to establish how many tests have to fail to meet specifications before

1. The work is considered unacceptable and the soil must be removed and replaced.
2. The contractor must accept a lower unit price for the work.

Formerly the acceptance criteria depended considerably on a "feel" for the project, which could be somewhat arbitrary, depending on the project engineer. Now it is more appropriate—especially for large projects—to quantify the "feel"

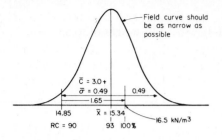

Figure 7-10 Normal distribution curve as would be obtained from plotting a number of field tests. A qualitative curve can be made to determine initial testing and for predetermining general compaction control requirements. Numbers shown correspond to text discussion.

using statistical quality control. This method is described in some additional detail by Turnbull et al. (1966) and will be briefly introduced here. Basically the procedure is as illustrated in Fig. 7-10. One develops the normal frequency distribution curve using either unit weight directly or relative compaction and estimates a reasonable standard deviation $\bar{\sigma}$. If this is estimated as, say, 2 percent relative compaction, then 98 percent of the field tests would be equal to or greater than the specified unit weight. Alternatively, the standard deviation might be, say, ±0.80 kN/m^3 (about 5 pcf), and any values of unit weight less than this value are unacceptable. In statistics terminology the 2 percent value is the coefficient of variation $\bar{C}$ for the unit weight. For a large project, a plot of all the field tests would result in a normal frequency distribution curve unless bias were introduced. Bias might develop from technician error, poorly calibrated equipment, or changes in compaction equipment. It might also develop from unnoticed changes in fill material and poor control of fill moisture content. The normal frequency distribution curve is, of course, symmetrical as in Fig. 1-3 and as in Fig. 7-10, which shows the curve that might be obtained from a reasonably large number of tests—more than 20 to 30.

Now if we assume that the standard deviation $\bar{\sigma} = 0.80$ kN/m^3 is to be based on this large number of tests, how many (or what percentage) can be allowed which fail to meet specifications? Note that unit weights that are too large are acceptable; it is only those too low which are "not to specification." The contractor might be concerned if a large number of tests are too large, since this represents a profit reduction. Whatever the statistical criteria, they must be reasonable, or earthwork costs would rapidly escalate because the contractor would lose money on the first job on which the criteria were applied, but word would get around and on the next job the earthwork costs would reflect the additional compaction effort. One report of what is "reasonable," reported by Sherman et al. (1967), indicates that a coefficient of variation of 2.5$^+$ is adequate, with somewhat larger values, around 3 to 3.5, for clayey soils with considerable gravel. A study of 29 projects gave a field average $\bar{C} = 3.3$ for all the projects. Thus, it appears that a coefficient of variation of 3 $\pm$ 0.3 will generally be adequate for all projects. Figure 7-10 illustrates the application of a coefficient of variation of 3$^+$ (actually $0.49/15.34 = 3.19$) to an example with numbers so that the principle can be more easily understood.

Shown on Fig. 7-10 is a situation where RC = 90 percent and the soil has a laboratory-determined standard compaction test dry unit weight of 16.50 kN/m³. The mean value $\bar{X}$ of a large number of field tests is determined (or estimated) to be 15.34 kN/m³. The relative locations of these several values are shown (in both RC and unit weight). Now the question is how many tests out of this large number could be less than 90 percent RC for "acceptable" work. We will assume that the project is one involving several hundred cubic meters of soil, so that at least 30 tests will be performed (in any case an estimate of the total number of tests should be made). Now from Fig. 7-10 we obtain the standard deviation as 15.34 − 0.9(16.50) = 0.49 as shown. From general statistics considerations we have

$$\bar{X} - 1.000\bar{\sigma} = 15.34 - 0.49$$

We need now to determine for these 30 (or other assumed number) tests the reliability which will produce the coefficient of 1.000 for $\bar{\sigma}$. From Table 1-2 at $N = 30$, we interpolate in the table for the coefficient:

Reliability percent	80	P	90
Coefficient	0.854	1.000	1.310

By proportion P is obtained as

$$P = 80 + \frac{1 - 0.854}{1.310 - 0.854}(10) = 83.2 \text{ percent}$$

With an 83.2 percent reliability, the number of failing tests (out of 30) would be

$$\tfrac{30}{100}(100 - 83.2) = 5.04 \qquad \text{or, say, 5 tests}$$

This means that if we do 30 tests, we would expect 1 test in 6 to be either larger than 15.34 + 0.49 = 15.83 kN/m³, which is also acceptable, or less than 15.34 − 0.49 = 14.85 (14.85 is the value for RC = 90 percent), which is unacceptable. The project engineer would now have to decide if this is satisfactory after consideration of all factors. These will include the fact the poorly compacted soil is surrounded by satisfactorily compacted soil, the test locations, and the intended use of the fill. On some projects, particularly small ones such as basement floors and around building walls, all the tests are required to meet specifications.

To obtain a smaller percentage of tests "failing" on any project, one can:

1. Increase the number of tests (but this is not of very much help. Inspection of Table 1-2 for $N = 100$ tests shows a reliability of only 83.5 percent).
2. Use a smaller RC, say 0.35 in this example (but we do not need statistics to tell us this!).
3. Reduce $\bar{\sigma}$ by exercising close field compaction control. If $\bar{\sigma}$ is reduced, more tests fall closer to $\bar{X}$, producing a narrow, peaked bell for the normal distribution curve.

7-11 SPECIAL PROBLEMS IN SOIL COMPACTION

Care must be taken that the fill being used is that specified. Stratified borrow should be used in fill in the same manner it was classified and laboratory tested, i.e., blended or on selected strata.

Compaction of sanitary landfills is difficult. Figure 7-6d illustrates a compactor used for landfill operations which can crush boxes, small metal cans, etc. Compaction is very haphazard in these sites. The primary purpose is to cover the material at the end of each day's operations with 15 to 30 cm of earth which has been compacted; this is primarily for rodent, insect, and odor control. It is recognized that over a period of time decay will produce a fill with large voids even if the earth layers are heavily compacted. Some advantage might be obtained by segregation of fill: papers and other organics in one location; tires, old refrigerators, hot water tanks, building rubble, etc., in another.

Placing logs, stumps, and large boulders in the bottom of fills as a means of disposal is not recommended. Unless wood is permanently below the water table, it will decay. It is difficult to compact the soil adjacent to any of these large objects, and local subsidence may be a problem after a period of time.

Replacing soil in roadway trenches such as those for water, power, or telephone replacement/repair is very difficult. Almost always the trench settles because the soil has not been backfilled properly. The result is an expense (tire and vehicle damage) and a nuisance to the road users. This problem can be avoided over 90 percent of the time by backfilling the full depth (not the top 15 cm) of the trench to a RC of 95^+. It is not necessary to make compaction tests; a pocket penetrometer (Fig. 13-25) can be used to determine resistance of the trench walls and the backfill can be compacted, using a hand-operated compactor on 7- to 10-cm lifts, to a value at least as great.

Frozen soil should not be compacted, as the lumps will thaw and the reduced water volume coupled with the looser initial soil condition in the lump will produce local soft spots. Additionally, even if the soil lumps are broken down, compaction at temperatures below $0°C$ produces lower unit weights than compaction above $0°$. Where it is absolutely essential to compact frozen soil, it may be treated with calcium chloride (this lowers the freezing point and melts the soil ice). As with other admixtures, the optimum percentage of $CaCl_2$ is determined by trial but will generally be on the order of 0.5 to 1.5 percent on a dry weight basis.

7-12 SUMMARY

This chapter has introduced the reader briefly to soil stabilization via the use of admixtures, with the optimum percent determined by trial.

We have considered both field and laboratory methods of compaction, or particle packing, in some detail. We note in particular the standardized laboratory values:

	Standard	Modified
Mold	0.000 942 2 m^3 (although a larger mold may be used)	
Hammer	24.5 N	44.5 N
Drop	0.305 m	0.457 m
Blows	25	25
Layers	3	5

Soil: $(-)$ No. 4 material commonly used; ASTM allows use of $(-)$ 19 mm material. The 152-mm-diameter mold should be used with the larger material.

The field compaction is nearly always accomplished by some type of static pressure or kneading action, depending heavily on the soil type. There are several types of equipment available for field compaction, but we may summarize then approximately as follows:

Method	Rollers	Soil type
Pressure-confining-vibration	Smooth-wheel, rubber tired	Cohesionless
Kneading	Sheepsfoot type, rubber tired	Cohesive

We note the use of destructive (digging holes) and nondestructive field testing and statistical concepts in field quality control of compaction. Particular emphasis was placed on using common sense and reasoning with the statistical methods.

Finally we briefly touched on some special problems, including use of large pieces of foreign material in the fill, compacting sanitary landfills, and compacting frozen ground.

HOMEWORK PROBLEMS

7-1 For the compaction curves of Fig. 7-1:
   (a) Estimate the maximum dry density and OMC for both curves.
   (b) Indicate the range of w for 95 percent RC for both curves.

7-2 The natural moisture content of a borrow material is 8 percent. Assuming 3000 g of moist soil for a standard compaction test, how much water is to be added to bring the sample to 11, 13, 15, 17, and 20 percent water contents?

7-3 For the soil shown in Fig. 7-1, a field unit weight test gave the following information:

$$w = 13.5 \text{ percent}$$

$$\text{Wet unit weight} = 20.09 \text{ kN/m}^3$$

Compute RC for both curves.
   Ans.: Curve 1—approx. 105 percent.

7-4 Approximately how many sacks of cement (42.7 kg per sack) are required to treat a road base (1.6 km × 8 m wide) to a depth of 15 cm (compacted) if the compacted dry unit weight is 19.5 kN/m^3? The in situ water content is 6.5 percent and OMC = 10.2 percent.

7-5 How much water must be added in Prob. 7-4, assuming 2 percent extra for atmospheric conditions and cement hydration?

7-6 Set up station numbers and distances (X and Y coordinates) from the left side of the road as field unit weight test locations to randomly sample the in situ compaction *and* cement contents for what you feel will be enough tests to obtain a 95 percent confidence in the work of Prob. 7-4.

7-7 Fifty-five field unit weight determinations were made as follows:

No. of tests:	16	5	4	6	2	10	7	5
RC:	98	102	99	101	100	97	94	90

(*a*) Is the job satisfactory if RC = 95 percent per specifications?

(*b*) What, if anything, will the contractor need to do to improve the efficiency of the operation?

7-8 Referring to Fig. 7-10 and Sec. 7-9, if all data are the same but $\bar{\sigma}$ is 0.39, what percent reliability do we obtain for 30 tests?

Hint: The new n is $0.49/0.39 = 1.256$.

Ans.: Approximately 89 percent.

7-9 Referring to Prob. 7-8, what $\bar{\sigma}$ will give only 1 "bad" test out of the 30? What RC would this correspond to?

Ans.: $\bar{\sigma} \cong 0.24$.

Chapter 8

Soil Hydraulics, Permeability, Capillarity, and Shrinkage

8-1 WATER IN SOIL

One of the most important considerations in soil mechanics is the effects of water in the soil on its engineering properties. The Atterberg limits test displays how soil may vary from a solid to a viscous fluid with water content. Individual observations of dry and wet soil on and around excavations, along roadways, and elsewhere indicate a considerable range in state. Cohesive soils are very hard, brittle, and tend to shrink when dry and are very soft, plastic, and tend to swell when wet. Cohesionless soils range from moldable to crumbly for the wet and dry states, respectively.

Chapter 5 introduced the concept of the ground water table and water flow from a higher to a lower energy potential. Wells as a source of water supply are intimately concerned with the flow of water through soils. Highway subdrainage is a water flow problem. Frost action in soils is a flow problem and also depends on capillary action. Chapter 2 introduced the concept of the buoyant, or submerged, unit weight and the loss of effective pressure which occurs due to soil pore water pressure. Seepage flow quantities, to be considered in Chap. 9, and consolidation settlements, in Chap. 11, are water-in-soil problems or conditions.

The following sections will present the general concepts and theory involving the effects of water on, and water flow in, soil. Many situations will occur where proper use of these concepts produces an adequate solution so that the soil and/or site can be used.

8-2 PERMEABILITY

The facility of fluid flow through any porous medium is an engineering property termed *permeability*. For geotechnical engineering problems the fluid is water and the porous medium is the soil mass. Any material with voids is porous and, if the voids are interconnected, possesses permeability. Thus, rock, concrete, soil, and many other materials are both porous and permeable. Materials with larger void spaces generally have larger void ratios, and thus, even the densest soils are more permeable than materials such as rocks and concrete. Materials such as clays and silts in natural deposits have large values of porosity (or void ratio) but are nearly impermeable, primarily because of their very small void sizes, although other factors may also contribute. The terms porosity n and void ratio e are both used to describe the voids in a soil mass.

The permeability of a soil mass is important in:

1. Evaluating the amount of seepage through or beneath dams and levees and into water wells
2. Evaluating the uplift or seepage forces beneath hydraulic structures for stability analyses
3. Providing control of seepage velocities so that fine-grained soil particles are not eroded from the soil mass
4. Rate of settlement (consolidation) studies where soil volume changes occur as water is expelled from the soil voids as a rate process under an energy gradient (Chap. 12).

To provide an appreciation of the flow of water through a soil mass, we will first develop the general equation of laminar flow through a capillary tube, as illustrated in Fig. 8-1a. During laminar flow the velocity varies across the tube diameter from zero at the tube walls—because of friction or viscosity effects—to a maximum value at the center.

From Fig. 8-1a, the velocity gradient at a radial distance r from the tube center is $-dv/dr$ and the unit shear force at this distance r is $-\eta(dv/dr)$, as shown in Fig. 8-1b. The forces on a free body of a cylinder of water with dimensions as shown in Fig. 8-1b at a radial distance r from the center of flow are also shown. Equating these forces, we obtain

$$-\eta(2\pi r L) \frac{dv}{dr} = \pi r^2 h_1 \gamma_w - \pi r^2 h_2 \gamma_w$$

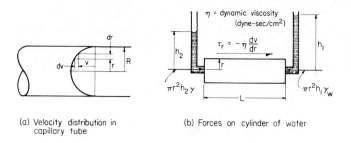

(a) Velocity distribution in capillary tube

(b) Forces on cylinder of water

Figure 8-1 Fluid flow through a capillary tube. $[\eta = \text{dynamic viscosity (dyne s/cm}^2)]$

Separating variables, combining terms, letting $i = (h_1 - h_2)/L$ be defined as the hydraulic gradient, and integrating, we obtain

$$v_r = -\frac{\gamma_w}{4\eta} r^2 i + C \qquad (a)$$

At $r = R$, $v_r = 0$, from which Eq. (a) becomes

$$v_r = i \frac{\gamma_w}{4\eta} (R^2 - r^2) \qquad (b)$$

The total flow in a unit time is obtained as the following integral:

$$q = i \frac{\gamma_w}{4\eta} 2\pi \int_0^R (R^2 - r^2) r \, dr$$

Integrating, we obtain

$$q = \frac{\gamma_w}{8\eta} \pi R^4 \qquad (c)$$

If the area of the tube is taken as $A = \pi R^2$, the average velocity of flow is

$$v = \frac{\gamma_w}{8\eta} R^2 i \qquad (8\text{-}1)$$

This expression for velocity is termed Hagen-Poiseuille's law since Hagen and Poiseuille, working independently, obtained experimental results almost simultaneously in 1839 and 1840.

Darcy (ca. 1856), in considering the flow of water through sand filters in France, proposed that flow of water through a soil could be expressed as

$$v = ki \qquad (8\text{-}2)$$

where $i = \Delta h/L =$ the head loss in a length of filter bed L (commonly referred to as the hydraulic gradient)

$k =$ coefficient of permeability with units of velocity.

Darcy's law is evidently a statistical representation of the average flow conditions in a porous medium. This equation is considered to be one of the most important equations in soil mechanics. It is generally considered valid for *laminar* flow but, as will be seen later, is applicable for any flow.

From a comparison of Eqs. (8-2) and (8-1) it is evident that

$$k \cong \frac{\gamma_w}{8\eta} R^2$$

thus depends on the unit weight and viscosity of the fluid—which is temperature-dependent—and on the tube radius. The tubes through a soil mass are of irregular shape, both in diameter and in the direction (longitudinal) of flow, and are dependent on the void ratio and, in particular, on the effective grain size. For this reason, k in coarse sands is larger by many orders of magnitude than in silts and clays even though the void ratio in the silts or clays may be as large as in sand, or often much larger.

Laminar flow occurs in smooth, straight pipes with Reynolds numbers N_R up to about 2100. The Reynolds number is defined as

$$N_R = \frac{v \, d\rho}{\eta} \tag{8-3}$$

where v = velocity, cm/s
d = tube diameter, cm
ρ = mass density = $\dfrac{\gamma}{g} \left(\text{units:} \dfrac{g \cdot s^2}{cm^4} \dfrac{980.7 \text{ dynes}}{980.7 \text{ g}} \right)$
η = dynamic viscosity (dynes $\cdot$ s/cm^2)

In soil it appears that turbulent flow occurs at much lower numbers, perhaps as low as 300 to 600. Darcy's law has been found experimentally not to be valid at N_R values smaller than this (see the discussion by Rumer, 1964), because of the discrepancy attributed to inertia forces developed in the water due to abrupt changes in direction of flow resulting from the irregularities in pores, pore sizes, and interconnections. Since these factors are not readily determinable, it follows that the best (and most valid) procedure is to obtain the permeability coefficient using a hydraulic gradient which is as close as practicable to the prototype. If this is done, it is academic whether the flow is laminar, turbulent, etc., since the k determined will be representative of flow conditions to be expected. A further mitigating factor is that for many fine-grained soils the flow velocity is so low (under the field hydraulic gradient) that the flow is laminar and the inertia forces are insignificant.

Inertia forces in near turbulent flow velocities may be a critical factor in controlling internal soil erosion. Inertia forces large enough to dislodge small silt or fine sand grains, or overcome interparticle attractive forces in clays, will result in material loss as these small particles are washed out of the soil matrix, with flow channel enlargement and further increased erosion. This phenomenon is called *piping*, and control measures are considered in Chap. 9.

8-3 SOIL WATER FLOW AND THE BERNOULLI ENERGY EQUATION

The Bernoulli equation is commonly used in pipe flow but also applies to flow of water through a soil mass. Figure 8-2 illustrates application of the Bernoulli equation for flow conditions at two points in a soil mass a length L apart. As in any fluid mechanics text, refer all heads to an arbitrary, but as convenient as possible, datum line.

At location B the total energy available, described as a measurable distance (called head) above the reference datum, is[†]

$$h_1 = Z_1 + \frac{p_1}{\gamma_w} + \frac{v_1^2}{2g} = \text{total energy} = \text{constant} \qquad (a)$$

At location C, a distance L downstream from point B,

$$h_2 = Z_2 + \frac{p_2}{\gamma_w} + \frac{v_2^2}{2g} \qquad (b)$$

The head loss between these two points is defined as $\Delta h = h_1 - h_2$.

For a constant area A of *saturated*, percolating soil from B to C, continuity of flow must exist, from which

$$q_{\text{in}} = q_{\text{out}} = Av[‡]$$

This requires that $v_1 = v_2 = $ constant, and equating Eqs. (a) and (b) for total energy, we have

$$\Delta h = Z_1 - Z_2 + \frac{p_1 - p_2}{\gamma_w} \qquad (8\text{-}4)$$

A critical evaluation of Eq. (8-4) shows that if $\Delta h = 0$, no flow can possibly take place.

[†] This distance can be measured by inserting a piezometer tube into the pipe as shown; the water level will rise to a level representing the current (at that point) energy head available.

[‡] This equation represents a *steady state* condition and, thus, the requirement of a saturated soil condition, since with a nonsaturated soil condition the voids could retain some of the entering water.

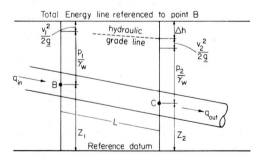

Figure 8-2 The Bernoulli energy equation for pipe flow.

In soil masses, p_1 and p_2 will be atmospheric pressures, unless we are investigating artesian conditions, resulting in the head loss being simply the change in free water surface elevation between any two points B and C. This elevation change is generally a very small value for in situ groundwater conditions. Since $i = \Delta h/L$ in Darcy's equation, if i is very small, it necessarily follows that the soil-water velocity v will also be very small.

With the rationale established for small velocities of flow in soil, it also follows that the velocity head $(h_v = v^2/2g)$ will be very small and can nearly always be neglected, e.g., for a velocity of 0.3 m/s the velocity head is only

$$h_v = \frac{0.3^2}{2(9.807)} = 0.0045 \text{ m (about 4.5 mm)}$$

In most soils the velocity is well under 0.3 m/s, resulting in velocity heads so small there would be some difficulty in accurately measuring them. For steady state flow, and a velocity head of zero, the piezometric head that is measured can only be the static or position head

$$h = \frac{p_{\text{static at piezometer point}}}{\gamma_{\text{water}}}$$

Considering Fig. 8-3, which represents a homogeneous, isotropic, confined soil mass, if piezometers are inserted at points A, B, C, and D, the water levels will be observed as shown. The piezometers shown are understood to be tubes of sufficient diameter that capillary effects (Sec. 8-7) are negligible. They are positioned to obtain the water level corresponding to the water pressure and would include any $v^2/2g$ effects (but we have already concluded that these are nearly zero) at the tip.

In geotechnical field work, a piezometer may consist of a 3- to 5-cm-diameter tube with the tip placed in a sand filter device and into a drill hole. The upper portion of the hole is plugged with a clay seal so that surface water is excluded and so that water pressures at the tip are observed. The piezometric head is measured by lowering a weighted tape or an electrical indicator which completes its circuit with water in the piezometer pipe. Some piezometer installations may use bourdon tube pressure gages or electronic pressure indicators, but these require a considerable installation expense.

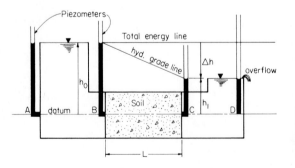

Figure 8-3 Flow of water through an isotropic, homogeneous soil mass of length L with the head conditions shown.

The piezometric heads of Fig. 8-3 are obtained as follows:

1. Take the datum along line $ABCD$ so that elevation head does not need to be considered. Also take $v^2/2g = 0$ at all locations.

2. At A: $h_A = h_0$ (d)

3. At B: $h_B = h_0$ (e)

4. At C: $h_C = h_1$ (f)

5. At D: $h_D = h_1$ (g)

With $h_A = h_B = h_0$, and $h_C = h_D = h_1$, it follows that the head loss shown occurs across the soil sample of length L. As we can only assume a uniform loss since the soil is considered to be homogeneous and isotropic, the slope of the energy line is as drawn on the figure. In this case the hydraulic gradient is

$$i = \frac{\Delta h}{L}$$

The hydraulic gradient is defined as the slope of the energy line defined by the free surface of flowing water in open channels, or the slope of the piezometric heads between two points in confined flow. It represents the head loss or energy loss per unit of length. It should again be noted that if $i = 0$ in Fig. 8-3, $h_1 = h_0$ and no flow can take place.

The low velocity of the pore fluid in a soil mass means that after sudden changes, as for example increasing h_0 or decreasing h_1 in Fig. 8-3, there will be a certain time lag before a new steady state condition is obtained. In earth dams or levees which become saturated during impoundment, or high water during floods, and then undergo a sudden to very rapid emptying, the time lag represents a time during which soil stability may be critical.

It may be apparent from Fig. 8-3 that the approach velocity v_i and the discharge velocity v_d are different from the actual velocity of the water through the soil pores v_a. From continuity, however, the flow rate must be constant; thus,

$$q_B = Av_i = Av_{s(\text{apparent})} = Av_d = q_C$$

And in the soil mass, referring to Fig. 8-4, we have

$$q = Av_s = A_v v_a \qquad (h)$$

But also from Fig. 8-4, we have

$$\frac{A_v}{A} = \frac{V_v}{V} = \text{porosity } n \qquad (i)$$

and substituting into Eq. (h),

$$Av_s = nAv_a$$

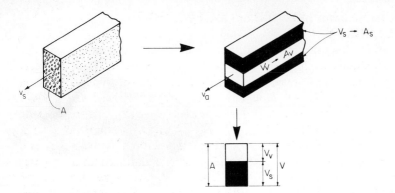

Figure 8-4 Distribution of the cross-sectional area of a percolating soil mass as voids and solids to approximate the true seepage velocity.

and since

$$n = \frac{e}{1+e}$$

the actual pore velocity in the soil mass is

$$v_a = \frac{1+e}{e} v_s \qquad (8\text{-}5)$$

This equation indicates that the actual seepage velocity in a soil mass may be substantially larger than the apparent seepage velocity v_s, which is usually computed when using Darcy's law.

Example 8-1

GIVEN A permeability test using a loose, coarse sand. $Q = 1650$ cm³ in a time of 15 min; void ratio $e = 0.65$; area of sample $= 45.4$ cm².

REQUIRED What is the actual water velocity through the sand in centimeters per second?

SOLUTION The nominal discharge velocity v_s is

$$v_s = \frac{1650}{45.4(15)(60)} = 0.0404 \text{ cm/s}$$

Using Eq. (8-5), the actual velocity is approximately

$$v_a = \frac{1.65}{0.65}(0.0404) = 0.102 \text{ cm/s (more than double the nominal value)}$$

This computation clearly illustrates that the void ratio or porosity of a soil affects the permeability [k as used in Eq. (8-2)], or how water flows through it.

8-4 DETERMINATION OF THE COEFFICIENT OF PERMEABILITY

The determination of the coefficient of permeability may be made in one of several ways. An approximate value may be obtained in the laboratory using either a *constant-head* or a *falling-head* permeability test. The falling-head test is more economical for tests of long duration, while the constant-head test is preferred for soils, such as sands or gravels, which have large void ratios and for which a large flow quantity is desired to improve computational precision. Figure 8-5 gives a schematic test setup for both tests. The test has been standardized for a temperature of 20°C as a convenience. Since the viscosity of water varies from 0.0157 dyne · s/cm^2 at 4°C to 0.008 97 at 25°C (the possible range of in situ soil temperatures of interest), the factor is 1.75; thus, a 175 percent difference in k can be obtained from two tests at 4 and 25°C.

For the constant-head permeability test (use the right side of Fig. 8-5):

$$Q = Avt = Akit$$

and by rearranging for k as the only unknown, obtain

$$k = \frac{QL}{Aht} \qquad \text{cm/s usual units} \qquad (8\text{-}6)$$

where Q = total discharge volume, cm^3, in time t s
   A = cross-sectional area of soil sample, cm^2
   h = differential head across sample, cm

For the falling-head permeability test (use the terms on the left side of Fig. 8-5):

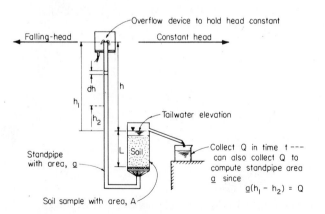

Figure 8-5 Line details of the laboratory test for determining the coefficient of permeability. Use the right side of the figure to identify terms used for the constant-head test, the left side for the falling-head test.

In the standpipe of cross-sectional area a cm^2,

$$v = -\frac{dh}{dt} \quad \text{(use minus since the head is decreasing)}$$

The resulting flow into the sample from the standpipe is

$$q_{in} = -a\frac{dh}{dt}$$

and the flow through and out of the sample is

$$q_{out} = Av = Aki$$

From continuity we can equate q_{in} and q_{out} to obtain

$$-a\frac{dh}{dt} = Ak\frac{h}{L}$$

Separating variables, integrating, and starting at time $t_1 = 0$ (which results in the constant of integration being 0) gives

$$k = \frac{aL}{At}\ln\frac{h_1}{h_2} \quad \text{cm/s} \tag{8-7}$$

when all dimensional units are in centimeters and seconds.

Example 8-2 A laboratory falling-head permeability test is performed on a light grey, gravelly, well-graded sand with the following test data obtained:

STAND PIPE $\quad a = 0.96$ cm^2 Soil $\quad A = 45.4$ cm^2 Flow DIST $\quad L = 20.0$ cm

Start water high $h_1 = 160.2$ cm End $\quad h_2 = 43$ cm $\qquad t = 65$ s for head to fall from h_1 to h_2

Water temperature of test $= 20°C$

REQUIRED Compute k.

SOLUTION Make a direct substitution into Eq. (8-7) to obtain

$$k = \frac{0.96(45.4)}{45.6(65)}\ln\frac{160.2}{43.0}$$

$$= 0.0085 \text{ cm/s}$$

Note
1. The hydraulic gradient is very large in this test and may be totally unrealistic (it also may produce turbulent flow).
2. If the water temperature had differed from 20°C, a temperature correction would be required.

Table 8-1 Temperature versus dynamic viscosity and surface tension for water

T, °C	γ, kN/m^3	η, dyne · s/cm^2	Surface tension, dynes/cm
4	9.807	0.015 67	75.6
16	9.7969	0.011 11	73.4
18	9.7935	0.010 56	73.1
20	9.7896	0.010 05	72.8
22	9.7854	0.009 58	72.4
24	9.7808	0.009 14	72.2
26	9.7758	0.008 74	71.8
28	9.7704	0.008 36	71.4
30	9.7646	0.008 01	71.2

When the test temperature differs from 20°C, the coefficient of permeability should be corrected to 20°C by recognizing that the value of k is inversely proportional to viscosity to obtain

$$k_{20} = k_T \frac{\eta_T}{\eta_{20}} \tag{8-8}$$

where k_T is the coefficient of permeability at any test temperature T. Table 8-1 gives several values of η versus T.

Figure 8-6 gives the approximate range of k values one may expect to obtain. The coefficient of permeability is plotted on a log scale, since the range of permeabilities is so large. No other engineering property of any material exhibits such a large range of values as does the permeability of soil.

An empirical equation relating the coefficient of permeability to the effective grain size (D_{10}) from a sieve analysis was reported by A. Hazen (ca. 1892) based on work with rapid sand filters in water treatment plants. He found that for sands

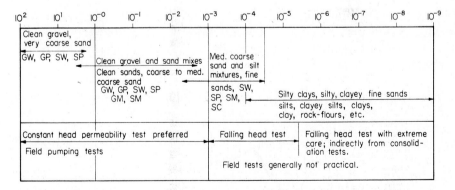

Figure 8-6 Typical ranges of permeability coefficients in cm/s and suggested test methods.

with D_{10} sizes between 0.1 and 3.0 mm, the coefficient of permeability could be expressed approximately† as

$$k = C(D_{10}^2) \qquad \text{cm/s} \tag{8-9}$$

Is this equation D_{10} is the effective grain size in centimeters, with C such that k is in centimeters per second. The coefficient C varies, according to Hazen, from about 40 to 150 and the values may be taken as follows:

C	Sand (any or all of the following applies)
40–80	Very fine, well graded or with appreciable fines [$(-)$ No. 200]
80–120	Medium coarse, poorly graded; clean, coarse but well graded
120–150	Very coarse, very poorly graded, gravelly, clean

One would expect that poorly graded sand would have a larger coefficient than well-graded materials, since the void spaces would be more ordered and larger with poorly graded soil.

An estimate of the permeability k_2 at a void ratio of e_2 when a test was performed with results of k_1 at void ratio e_1 may be made as

$$k_2 = k_1 \left(\frac{e_2}{e_1}\right)^2 \tag{8-9}$$

Other equations more complicated than this have been suggested, but in the range of void ratios (0.5 to 1.1) likely to be used, this equation is considerably simpler and the results are sufficiently precise considering the precision with which k_1 can be determined.

8-5 LIMITATIONS AND OTHER CONSIDERATIONS IN DETERMINING k

The laboratory test for determination of k is very unreliable, with considerable attention to test procedure and equipment design necessary to obtain even the correct order of magnitude of the permeability coefficient. Some of the factors producing this unhappy condition are:

1. Soil in situ is generally stratified, and it is hard to duplicate in situ conditions in the laboratory test. The horizontal value k_h is usually needed, but tube samples are likely to be tested, with vertical values k_v obtained.
2. In sand the values of k_v and k_h are considerably different, often on the order of

---

† Hazen (1911), which is commonly cited as the reference of origin, does not give this form of the equation; instead, $v = cd^2h/L(0.73 - 0.03T)$ is given, where T is the temperature of the water.

$k_h = 10$ to $1000k_v$, due to the sedimentation process of soil deposit formation. The field soil structure is invariably lost in the laboratory because an undisturbed sample cannot be tested (even if it were possible to obtain one), since it would have to be transferred from the recovery device to the permeameter.

3. The small size of laboratory samples leads to effects of boundary conditions, such as smooth sides of the test chamber affecting flow and air bubbles either in the water or trapped in the test sample affecting the test results.

4. No method is available to evaluate k for other than saturated steady state soil conditions, yet many flow problems will involve partially saturated soil-water flow. Where k is very small, as in clays and fine silts, it may be difficult to determine when a steady state has been obtained.

5. When k is very small, say from 10^{-5} to 10^{-9} cm/s, the time necessary to perform the test will cause evaporation and equipment leaks to become very significant factors. Leaks which are not visible to the eye can leak (with continual evaporation) with no accumulation of water ever taking place yet be of sufficient size to affect k by several orders of magnitude.

6. In the interests of time, the laboratory hydraulic gradient $\Delta h/L$ is often made 5 or more, whereas in the field more realistic values may be on the order of 0.1 to 2.0^-.

    (a) In clays a threshold i of 2 to 4 may be necessary to produce any flow (and an apparent k); thus, a totally unrealistic flow may be obtained if the field hydraulic gradient is not as large as the threshold gradient.

    (b) In sands, unrealistically high i values may produce turbulent conditions, indicating a flow condition different from the field i, which may indeed produce laminar flow.

    (c) Unrealistically high i values may produce sample packing and a void ratio different from that in the field. This may be very critical when testing loose sands.

The coefficient of permeability is sometimes obtained from the one-dimensional compression test (the consolidation test of Chap. 10). Sometimes the samples are tested in a triaxial cell (see Chap. 13) to obtain k.

In the field, the permeability may be evaluated on the basis of observing the length of time it takes dye, salt, or radioactive tracers to travel between two wells, wellpoints, or borings in which the differential head can also be measured. Field pumping tests and the well equations given in Chap. 9 can be used to compute the coefficient of permeability. The permeability may be computed from the rate of water fall in a borehole filled with water (above the water table) or from its rising some height in the borehole after bailing (below the water table). Procedures and computations to obtain the coefficient of permeability using the several methods cited above can be found in references such as Cedergren (1977), USBR (1968), and NAFAC (1971). Almost any well-performed field test gives a more reliable value of the coefficient of permeability than a laboratory test, except on remolded soil used for compacted fills, but is considerably more expensive and time consuming.

8-6 EFFECTIVE COEFFICIENT OF PERMEABILITY OF STRATIFIED SOILS

Figure 8-7 illustrates a stratified soil condition where it may be convenient to replace the stratified system by an equivalent soil mass with a single effective thickness $L = \sum H_i$ and a single value of k—either k'_v or k'_h, depending on the direction of flow being considered. There is a direct analogy between a simple electric circuit containing only resistors in either series or parallel and water flow through the stratified soil mass of Fig. 8-7. Flow perpendicular to parallel-bedded strata is analogous to series resistor networks. When resistors are in series, the *largest* resistance controls the current flow (analogy = fluid), and with flow perpendicular to the bedding planes, as k_v of Fig. 8-7, the stratum with the *smallest* permeability essentially controls the flow quantity. When flow is parallel to the bedding planes, as k_h of Fig. 8-7, the flow quantity is controlled by the stratum with the largest coefficient of permeability (and in electric circuits by the smallest resistor). With these concepts in mind, we will now develop equations for the equivalent k's for the stratified deposit. For equivalent k'_v, we have

$$q_{in} = q_{out}$$

from continuity; therefore $v = $ constant, and

$$v = k'_v i = k_1 \frac{h_1}{H_1} = k_2 \frac{h_2}{H_2} = k_3 \frac{h_3}{H_3} = \cdots = k_n \frac{h_n}{H_n}$$

Rearranging,

$$\frac{H_1}{k_1} = \frac{h_1}{v}$$

$$\frac{H_2}{k_2} = \frac{h_2}{v}$$

$$\frac{H_3}{k_3} = \frac{h_3}{v}$$

$$\cdots\cdots\cdots\cdots$$

$$\frac{H_n}{k_n} = \frac{h_n}{v}$$

Adding, we obtain

$$\frac{h_1}{v} + \frac{h_2}{v} + \frac{h_3}{v} + \cdots + \frac{h_n}{v} = \frac{H_1}{k_1} + \frac{H_2}{k_2} + \frac{H_3}{k_3} + \cdots + \frac{H_n}{k_n}$$

Factoring the left side and recognizing that $H_1 + H_2 + H_3 + \cdots + H_n = L$ and $v = k'_v(h/L)$, with some rearranging to solve for k'_v we obtain

$$k'_v = \frac{L}{H_1/k_1 + H_2/k_2 + H_3/k_3 + \cdots + H_n/k_n} \qquad (8\text{-}10)$$

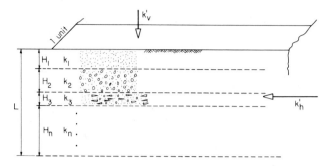

Figure 8-7 Stratified soil system with permeability as shown.

The equivalent k'_h can be obtained as

$$q = A v_{\text{average}} = L(k'_h)i$$

which is also the sum of the flow in each stratum:

$$L(k'_h)i = k_1 H_1 i + k_2 H_2 i + k_3 H_3 i + \cdots + k_n H_n i$$

Cancelling i and solving for k'_h, we obtain

$$k'_h = \frac{k_1 H_1 + k_2 H_2 + k_3 H_3 + \cdots + k_n H_n}{L} \tag{8-11}$$

8-7 CAPILLARITY AND CAPILLARY EFFECTS IN SOIL

All materials possess intermolecular forces. These may be termed *cohesion* for the internal molecular forces and *adhesion* for the attraction between molecules of dissimilar materials, such as water and glass. If the adhesion forces between a liquid and any other material are larger than the intermolecular attraction of the liquid, the surface of the dissimilar material will be "wetted" by the liquid. Mercury, for example, has a very considerable cohesion; thus, it will wet only a limited number of dissimilar materials. Water, on the other hand, with little internal cohesion, will wet almost all materials it contacts. Wetting agents may be used to increase the adhesion effects between liquids and solids.

Any quantity of liquid will behave as though the surface were a tightly stretched skin due to the intermolecular attractive forces in the interior. This phenomenon is termed *surface tension*. This liquid material property accounts for the spherical shape of water drops on oily dust and the nearly true spheres of mercury drops on glass plates. Wetting agents (soaps and detergents) tend to reduce surface tension (the internal cohesion) and increase adhesion of the material to the foreign surface (or make it wet).

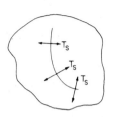

(a) Surface tension perpendicular to any line of interest on surface of liquid.

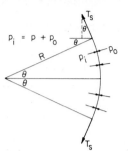

$p_i = p + p_o$

(b) Surface tension on a portion of 3-dimensional curved surface.

Figure 8-8 Surface tension.

Since surface tension is a material property of liquids and depends on intermolecular attraction, it will be temperature-dependent (below some critical temperature, which for water is 100°C, the material is a liquid, above it is a gas). Table 8-1 also gives the surface tension value for water at several values of temperature.

Surface tension allows a razor blade or needle carefully placed on the surface of water to float and produces the rise above the static surface level of water in a small glass tube placed in a container of water.

Since surface skin tension effects are caused by intermolecular attraction, it follows that at the interface with air the skin is in vertical equilibrium and that over the surface the pull must be equal in all directions, or perpendicular to any line of interest as in Fig. 8-8a. With a skin thickness of one molecule, the units of surface tension must be force/length.

Now let us investigate the effects of surface tension on a curved surface, as shown in Fig. 8-8b.

For equilibrium, $\sum F_h = 0$, or

$$T_s \sin \theta (2\pi \sin \theta) = \frac{p}{4} \pi (2R \sin \theta)^2$$

from which

$$p = \frac{2T_s}{R} \tag{8-12}$$

This equation states that the pressure difference ($p_i - p_o$) inside the curved surface is directly proportional to the surface tension T_s and inversely proportional to the radius of curvature R. It is well known that water vaporizes (or boils) at temperatures less than 100°C when the pressure is less than atmospheric (101.3 kPa or 1 bar). Equation (8-12) indicates that if R is sufficiently small, a higher temperature will be necessary to cause vaporization due to the increased pressure.

A Height of Water Rise in Capillary Tubes

When a hollow, open-ended tube is inserted into a container of liquid, and if the liquid wets the contact surface, it will climb the inside walls of the tube because of surface tension effects, producing a concave spherical upper surface, as shown in Fig. 8-9a. This is both a theoretical and readily observed phenomenon.

For clean glass, the angle α of the concave surface film with the tube walls is:

Liquid to clean glass	α, degrees
Water	0
Mercury	139

The height of rise h_c can be computed as $\sum F_v = 0$, from which in Fig. 8-9a

$$\frac{\pi d^2}{4} h_c \gamma = \pi d T_s \cos \alpha$$

and solving for h_c,

$$h_c = \frac{4 T_s \cos \alpha}{\gamma d} \qquad (8\text{-}13)$$

where T_s = surface tension
γ = unit weight of fluid
d = tube diameter, all in consistent units

For water at 20°C; $T_s \cong 72.8$ dynes/cm; $\gamma = 9.7896$ kN/m³. Since 10^8 dynes = 1 kN and 10^6 cm³ = 1 m³, we obtain

$$h_c = \frac{4(72.8 \text{ dynes/cm})(10^6 \text{ cm}^3/\text{m}^3)}{9.7896 \text{ kN/m}^3(10^8 \text{ dynes/kN})(d \text{ cm})}$$

From which

$$h_c = \frac{0.297\,46}{d \text{ (cm)}} \qquad \text{units of cm} \qquad (8\text{-}13a)$$

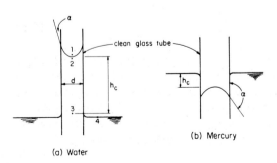

(b) Mercury

(a) Water

Figure 8-9 Capillary rise.

Example 8-3

GIVEN $d = 0.001$ cm, $T = 20°C$.

REQUIRED Height of capillary rise h_c.

SOLUTION Using Eq. (8-13a),

$$h_c = \frac{0.297\,46}{0.001} = 297.46 \text{ cm} = 2.97 \text{ m}$$

Example 8-4

GIVEN $d = 0.05$ cm

Mercury with $\alpha = 139°$

$$\gamma_{Hg} = 13.6 \times 9.807 = 133.4 \text{ kN/m}^3$$

$$T_s = 473 \text{ dynes/cm}$$

REQUIRED Height of capillary rise.

SOLUTION Use Eq. (8-13).

$$h_c = \frac{4(473)(10^{-2})(\cos 139°)}{133.4(0.05)} = -2.14 \text{ cm (below liquid surface)}$$

Note that the sign of the angle allows Eq. (8-13) to be used for all liquids, and that $\alpha > 90°$ will produce a depressed height of capillary rise.

Referring to Fig. 8-9a, what is the stress in the water at points 2 and 3? From fluid statics, the stress at point 3 is atmospheric, since it is in equilibrium with point 4, or

$$\sigma_3 = \text{atmospheric} = 0 \text{ gage pressure (or 101.3 kPa absolute pressure)}$$

At point 2, the stress must be tensile from Eq. (8-12), and from Fig. 8-8b $p_i = p + p_o$, but $p_i =$ atmospheric or 0 gage pressure, and therefore, $p = -p_o =$ water pressure; $(-)$ must be tension from the given sign convention, and since

$$\alpha = 0 \qquad R = \frac{d}{2}$$

we obtain

$$-p_o = \sigma_2 = \frac{2T_s}{R} = \frac{4T_s}{d} \qquad \text{(tension)}$$

Now since $4T_s \cong 291.2$ at $T = 20°C$, say 300 dynes/cm (for all temperatures from 0 to about 25°C), the stress is approximately

$$\sigma_t = \frac{0.03}{d \text{ (cm)}} \quad \text{kPa} \tag{8-14}$$

And from Ex. (8-3) the tension stress is (exactly)

$$p_o = -\frac{0.029\,746}{0.001} = -29.75 \text{ kPa} \quad \text{(tension)}$$

Since the tension stress should be sufficient to "pull" the column of water to this height (2.97 m), this should provide an alternative means of computing the tension stress as follows:

$$\sigma_2 A = \text{weight of column of water 2.97 m high} = W_c$$

The weight of the column of water is

$$W_c = \gamma_w h_c A$$

from which the tension stress is always

$$\sigma_t = -\gamma_w h_c$$

and in Ex. 8-3, the tension stress at $T = 20°C$, using Table 8-1, is

$$\sigma_t = -9.7896(2.97) = -29.12 \text{ kPa} \quad \text{(vs. 29.75 previously computed)}$$

The small discrepancies between the two computed values of tension stress are due to small inconsistencies between the tabulated values of surface tension and the unit weight of water in Table 8-1. Using Eq. (8-14), the tension stress is approximately

$$\sigma_t = \frac{0.03}{0.001} = 30.0 \text{ kPa}$$

For all practical purposes, the tension stress in water in the capillary zone at any height h_i is

$$\sigma_t = -\gamma_w h_i \tag{8-15}$$

where terms are identified in Fig. 8-10a. Again it should be noted that with a tube diameter of $d \cong 0.0003$ cm, the maximum tension stress $\sigma_t \cong 101.3$ kPa $= 1$ atmosphere tension (vacuum), which would be a condition of vaporization or "boiling" of the water at any ambient temperature at point 3 of Fig. 8-9a. According to Eq. (8-14), this "boiling" may not be possible due to surface tension effects. It may, however, be that with a tube of 0.0003 cm diameter, flow will be so slow that evaporation and/or vapor pressure above the capillary water may invalidate this concept.

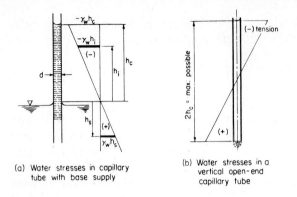

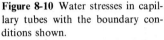

(a) Water stresses in capillary
tube with base supply

(b) Water stresses in a
vertical open-end
capillary tube

Figure 8-10 Water stresses in capillary tubes with the boundary conditions shown.

B Capillary Tubes with Variable Radius

Figure 8-11 illustrates the rise which can be obtained in capillary tubes of variable cross section. Figure 8-11a and b illustrates that the capillary rise h_c depends on tube diameter, and Fig. 8-11a further illustrates that the meniscus may not fully develop if the tube height is less than h_c. Figure 8-11c illustrates that a sudden enlargement may halt the capillary rise unless it is located at a height which corresponds to diameter d_2. Figure 8-11d illustrates how capillary rise can bypass a tube enlargement if a water supply from above flows downward and fills the tube to above the enlargement. The enlargement can be bypassed by raising the water supply so that h_c for diameter d_2 is above the enlargement; after the tube is filled, the water surface can be lowered. This condition corresponds to rainfall infiltration and/or a temporary rise in the ground water table. In Fig. 8-11e, an open based box with a capillary tube in the top is lowered into the water until the capillary tube has water in the zone h_c for that tube diameter. If the box is slowly raised, sufficient capillary tension stresses develop (at least in theory) throughout the water to hold it in the box. Figure 8-11f is a large tube with soil particles such that the pores make capillary tubes. Note that the large void area in the center at

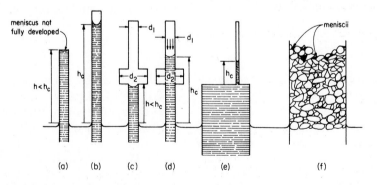

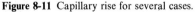

Figure 8-11 Capillary rise for several cases.

the top produces an irregular h_c level. In this case the tensile stresses in the water produce compressive interparticle stresses, resulting in an increase in the intergranular pressure.

C Capillarity in Soils

Figure 8-12 represents several soil particles with the voids filled with water and evaporation taking place. The soil "tubes" will vary in cleanliness and the water-soil interface will range from smooth to rough; however, the upper limit of intergranular stress due to compression of the tube walls as tension stresses develop in the pore water (refer to Fig. 8-12c) will be on the order of 80 to 100 percent of the maximum possible capillary tension stresses computed using Eq. (8-14). The pore tubes are quite irregular and variable in diameter, but it is still convenient, and a satisfactory approximation, to use Eq. (8-13a) to obtain h_c.

The pore diameter is impossible to measure and must be approximated. A commonly used approximation is

$$d \cong \tfrac{1}{5}D_{10} \qquad \text{mm}$$

Particle packing may influence the pore diameter(s) considerably, but this is not reflected in this equation. In clayey soils where D_{10} may be on the order of 0.0015 mm, the tube diameter is approximately

$$d = \frac{0.0015}{5} = 0.003 \text{ mm}$$

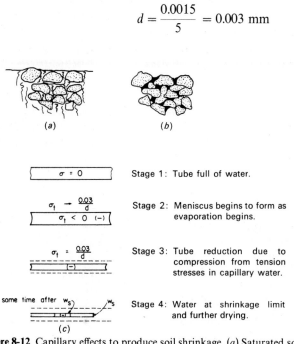

(a)   (b)

$\sigma = 0$	Stage 1: Tube full of water.
$\sigma_t \rightarrow \dfrac{0.03}{d}$ $\sigma_t < 0 \ (-)$	Stage 2: Meniscus begins to form as evaporation begins.
$\sigma_t = \dfrac{0.03}{d}$ $(-)$	Stage 3: Tube reduction due to compression from tension stresses in capillary water.
w at some time after w_s w_s	Stage 4: Water at shrinkage limit and further drying.

(c)

Figure 8-12 Capillary effects to produce soil shrinkage. (a) Saturated soil. (b) Soil after partial drying, with meniscii, water tension, and grain compressive stresses being developed. (c) Evaporation stresses in a horizontal capillary tube.

and the corresponding height of capillary rise is

$$h_c = \frac{0.03}{0.003} = 10 \text{ m}$$

It is very doubtful if an observer would be able to detect a capillary zone of over 1 to 2 m in this (or any) soil because evaporation removes the water as fast as it is pulled to higher elevations.

D Capillarity and Shrinkage

The shrinkage limit was defined in Chap. 2 as the water content ($S = 100$ percent) below which no further volume change occurs. Figure 8-12c illustrates the general conditions of shrinkage; at stage 1 the tube is full of water and no meniscus radius can form. As evaporation takes place, the meniscus forms and water tension stresses develop; this squeezes water out of the pores, reducing pore sizes, tube diameter, and radius, and more water evaporates. The limiting case occurs when the soil matrix will reduce in volume no further from the water tension stresses. This is the shrinkage limit, and from this point on the water simply evaporates with no further soil structure change. As Eq. (8-12) indicates, very high compressive stresses can be developed during drying due to the large number of soil pores simultaneously being subjected to water tension. It is not possible to evaluate the magnitude of the compressive stresses developed; however, the effects have been observed. For many cohesive soils the drying (often several times over long geological periods and perhaps aided by pore and pore-water contamination as in marine deposits) has altered the soil structure just as if large depths of overburden had been applied and later eroded. In many parts of the United States the drying has loaded the soil (termed *preconsolidated* in Chap. 11) to apparent overburden loads on the order of 200 to 800 kPa. Tschebotarioff (1936) and others found shrinkage values on the order of 900 kPa in the Nile valley of Egypt. Large values of shrinkage stresses have also been found in Australia, India, and the Middle East.

Capillarity is a significant factor in sands (gravel is too large to be much affected), especially in fine to medium-fine sands. When the sand is fully saturated or completely dry, capillarity is not present (Fig. 8-12c, stage 1 for the saturated case) and the sand grains are easily displaced. At intermediate water contents, capillary effects are present, and due to the many particles the cumulative effect is great enough to allow sand to stand on vertical cuts and to be molded when damp. It is well known and easily observed that it is much easier to walk or drive a wheeled vehicle on damp than on dry sand due to grain displacement. Another practical consideration with damp sand is *bulking*, which was considered in Sec. 6-4. When sand is dry, it is totally impossible to make a vertical excavation or to mold it. The capillary effect in damp sand is termed *apparent cohesion* since it disappears when the sand is either dry or saturated.

Slaking is the rapid, almost explosive, disintegration of dry, or nearly dry, cohesive soil lumps when they are immersed in water; it is a capillary phenomenon. Soil that is not very nearly dry will only swell when immersed in water; however, if the water content is well below the shrinkage limit, the soil capillaries contain air. When the soil is placed in water, surface tension pulls water into the capillary tubes, confining and compressing the air. As the air becomes more compressed, the interparticle tension stresses increase, and when they become larger than the particle attractions, the soil slakes. Only the clayey soils and some clayey shales, where the material in the elemental state has a shrinkage limit, will slake. Other soils and rocks without a shrinkage limit do not exhibit this phenomenon.

E Shrinkage and Volume Change

Volume change is a serious problem in shrinkage-susceptible soils all over the world. It can involve almost any cohesive soil, but is more pronounced in arid to semiarid areas, or where the more active montmorillonite, or bentonite, clay minerals have not sufficiently weathered to a less active state. Large areas of the western and southwestern United States, Australia, India, the Middle East, and southern Africa are covered with soils subject to large volume changes which cause serious engineering problems. Cohesive soils in other areas are often susceptible to lesser volume changes, but these are usually more a nuisance than a serious problem. Jones and Holtz (1973) estimate the economic damage just in the United States at over $2\frac{1}{4}$ billion dollars annually and indicate that some 20 to 25 percent of the land area in the United States is covered with soil susceptible to volume change.

(a)   (b)

Figure 8-13 Shrinkage cracks. (a) Typical for clayey soil. The crack is about 2 cm wide and about 1.5 m deep, determined by inserting a small wire. (b) Cracks in a lake bed deposit. They are about 4 cm wide and at least 0.3 m deep.

Expansive clay soils are dense and very hard in the dry state due to shrinkage stresses. Even at small water contents the soil is quite dense and hard—so hard that obtaining thin-walled tube samples for laboratory testing is often difficult or nearly impossible. Often these soils will contain a maze of shrinkage cracks ranging from hairline to 2 or 3 cm wide. Figure 8-13 illustrates shrinkage cracks in a grass-covered area and in a dried lake bed, indicating that the phenomenon is not limited to freshly deposited water deposits. Shrinkage cracks similar to those shown for the grassy area can considerably influence the rate of soil saturation from surface infiltration. Laboratory strength test values are considerably affected by cracks and/or cracks which have "silted up" with foreign windblown or water-borne material. If the test samples are badly fissured, strength tests may not be practical.

The expansion problem can be avoided only in cases where the soil can be reliably protected against water infiltration by using surface or subsurface drainage, landscaping, or using impervious membranes (asphalt or plastic cloth). The only other recourse is altering the clay by chemical additives such as cement, lime, lime–fly ash, cement–fly ash, calcium chloride, etc. Simply covering an area with a floor slab or pavement does not control water infiltration in the zone of importance, since water vapor tends to condense on the under side over a period of several years and will saturate the soil. This can be readily observed by turning over a small rock in the summer; the underside will be damp, even if there has been a prolonged dry period. Unfortunately, near the edges of a floor slab or pavement the water content will be less than in the interior due to perimeter evaporation; thus the volume change will be differential rather than uniform.

Volume change is directly related to the shrinkage limit and somewhat related to the liquid and plastic limits. Table 8-2 gives an approximate relationship which has been found to be reasonably reliable in predicting the occurrence of volume change. Unfortunately, there is no reliable way to numerically quantify volume change. One would expect a "little" potential to be less than a "high" potential, but for design it would be of considerable benefit to be able to say that a "little" is, say, 2 or 3 percent or less, and "high" corresponds to 8 or 10 percent or more. At this point there is no means of quantifying the potentials given in the table, and considerable research needs to be done before they can be quantified. Factors such

Table 8-2 Relationship between Atterberg limits and volume change potential†

Volume change potential	Plasticity Index I_p		Shrinkage limit w_s
	Arid area	Humid area	
Little	0–15	0–30	> 12
Moderate	15–30	30–50	10–12
High	> 30	> 50	< 10

† From Holtz and Gibbs (1956).

as type of clay, surcharge load, void ratio, method of saturation, and general environment produce too wide a range of problem parameters to provide a ready and direct answer.

Since volume change is a result of the presence of cohesive soils and water, the problem may develop also in less arid areas as a result of tree and shrub growth. In the growing season tree roots may temporarily desiccate the soil, but when the tree is in the dormant season the water content builds up, resulting in a seasonal volume change. This may develop into an environmental problem: how to avoid the volume change and at the same time avoid destroying the trees or shrubs.

Example 8-5

GIVEN The shrinkage limit of a very clayey soil is 9.6 percent. $G_s = 2.60$, which is a low value for clay. Compaction test data gives $\gamma_d = 16.70$ kN/m³ at OMC = 16.5 percent.

REQUIRED Estimate the dry unit weight of the soil at the shrinkage limit and compare with the standard compaction test value.

SOLUTION By definition, at the shrinkage limit $w_s = 9.6$ percent, the degree of saturation $S = 100$ percent. From Eq. (2-12) we have

$$Se = wG_s \quad \text{and} \quad e = \frac{0.096(2.60)}{1.00} = 0.2496$$

From Eq. (2-11) the dry unit weight at the shrinkage limit is

$$\gamma_d = \frac{G_s \gamma_w}{1 + e} = \frac{2.60(9.807)}{1 + 0.25} = 20.40 \text{ kN/m}^3$$

The percent improvement is

$$\text{Percent improvement} = \frac{20.40}{16.70}(100) = 122 \text{ percent}$$

This computation indicates that the capillary stresses during drying produce very high compressive stresses in the soil structure and are considerably more efficient in producing a dense soil matrix than the standard compaction test. Note, however, that this unit weight was computed and not actually measured; the field value might be somewhat less, but could be even more if a large enough number of drying cycles had been applied.

8-8 SEEPAGE FORCES AND QUICK CONDITIONS

From Eq. (2-21) the intergranular or *effective* stress is

$$\sigma' = \sigma - u \qquad (2\text{-}21)$$

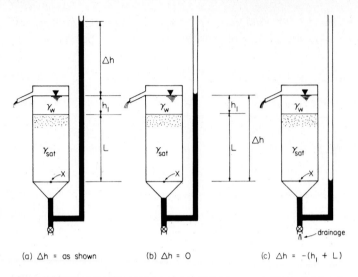

(a) Δh = as shown   (b) Δh = 0   (c) Δh = -(h₁ + L)

Figure 8-14 Intergranular pressure at point X for several differential heads.

A critical inspection of this equation shows that both the total pressure on the plane of interest and the pore water pressure can be measured and/or calculated. The pore water pressure can be measured using a piezometer with little difficulty (see discussion in Sec. 8-3). Note, however, that the effective stress can only be calculated.

When water flows in soil under a hydraulic gradient as considered in Secs. 8-3 and 8-4, the differential pressure head produces a force on the soil grains in the direction of flow. This pressure accumulation will be termed a *seepage force*. Consider the intergranular pressure at point X of Fig. 8-14 for the cases shown in (a), (b), and (c).

In Fig. 8-14, at the instant the differential head is as shown, the following are obtained for point X:

$$\sigma = h_1 \gamma_w + L\gamma_{sat} = h_1 \gamma_w + \frac{G_s \gamma_w}{1 + e} + \frac{\gamma_w e}{1 + e} = h_1 \gamma_w + \frac{(G_s + e)\gamma_w}{1 + e}$$

$$\gamma' = \gamma_{sat} - \gamma_w$$

$$u = L\gamma_w + (h_1 + \Delta h)\gamma_w$$

In Fig. 8-14a, the intergranular pressure is

$$\sigma' = h_1 \gamma_w + L\gamma_{sat} - L\gamma_w - (h_1 + \Delta h)\gamma_w$$

or

$$\sigma' = L\gamma' - (\Delta h)\gamma_w$$

and flow is upward through the sample. In Fig. 8-14b, the intergranular pressure at point X is

$$\sigma' = L\gamma'$$

and no flow is taking place, since $\Delta h = 0$. In Fig. 8-14c, the intergranular pressure at point X is

$$\sigma' = h_1 \gamma_w + L\gamma_{sat}$$

Alternatively, the intergranular pressure in this case may be computed as

$$\sigma' = L\gamma' - (-h_1 - L)\gamma_w = h_1 \gamma_w + L\gamma_{sat}$$

but note that the flow has reversed from upward to downward. This flow reversal effectively increases the intergranular pressure of the water in the soil mass as shown.

If we sum forces (pressure $\times$ area $= \sigma A$) in the three cases, we obtain

Case a: $\qquad \sigma' A = L\gamma' A - \Delta h \gamma_w A = $ intergranular force $-$ seepage force

Case b: $\qquad \sigma' A = L\gamma' A - 0xA = L\gamma' A \qquad$ (seepage force $= 0$)

Case c: $\qquad \sigma' A = L\gamma' A + (h_1 + L)\gamma_w A = L\gamma' A + \Delta h' A$

From inspection of these equations we can interpret the *seepage force* as

$$\text{Seepage force} = \text{excess}\dagger \text{ pore pressure} \times \text{area} \qquad (8\text{-}17)$$

Note carefully that a seepage force is always obtained when there is a differential pressure head Δh across a flow path of length L and will vary from point to point within the length of soil under consideration as shown in Fig. 8-15. The seepage force can be computed as $\Delta h \gamma_w A$. The hydraulic gradient is dissipated as in the figure:

$$d\,\Delta h = i\,dy$$

$$\Delta h = \int_0^L \frac{\Delta h}{L}\,dy$$

$\dagger$ Excess pore pressure is that due to Δh of Fig. 8-14, or the value over the static head at the point, as used in Sec. 2-13.

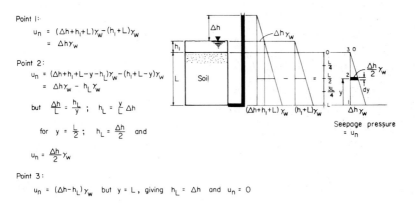

Point 1:

$$u_n = (\Delta h + h_1 + L)\gamma_w - (h_1 + L)\gamma_w$$
$$= \Delta h \gamma_w$$

Point 2:

$$u_n = (\Delta h + h_1 + L - y - h_L)\gamma_w - (h_1 + L - y)\gamma_w$$
$$= \Delta h \gamma_w - h_L \gamma_w$$

but $\quad \dfrac{\Delta h}{L} = \dfrac{h_L}{y}; \quad h_L = \dfrac{y}{L}\Delta h$

for $\quad y = \dfrac{L}{2}; \quad h_L = \dfrac{\Delta h}{2} \quad$ and

$$u_n = \frac{\Delta h}{2}\gamma_w$$

Point 3:

$$u_n = (\Delta h - h_L)\gamma_w \quad \text{but } y = L, \text{ giving } h_L = \Delta h \text{ and } u_n = 0$$

Figure 8-15 Method of computing seepage pressure using piezometric relationships and the definition of the hydraulic gradient.

The loss of Δh is an energy loss from viscosity, friction, and inertia effects as the water flows through the voids and around the soil grains making up the irregular and rough pore channels.

When the seepage force (or excess pore pressure) is sufficiently large, the individual soil grains will be suspended in the upward flowing water—a condition visibly resembling boiling. This phenomenon can be readily observed both in the laboratory and in field locations (on the land side of levees during floods) and can be approximately evaluated for sands. With cohesive soils, inter-particle attractive forces produce a condition in which a mass of soil may be lifted rather than individual grains. These conditions will be considered separately.

A Quick Conditions or Sands

When the intergranular pressure is zero for sand, the soil grains just touch with no friction resistance $(\sigma_f = f\sigma_n)$ available. This state is termed a *quick* or *quicksand* condition. Intergranular stresses less than zero are not possible in sand, since this would be a state of tension. At a quick condition,

$$\sigma' = 0 = \sigma - u$$

thus, when the pore pressure equals the total pressure on the plane, a quick condition exists, and the pore pressure can only equal the total pressure when $\Delta h > 0$, which is a flow condition.

Example 8-6

GIVEN Void ratios of $e = 0.5$, 0.8, and 1.0 for a sand with $G_s = 2.67$.

REQUIRED What is the critical hydraulic gradient i_c for these void ratios?

SOLUTION Noting from the just completed discussion that a quick condition is a flow condition, the critical hydraulic gradient will be taken as that hydraulic gradient causing a quick condition,

$$\sigma = u$$

For any tailwater h_i and differential head Δh sufficiently large to produce i_c, we have

$$L\gamma_{sat} + h_i\gamma_w = (L + h_i + \Delta h)\gamma_w$$

and simplifying,

$$(\gamma_{sat} - \gamma_w)L = \Delta h\gamma_w$$

$$L\gamma' = \Delta h\gamma_w$$

Rearranging,

$$\frac{\Delta h}{L} = i_c = \frac{\gamma'}{\gamma_w}$$

or the critical hydraulic gradient i_c is

$$i_c = \frac{\gamma'}{\gamma_w} \tag{8-18}$$

Next we will develop a general expression for γ' using Eqs. (2-10) and (2-12) to obtain

$$\gamma_{sat} = \frac{\gamma_w(G_s + e)}{1 + e}$$

$$\gamma' = \frac{\gamma_w(G_s + e)}{1 + e} - \gamma_w$$

and the desired relationship is

$$\gamma' = \frac{\gamma_w(G_s - 1)}{1 + e} \tag{8-19}$$

Now from Eqs. (8-18) and (8-19) the critical hydraulic gradient is

$$i_c = \frac{G_s - 1}{1 + e} \tag{8-20}$$

and for this example,

$$e = 0.5: \quad i_c = \frac{2.67 - 1}{1 + 0.5} = 1.11$$

$$e = 0.8: \quad i_c = 0.93$$

$$e = 1.0: \quad i_c = 0.835$$

The maximum range of i_c for any sand is on the order of

e	i_c	
0.3	1.3	(No sand is likely to have a void ratio much less than this value)
1.2	0.76	(No sand is likely to have a void ratio much larger than this value)

The likely range of e is from about 0.45 to 0.7; thus $i_c \cong 1.0$ for any practical G_s and sand deposit.

B Seepage Uplift Pressure on Clay Strata

The interparticle attractive forces in clays are such that it is more likely that the mass, rather than individual particles, may float or isolated channels may erode (caused by defects from worm or animal burrows and root decay). Consider

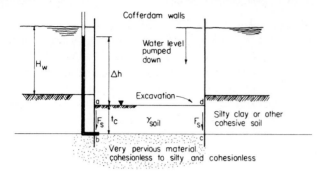

Figure 8-16 Excavation into a silty clay deposit inside a cofferdam.

Fig. 8-16 as a typical condition where seepage forces would require investigation of stability. Assume that a piezometer located on the bottom of the silty clay indicates the piezometric head shown. One would be concerned with the amount of excavation to produce t_c which can be safely done. Neglecting any side friction resistance F_s, and with the excavation pumped dry as shown, $\sum F_v = 0$ gives the critical thickness t_c where the clay is on the verge of floating (safety factor $F = 1$). There are no seepage forces to consider within the clay, since the coefficient of permeability is so low that the seepage quantity is very nearly zero. With $\sum F_v = 0$ on block $abcd$, obtain

$$t_c \gamma_{\text{soil}} = (\Delta h + t_c)\gamma_w$$

from which

$$t_c(\gamma_{\text{soil}} - \gamma_w) = \Delta h \gamma_w$$

where γ_{soil} is probably (but not necessarily) the saturated unit weight. Rearranging, we obtain

$$t_c = \frac{\Delta h \gamma_w}{\gamma_{\text{soil}} - \gamma_w} \qquad F = 1$$

and any $t > t_c$ gives a safety factor $F > 1$.

8-9 SUMMARY

This chapter has developed the concept of the hydraulic gradient i as an energy loss occurring when water flows in a soil mass. The flow has been analyzed in a manner similar to any other fluid flow, and the Bernoulli equation has been used together with the Darcy coefficient of permeability to obtain the flow quantity. The velocity head in soil flow problems has been shown to be negligible using the Bernoulli equation. Means of obtaining the coefficient of permeability and consideration of laminar vs. turbulent flow and the effects of these flows on the validity of Darcy's law have been critically evaluated.

The concept of soil capillarity has been considered in some detail. This includes both the simple rise in a tube of a column of water above the water surface and the effects of capillary tension on soil shrinkage and the shrinkage limit.

Conceptual relationships have been made and/or developed between:

1. Hydraulic gradient and seepage forces
2. Hydraulic gradient and quick conditions for cohesionless soils
3. Hydraulic gradient and mass uplift for cohesive soils

HOMEWORK PROBLEMS

8-1 A sample of soil for a constant-head permeability test provides the following data (refer to Fig. P8-1): Diameter = 7.6 cm; L = 20.0 cm; Δh = 15.0 cm; e = 0.55; γ_{sat} = 2.08 g/cm^3; time of test duration = 8 min; Q = 1200 cm^3.

REQUIRED: Sketch the test setup and compute
(a) Coefficient of permeability k, cm/s
(b) Nominal seepage velocity, cm/s
(c) Approximate actual discharge velocity
Ans.: (c) v_a = 0.155 cm/s.

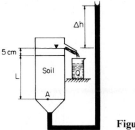

Figure P8-1

8-2 Assume the sample of Prob. 8-1 is vertical as in the sketch.
REQUIRED: What differential head Δh will cause a quick condition at point A?
Ans.: 21.6 cm.

8-3 Given are the following data from a falling-head permeability test: Diameter = 7.62 cm: L = 20.3 cm; h_1 = 50 cm; h_2 = 20 cm; Time = 82 s; Q = 34 cm^3, and T = 28°C.
REQUIRED: Sketch the test setup and find k both for test temperature and for 20°C.
Partial ans.: 5.62 × 10^{-3} cm/s at 28°C.

8-4 Refer to Fig. 8-7 and take:

Stratum	k, cm/s	H_i, m
1	1 × 10^{-2}	1.6
2	2 × 10^{-4}	2.5
3	1 × 10^{-1}	1.5
4	6 × 10^{-3}	0.9

REQUIRED: Find the equivalent k_v and k_h for the soil mass.
Ans.: k_v = 5.07 × 10^{-4}; k_h = 2.64 × 10^{-2} cm/s.

8-5 A falling-head permeability test was performed with data as follows: $h_1 = 150$ cm; $h_2 = 50$ cm; $L = 15$ cm; $A = 100$ cm^2; time $= 10$ min; area of standpipe $a = 0.959$ cm^2; $T = 25°C$.

REQUIRED: Compute the coefficient of permeability at the temperature of the test and at 20°C.

Ans.: $k_{20} = 2.3 \times 10^{-4}$ cm/s.

8-6 What is the expected height of capillary rise of the soil of curve A of Fig. 3-2?

8-7 Assuming the shrinkage limit of the soil of Fig. 7-1 $w_S = 9.4$ percent, what is the density at the shrinkage limit and how does this compare to the density values shown?

8-8 A sand soil has $G_s = 2.68$. Make a plot of critical hydraulic gradient i_c versus void ratio for the likely range of realistic void ratios.

8-9 Referring to Fig. 8-16, if $\gamma_{\text{soil}} = 18.2$ kN/m^3 and $\Delta h = 6.3$ m and t must be excavated to a thickness of 5.4 m, draw a neat working drawing of the problem conditions and compute the safety factor F; if it is less than 1.15, state what can be done to allow the excavation to safely proceed.

8-10 Referring to Fig. 8-14, verify that the plot of seepage pressure is linear by computing the seepage pressures at $y = 0.25L$ and $0.75L$.

Chapter 9

Seepage
and Flow Net Theory

9-1 INTRODUCTION

Chapter 8 introduced the concept of permeability as the facility of fluid flow through a porous medium. This chapter will be concerned with using the coefficient of permeability to evaluate the amount of seepage flow through a soil mass.

9-2 TWO-DIMENSIONAL SEEPAGE FLOW THEORY

Seepage analysis is not a very exact procedure, with computed results often not much more than estimates. This unhappy situation may be improved somewhat if steady state flow is considered. Steady state flow would be obtained when the soil is fully saturated, the pressure gradient is unchanging, a constant soil mass is involved, and the flow rate is constant. We will consider only two-dimensional flow (in the XY plane), since a very large number of problems are, or can be, treated as two-dimensional flow by a transformation (or rotation) of axis. The soil is often assumed to be homogeneous but may be anisotropic ($k_x \neq k_y$).

With these ground rules established, and referring to Fig. 9-1, for any typical differential element,

$$q_{in} = q_{out}$$

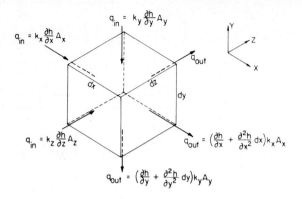

Figure 9-1 Flow of water through a soil element of dimensions dx, dy, and dz as shown.

At the entrance faces of the element and parallel to the X and Y axes, the hydraulic gradients are

$$i_x = \frac{\partial h}{\partial x} \qquad i_y = \frac{\partial h}{\partial y}$$

respectively, and at the exit faces the hydraulic gradients are

$$i_x + \frac{\partial i_x}{\partial x}\, dx = \frac{\partial h}{\partial x} + \frac{\partial}{\partial x}\left(\frac{\partial h}{\partial x}\right) dx = \frac{\partial h}{\partial x} + \frac{\partial^2 h}{\partial x^2}\, dx$$

And for flow parallel to the Y axis we have

$$i_y + \frac{\partial i_y}{\partial y} = \frac{\partial h}{\partial y} + \frac{\partial^2 h}{\partial y^2}\, dy$$

For the entering flow quantities, using Darcy's law for $v = ki$ and a flow rate of $q = Av$, we obtain the following:

$$q_{x,\,\text{in}} = dy\, dz\, k_x\, \frac{\partial h}{\partial x}$$

$$q_{y,\,\text{in}} = dx\, dz\, k_y\, \frac{\partial h}{\partial y}$$

The exit flow quantities are

$$q_{x,\,\text{out}} = dy\, dz\, k_x \left(\frac{\partial h}{\partial x} + \frac{\partial^2 h}{\partial x^2}\, dx\right)$$

$$q_{y,\,\text{out}} = dx\, dz\, k_y \left(\frac{\partial h}{\partial y} + \frac{\partial^2 h}{\partial y^2}\, dy\right)$$

Equating q_{in} to q_{out} and canceling appropriate terms, we obtain the Laplace equation for two-dimensional flow as

$$k_x \frac{\partial^2 h}{\partial x^2} + k_y \frac{\partial^2 h}{\partial y^2} = 0 \qquad\qquad (9\text{-}1)$$

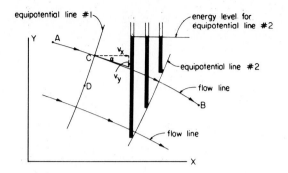

Figure 9-2 Intersection of a flow line with an equipotential line.

This is the equation of families of curves intersecting in the XY plane. Let us investigate if these curves have any special properties. Figure 9-2 is a physical interpretation of two of these curves. One of the curves is the *flow path* of a particle of water from A to B, and the other curve is a line of constant pressure head h, termed an *equipotential line*. At point C the slope of curve AB is α, computed as follows:

The velocity vectors are (noting that a sign will go with the derivatives)

$$v_x = k_x \frac{\partial h}{\partial x} \qquad v_y = k_y \frac{\partial h}{\partial y}$$

By inspection of the figure,

$$\tan \alpha = \frac{v_y}{v_x} = \frac{k_y \, \partial h/\partial y}{k_x \, \partial h/\partial x} \tag{a}$$

Now along the line of constant pressure (or head), say from C to D, $h =$ constant and therefore the derivative $dh = 0$, but the derivative dh is

$$k_x \frac{\partial h}{\partial x} dx + k_y \frac{\partial h}{\partial y} dy = dh = 0$$

Dividing by dx and solving for dy/dx, we obtain

$$\frac{dy}{dx} = -\frac{k_x \, \partial h/\partial x}{k_y \, \partial h/\partial y}$$

which is the negative reciprocal of $\tan \alpha$; thus, the families of curves defined by the Laplace equation always intersect at right angles.

The lines along which the velocity vectors were considered are called stream or *flow lines*. The lines along which the total energy or head = constant are called *equipotential lines*. Note that along any equipotential line the total head is the sum of

Total head $h =$ static (elevation) head + pressure head

+ velocity head (negligible quantity)

thus, the static and pressure heads vary with the Y coordinate. The pressure head and head differences can be measured in the field with piezometers.

9-3 FLOW NETS

Figure 9-3 represents a layer of soil one unit in width confined between two impervious plates (or soil layers). The soil will be divided into sections by drawing a series of flow lines and equipotential lines which intersect at as close to right angles as it is practicable to draw. The channel formed by any two adjacent flow lines will be called a *flow path*. The difference in energy head represented by any two adjacent equipotential lines is a head loss, defined by Δh_i, and it is evident that the total head loss between two points, say, B and C, is $\sum (\Delta h)$. The dimensions of the segments in Fig. 9-3 are a and b as shown on the figure. From this, the area of element 1 through which water flows is

$$A = a(1)$$

The total area is obtained from the three flow paths $(n_f = 3)$ as

$$A_{total} = n_f(a)$$

where n_f = number of flow paths and may be an integer or decimal, as 5, 7, 9, 9.2, 10.7, 4.3, etc.

The length L of soil across which a total head loss h occurs is, by inspection of Fig. 9-3,

$$L = n_d(b)$$

where n_d = number of equipotential drops (spaces, not lines) between the two boundary points where h is developed and must be an integer since Δh = constant

The flow quantity per unit of width can be computed as

$$Q = kiA = k\frac{h}{L}A = k(h)\frac{n_f}{n_d}\frac{a}{b}$$

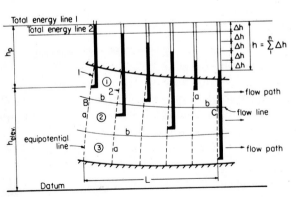

Figure 9-3 A percolating soil mass subdivided into a grid or flow net using flow lines and equipotential lines.

Now if squares are drawn, then $a = b$ for any possible quadrilateral, and one obtains

$$Q = k(h)\frac{n_f}{n_d} \tag{9-2}$$

Squares can be drawn when the soil is isotropic ($k_x = k_y$), as is evident by inspection of the Laplace equation [Eq. (9-1)]. By use of the method to be considered in Sec. 9-8, squares may be obtained when $k_x \neq k_y$.

Equation (9-2) can be used to obtain seepage quantities by using a graphical solution of Eq. (9-1). This is especially useful since an analytical solution of the Laplace equation for seepage can be obtained only for very simple cases, and even then the mathematics generally is prohibitive. Flow nets can be rapidly sketched, and they provide answers well within the precision with which the coefficient of permeability is likely to have been determined. Approximate flow nets can be obtained for very complicated boundary conditions for which analytical solutions would be very nearly impossible, and these approximations will provide seepage quantities with an accuracy that is compatible with the determination of the coefficient of permeability.

One of the principal drawbacks of evaluating seepage quantity is that k_x is very seldom equal to k_y. Laboratory tests often do not reveal this situation—it usually requires field testing to obtain reliable values for the two values.

9-4 FLOW NETS FOR EARTH DAMS

Flow through earth dams was one of the first applications of flow net theory. Consider an idealized earth dam as shown in Fig. 9-4. The equation of the top flow line (the upper seepage boundary and the *phreatic* surface) can be obtained as follows:

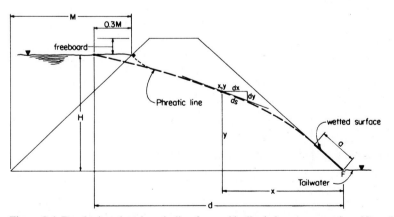

Figure 9-4 Developing the phreatic line for an idealized dam cross section. Note that point F is always taken at the intersection of the tailwater and the downstream face. In the case shown here, the tailwater is coincident with the "original" ground line.

At a point of coordinates (x, y) measured from the downstream toe as shown we have

$$i = \frac{dy}{ds} \qquad v = ki = k\frac{dy}{ds} \qquad A = y \text{ (width)}$$

For cases of small β angles (generally $\beta < 30°$), one may replace dy/ds with $i = dy/dx$. Making this substitution and solving for flow rate,

$$q = Av = k\frac{dy}{dx}(y)(1) \qquad (a)$$

Separating variables, obtain

$$q(dx) = k(y)\,dy$$

This differential equation assumes that either $k = k_x = k_y$ or, if $k_x \neq k_y$, a linear transformation, as in Sec. 9-9, has been used. Integrating this differential equation, one obtains

$$q(x) = k\frac{y^2}{2} + C \qquad (b)$$

At $x = d$, $y = H$ and we obtain

$$C = q(d) - k\frac{H^2}{2}$$

and substituting this value of C into Eq. (b), we obtain

$$q(x - d) = \frac{k}{2}(y^2 - H^2) \qquad (9\text{-}3)$$

Equation (9-3) shows that the equation of the phreatic (sometimes called the saturation) line for this problem geometry is a parabola.

Angle of Exit of the Phreatic Line

The angle of exit of the phreatic line can be developed by referring to Fig. 9-5 as follows:

At point 1 along the downstream face,

$$\frac{\Delta h}{c} = \sin(\beta - \alpha) \qquad (c)$$

Since we will always use squares, inspection of the figure gives $b = c$; thus,

$$\Delta h = b \sin(\beta - \alpha) \qquad (d)$$

Also from the figure,

$$\frac{\Delta h}{d} = \sin\beta \qquad (e)$$

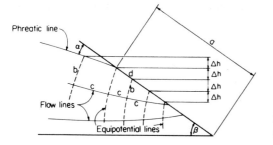

Figure 9-5 Exit of phreatic line at downstream face of an earth dam.

At point 2 we have

$$\frac{c}{d} = \cos \alpha \qquad (f)$$

Now dividing (e) by (f), we obtain

$$\frac{\Delta h}{c} = \frac{\sin \beta}{\cos \alpha} \qquad (g)$$

Finally, equating Eqs. (c) and (g),

$$\sin (\beta - \alpha) = \frac{\sin \beta}{\cos \alpha} \qquad (h)$$

This equality can only be obtained if $\alpha = 0$, since $\cos \alpha = \cos 0 = 1$ and $\sin (\beta - 0) = \sin \beta$. With the angle of exit of the phreatic surface $= 0$, the exit is parallel and coincident with the downstream face of the dam at the top of the wet zone a of Fig. 9-6.

9-5 DRAWING THE PHREATIC LINE

The downstream wetted surface a of Fig. 9-6 can be computed as follows:
Rearrange Eq. (9-3) to obtain

$$q = \frac{k}{2} \frac{y^2 - H^2}{x - d} \qquad (i)$$

Next consider the following to obtain q:

$$i = \frac{dy}{ds} \cong \frac{dy}{dx} \qquad \text{for small } \beta \text{ angles}$$

but $\tan \beta = dy/dx$ at the exit of the phreatic line. From Fig. 9-6, $y = a \sin \beta$, and since fluid flow $q = Av$, we have

$$q = kiA = k(\tan \beta)(a \sin \beta)$$

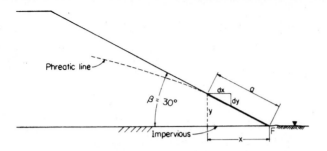

Figure 9-6 Computation of the wetted downstream face distance a and the direct computation of seepage quantity for small β angles.

or, rearranging,

$$q = k(a) \sin \beta \tan \beta \qquad (9\text{-}4)$$

Now if we equate Eq. (9-4) and Eq. (i) and again substitute $y = a \sin \beta$ and $x = a \cos \beta$, we obtain

$$a = \frac{d}{\cos \beta} - \sqrt{\frac{d^2}{\cos^2 \beta} - \frac{H^2}{\sin^2 \beta}} \qquad \text{for } \beta \leq 30° \qquad (9\text{-}5)$$

where all terms are identified on Fig. 9-4. This equation allows a direct computation of the wet distance a for a graphical solution or for an analytical solution using Eq. (9-4) with the limitations indicated.

9-6 DIRECT COMPUTATION OF SEEPAGE QUANTITY

Equation (9-4) can be used for a direct computation of seepage quantity and is generally recommended to obtain a rapid solution when $\beta \leq 30°$. When $\beta > 30°$, the use of Eq. (9-4) may give satisfactory estimates of the seepage quantity in many cases.

Example 9-1 Compute the estimated seepage quantity for the dam shown in Ex. 9-2 (Fig. E9-2a).

SOLUTION Obtain dimensions and data from Fig. E9-2a as follows:

$$H = 18.5 \text{ m} \qquad d = 40 + 13.8 + 46(0.3) = 67.6 \text{ m}$$

$$k = 4 \times 10^{-4} \text{ m/min} \qquad \beta = \tan^{-1} \tfrac{20}{40} = 26.6°$$

Step 1 From Eq. (9-5) compute the wet face distance a as

$$a = \frac{67.6}{\cos 26.6} - \sqrt{\frac{67.6^2}{\cos^2 26.6} - \frac{18.5^2}{\sin^2 26.6}}$$

$$= 75.6 - \sqrt{5715.7 - 1707.1}$$

$$= 75.6 - \sqrt{4008.6}$$

$$= 75.6 - 63.3 = 12.3 \text{ m}$$

Step 2 From Eq. (9-4),

$$q = k(a) \sin \beta \tan \beta$$

$$= (4 \times 10^{-4})(12.3)(\sin 26.6)(\tan 26.6)$$

$$= 11.01 \times 10^{-4} \text{ m}^3/\text{min/m of width}$$

9-7 METHODS FOR OBTAINING THE PHREATIC LINE FOR EARTH DAMS

The phreatic surface must be considered separately for cases of $\beta < 30°$ and for larger downstream slope angles.

Case I $\beta \leq 30°$

Draw the earth dam to scale as in Fig. 9-7. Calculate the wetted downstream face using Eq. (9-5) and lay off as shown. Observations of model flow nets have indicated that the apparent origin of the parabola is located a distance of 0.3 m upstream, as shown in the figure.

Since the phreatic line is a parabola, we may use the simplest form of the equation,

$$y = Kx^2$$

and at x_o, $y = y_o$, which gives $K = y_o/x_o^2$.

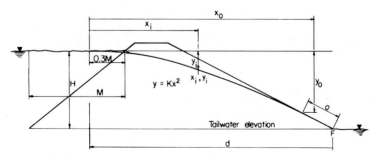

Figure 9-7 Locating the phreatic line when $\beta \leq 30°$.

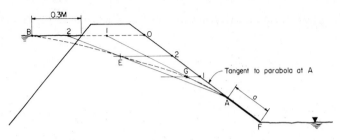

Figure 9-8 Rapid means of drawing an approximate phreatic line. Note that four points, such as D, E, G, and A with the slope at A known, are generally sufficient to draw a parabola.

Scale several distances x_i, compute the corresponding offsets y_i, and, using a french curve, draw a smooth curve through the locus of points thus obtained. Note that the parabola is tangent to the downstream face at the top of the wetted face (point A) and is faired into a perpendicular to the upsteam face at the water line. This is necessary, since the upstream face is an equipotential line and the phreatic line is a flow line.

Figure 9-8 illustrates a rapid alternative means of establishing an approximate phreatic line. The steps are as follows:

1. Establish a *tangent* to the parabola at some point. Use the downstream face when $\beta \leq 30°$. Strictly, OA should be a vertical tangent through point 1 of Fig. 9-9.
2. Extend this tangent from point A to point O and draw a line parallel to the horizontal axis from point B at the upstream face to point O. Divide OA and OB into the same number of equal parts (not more than 4 or 5) and mark locations.
3. On line OA, through the marked points from step 2, lightly draw lines parallel to the horizontal axis.
4. From A as the origin lightly draw rays to the bisector points marked on line OB. At the intersection of the corresponding ray and the horizontal line of step 3, mark the point.
5. Through the locus of points obtained from step 4, draw the phreatic line and adjust the entrance as previously outlined.

Case II $\beta > 30°$

An alternative means of establishing the phreatic line may be required when $\beta > 30°$. The reason is that the method of Case I is only valid as long as $dy/ds \cong dy/dx$, with the break point occurring at approximately 30°. How much larger than 30° one may continue to use Case I is an exercise in engineering judgment, taking into account the order of magnitude of k and how accurately k can be determined.

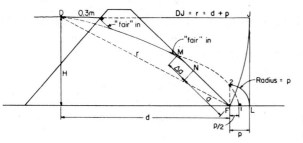

Figure 9-9 Locating the phreatic surface when $\beta > 30°$.

To obtain the phreatic line when it is necessary to account for $dy/ds \neq dy/dx$, it is necessary to obtain the parameter distance p of a parabola as measured from the focus F (see Fig. 9-9 and any text on analytical geometry). Once the half-parameter distance p is established, this locates a point on the phreatic line. An arc from F as shown on Fig. 9-9 locates a second point on the parabola along a perpendicular from F to the arc. Since the upstream starting point is known, this may give sufficient points to draw the parabolic phreatic line. If not, other offsets from the horizontal line DJ may be computed using

$$y = \frac{y_0}{x_0^2} x^2$$

as for Case I.

The procedure for obtaining the distance p for the parabola is as follows:

1. Let
   p = half-parameter of any parabola
   p = perpendicular distance from F both to the parabolic curve at point 2 and to the directrix (point K)
   F = focus of the parabola, the intersection of the tailwater elevation and the downstream slope
2. From the dimensions d, H, and r of Fig. 9-9, one can compute

$$p + d = \sqrt{H^2 + d^2} = r$$

and rearranging we obtain

$$p = \sqrt{H^2 + d^2} - d \qquad (9\text{-}6)$$

3. Next let
   M = point where the theoretical parabola outcrops the downstream face using the p values computed from Eq. (9-6)
   N = point where the actual seepage line outcrops the downstream face a distance of Δa from M (to be found)
   FN = the wetted distance a (also to be found)

It has been found (Casagrande, 1937) that the ratio $\Delta a/(a + \Delta a)$ (refer to Fig. 9-9) is a special scalar which we may call ρ'. From theoretical work cited by Casagrande, this scalar is related to the downstream slope angle β as follows:

β	ρ'
30°	0.375
60	0.320
90	0.260
120	0.185
150	0.105
180	0.000

Steps in computing the seepage quantity for $\beta > 30°$ are as follows:

1. Compute p using Eq. (9-6), and note that d includes the distance of 0.3 m as shown in Fig. 9-4. Alternatively, initially locate the point D on the upstream surface, then with a compass scribe an arc to the horizontal DJ using DF as a radius, as shown in Fig. 9-9. Drop a vertical JK as shown, measure p, and locate the two points (1 and 2) from F using the radius distances of p and $p/2$, also as shown.
2. Using any accepted procedure, construct the theoretical parabola from D through M to point 1.
3. Locate and scale the distance $FM = a + \Delta a$.
4. Compute Δa using the value of ρ' from the table of values depending on the value of β. Interpolate for intermediate values of β. Then

$$\Delta a = FM(\rho')$$

5. Lay off Δa from point M and fair in the parabola so that it is tangent to the downstream face for all $\beta \leq 90°$ and is *vertical* for all $\beta > 90°$.

Other Cases When $\beta > 30°$

The odd downstream face slopes shown in Fig. 9-10 are attempts to depress the seepage outcrop to the interior of the dam so that the downstream face is dry (more confidence-inspiring), to control seepage pressures, and to reduce the possibility of piping (Sec. 9-15 will consider piping in some detail).

9-8 FLOW NET CONSTRUCTION

Figure 9-11 illustrates several flow nets for indicated flow geometry. The general shape of a flow net will be determined by the boundary conditions in most cases. Exceptions are at singular points, where the flow net may determine the boundary conditions. General requirements for the flow net boundary conditions are:

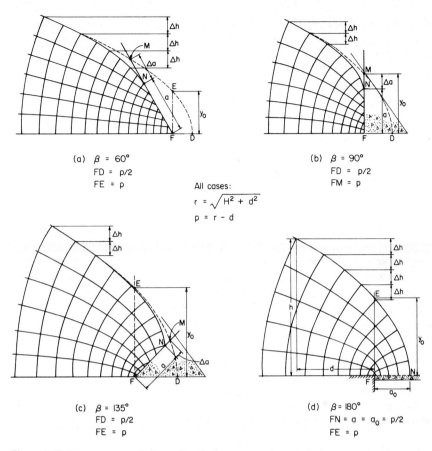

Figure 9-10 Downstream exit faces. In all these cases, the exit is into granular (or rock toe or underdrain) material with $k_2 \gg k_{dam}$. (*Casagrande, 1937.*)

1. The flow net intersects with equipotential lines at right angles, except at singular points where the velocity is zero (stagnation) or $v \to \infty$, as occurs at corners or tips of impervious cutoff walls.
2. From the definition of q, Δh must be the same amount for each equipotential line.
3. The pressure head at the intersection of the phreatic line and any equipotential line is zero.
4. All flow paths must have continuity, so that $q_{in} = q_{out}$.

In constructing flow nets the following guidelines may prove helpful.

1. Always draw squares which intersect at right angles as nearly as it is possible to draw them (exceptions are at singular points, such as corners). When condi-

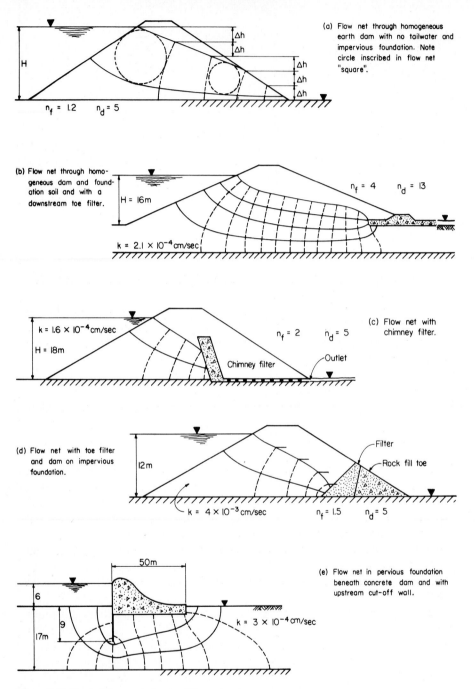

(a) Flow net through homogeneous earth dam with no tailwater and impervious foundation. Note circle inscribed in flow net "square".

$n_f = 1.2$ $n_d = 5$

(b) Flow net through homogeneous dam and foundation soil and with a downstream toe filter.

$H = 16m$

$k = 2.1 \times 10^{-4}$ cm/sec

$n_f = 4$ $n_d = 13$

(c) Flow net with chimney filter.

$k = 1.6 \times 10^{-4}$ cm/sec

$H = 18m$

$n_f = 2$ $n_d = 5$

Chimney filter

Outlet

(d) Flow net with toe filter and dam on impervious foundation.

$12m$

Filter

Rock fill toe

$k = 4 \times 10^{-3}$ cm/sec $n_f = 1.5$ $n_d = 5$

(e) Flow net in pervious foundation beneath concrete dam and with upstream cut-off wall.

50m

6

9

17m

$k = 3 \times 10^{-4}$ cm/sec

Figure 9-11 Typical flow nets for conditions shown.

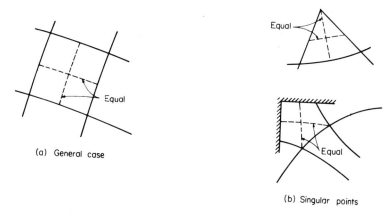

Figure 9-12 Definition of a "square" for constructing flow nets.

tions exist which require the use of rectangles, refer to Sec. 9-9 for modifications to allow the use of squares.

2. Use as few flow paths (and resulting equipotential drops) as possible while maintaining square sections. Generally four to six flow paths will be sufficient, used with a modest plotting scale so that the drawing does not tend to be more precise than soil data can justify.

3. Check the accuracy of squares by adding selected lines and observing if they subdivide large squares into recognizable smaller ones.

4. Use a pair of dividers to measure the square dimensions as illustrated in Fig. 9-12.

5. Always watch the appearance of the entire flow net. Do not make fine detail adjustments until the entire flow net is approximately correct.

6. Take advantage of symmetry where possible. Symmetrical geometry may result in portions of the net being exact squares; if so, develop this area first, then extend the net into the adjacent zones.

7. Use smooth transitions around corners and reentrant corners. Use gradual transitions from small to larger "squares."

8. A discharge face in contact with air is neither a flow line nor an equipotential line; however, such a boundary must fulfill the same conditions of equal drops in potential where the equipotential lines intersect it (a concept used in Fig. 9-5).

9. To get very good results a roughed-in flow net is adequate. While rough nets are usually not acceptable for academic work, the student should be aware that the ratio of n_f/n_d will be relatively unchanged in a very precise, right-angle-intersection flow net from a very rapidly drawn and uncorrected flow net. Keep in mind that the coefficient of permeability may be correct to the order of magnitude (power of 10), which is far less precise than the n_f/n_d ratio.

Example 9-2

GIVEN The earth dam cross sections shown in Fig. E9-2. The dam section of Fig. E9-2b has a toe filter so that the phreatic line is entirely inside the dam.

REQUIRED Compute the expected seepage in cubic meters per second per meter of dam width.

SOLUTION Superimpose the flow nets directly on the profile section after drawing to scale to save text space and obtain the flow nets shown.

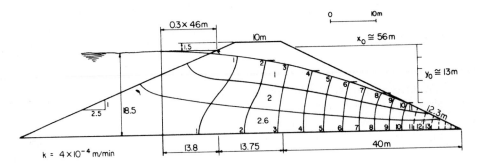

Figure E9-2a

For Fig. E9-2a the flow rate is

$$q = (4 \times 10^{-4})(18.5)(1)\left(\frac{2.6}{15}\right) = 12.8 \times 10^{-4} \ \mathrm{m^3/(min \cdot m)}$$

The number of drops, 15, is averaged from the 2.6 flow paths. Alternatively, each flow path in turn could be computed and the results summed to obtain the total flow rate q. This value compares to $q = 11.01 \times 10^{-4} \ \mathrm{m^3/(min \cdot m)}$ directly computed in Ex. 9-1.

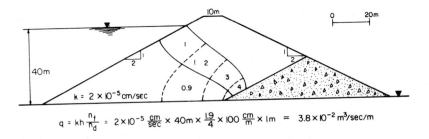

Figure E9-2b

9-9 THE FLOW NET WHEN $k_x \neq k_y$

When the coefficient of permeability in the horizontal direction (usually k_x) is not the same as in the vertical direction, which occurs most of the time due to the process of stratification resulting from sedimentation, the Laplace equation

$$k_x \frac{\partial^2 h}{\partial x^2} + k_y \frac{\partial^2 h}{\partial y^2} = 0$$

indicates that rectangles may result, the side ratio being $\sqrt{k_x/k_y}$. Therefore let

$$u^2 = \frac{k_x}{k_y}$$

from which $k_x = u^2 k_y$, and substituting this into the Laplace equation obtain

$$u k_y \frac{\partial^2 h}{\partial x^2} + k_y \frac{\partial^2 h}{\partial y^2} = 0$$

which can be rewritten as

$$\frac{\partial^2 h}{\partial (x/u)^2} + \frac{\partial^2 h}{\partial y^2} = 0 \tag{9-7}$$

A similar approach using

$$v^2 = \frac{k_y}{k_x}$$

can also be made.

From Eq. (9-7) above, we have

$$\frac{x}{u} = \frac{x}{\sqrt{k_x/k_y}}$$

but the value of x/u is simply a transformed x distance we may call x'. Once the transformed distance x' is used, Eq. (9-7) indicates that the coefficient of permeability is no longer a factor. Since this is so, it follows that if we use transformed distances when $k_x \neq k_y$, then the resulting flow net is the same as when $k_x = k_y$, and we always use squares for that case. The use of "squares" results in a net that is easy to check with a pair of dividers and easier to draw since the sides are the same.

Example 9-3

GIVEN $k_x = 12.1 \times 10^{-3}$ cm/s; $k_y = 3.0 \times 10^{-3}$ cm/s.

REQUIRED Set up a table of several transformed x' distances for drawing the outlines of an earth dam.

SOLUTION Let

$$u^2 = \frac{k_x}{k_y} = \frac{12.1 \times 10^{-3}}{3 \times 10^{-3}} = 4.0$$

$$u = \sqrt{\frac{k_x}{k_y}} = \sqrt{4} = 2.0$$

$$x' = \frac{x}{u} = \frac{x}{2}$$

For:

x, m	x', m	y, m
100	50	25
50	25	10
25	12.5	5

These distances would be plotted to obtain the shape of the dam and various parts of the dam so that a flow net could be sketched using squares.

If this transformation had not been made, the resulting flow net would be rectangular with dimension $b = u(a) = 2a$. This would be hard to check using dividers, as they would have to be reset each time for each flow net square.

Computation of Seepage Quantity When $k_x \neq k_y$

When the flow net is completed using the transformed dimensions for x, the flow rate is computed as

$$q = k'(h)\frac{n_f}{n_d}$$

where k' = the effective coefficient of permeability, obtained as follows (refer to Fig. 9-13)

Since the quantities of flow in Fig. 9-13a and 9-13b are the same, we may equate the flow quantities shown on the figure to obtain

$$k'(h)a/b = k_x(h)\frac{a}{a\sqrt{k_x/k_y}}$$

Simplifying and rationalizing the denominator, we obtain the equivalent coefficient of permeability as

$$k' = \sqrt{k_x k_y} \qquad (9\text{-}9)$$

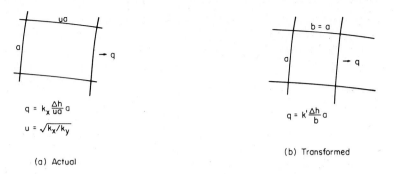

$$q = k_x \frac{\Delta h}{ua} a$$
$$u = \sqrt{k_x/k_y}$$

(a) Actual

$$q = k' \frac{\Delta h}{b} a$$

(b) Transformed

Figure 9-13 Obtaining the effective coefficient of permeability when $k_x \neq k_y$.

9-10 FLOW NETS FOR OTHER STRUCTURES

Flow nets can be drawn for many seepage problems other than flow through earth dams. Figure 9-14 illustrates several problems where flow nets can be constructed to evaluate seepage quantities.

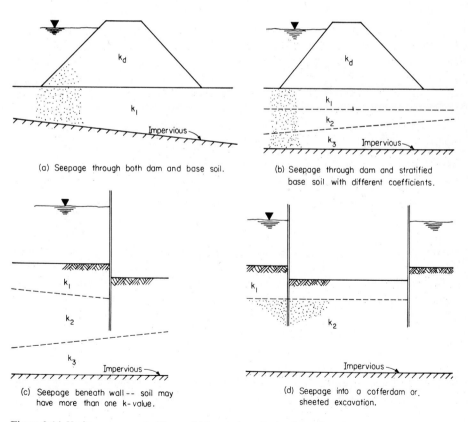

(a) Seepage through both dam and base soil.

(b) Seepage through dam and stratified base soil with different coefficients.

(c) Seepage beneath wall -- soil may have more than one k-value.

(d) Seepage into a cofferdam or sheeted excavation.

Figure 9-14 Various seepage problems which may be solved using a flow net.

In all the illustrated cases, it is necessary to define the boundary conditions, recognizing that abrupt changes in the flow paths do not exist and that equipotential and flow lines intersect at right angles and form squares. Singular points produce non-right angle intersections. When $k_x \neq k_y$, it is necessary to use transformed dimensions for one set of values in order to develop a net with squares.

Example 9-4

GIVEN The flow net shown for a sheet-pile cutoff wall. Assume the sheet-pile cutoff is impervious (although in practice it seldom is).

REQUIRED Find the effective pressure at point A (2 m below the downstream ground surface) and compute the seepage quantity per meter of wall width per day based on $k = 4 \times 10^{-2}$ cm/s.

SOLUTION

Step 1   Draw the sheet-pile wall-soil system to scale and sketch an acceptable flow net, as shown in Fig. E9-4.

Step 2   Scale the distance from the ground surface to point A, which is the next-to-last equipotential line downstream. The downstream ground line is the last equipotential line.

Step 3   Find the remaining seepage head at point A. Since Δh is defined as constant between any two consecutive equipotential lines and the total head

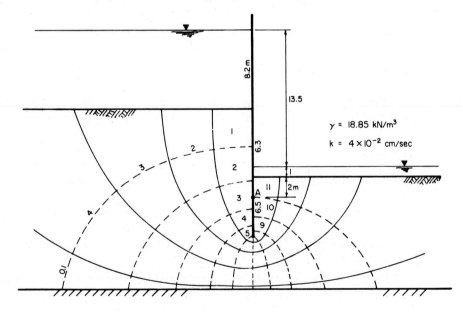

Figure E9-4

loss for 11 drops is 13.5 m, by proportion the remaining seepage head at A is

$$\Delta h = \tfrac{1}{11}(13.5) = 1.23 \text{ m}$$

Step 4 The total equivalent static pressure head at A is

$$h = 1.23 + 1.0 + 2.0 = 4.23 \text{ m}$$

The total vertical pressure due to the weight of material overlying a plane through point A is

$$\sigma_{\text{total}} = 1.0(9.807) + 18.85(2) = 47.51 \text{ kPa}$$

Step 5 Compute the effective pressure as

$$\sigma' = \sigma_{\text{total}} - u$$

$$\sigma' = 47.51 - 4.23(9.807) = 47.51 - 41.48 = 6.03 \text{ kPa}$$

Since $\sigma' > 0$, point A is not "quick," but 6.03 kPa is not large enough to be very confidence-inspiring.

Step 6 Compute the seepage quantity. Use Eq. (9-2) and count $n_f = 4.1$, $n_d = 11$; 1 day $= 86,400$ s, and 1 m $= 100$ cm. Substituting into Eq. (9-2),

$$Q = k(h)\frac{n_f}{n_d} \times \text{width} = (4 \times 10^{-2})(13.5)\frac{4.1}{11}\frac{86,400}{100}$$

$$= 173.9 \text{ m}^3/\text{day/m of wall}$$

Example 9-5

GIVEN The flow net of Ex. 9-4 (Fig. E9-4).

REQUIRED What depth of water outside the sheet-pile wall will cause instability at point A, i.e., a "quick" condition?

SOLUTION

Step 1 Equate the vertical pressures at point A to 0 as follows:

$$\sigma_{\text{total}} \text{ downward} = 47.51 \text{ kPa} \qquad \text{from step 4 of Ex. 9-4}$$

Finding the piezometric head of water at A which gives 47.51 kPa, we have $h' = 47.51/9.807 = 4.84$ m

Since 3.0 m of water is caused by tailwater (1 m) and location of A (2 m), it follows that

$$\Delta h' = 4.84 - 3.0 = 1.84 \text{ m}$$

Step 2 Convert $h = 1.84$ m to the equivalent depth of water. Since n_f is unchanged, we can back compute the h' to produce 1.84 m as

$$h' = 1.84 \tfrac{11}{1} = 20.24 \text{ m}$$

Since there is $6.3 - 1 = 5.3$ m of free excavation depth, the new water depth is

$$20.24 - 5.3 = 14.94 \quad \text{(versus 8.2 m previously)}$$

Example 9-6

GIVEN The flow net of Ex. 9-4 (Fig. E9-4).

REQUIRED What is the effective pressure at A if the water level stabilizes on both sides of the wall at 8.2 m, as shown in Fig. E9-6?

SOLUTION The total pressure on A is

$$2(18.85) + 14.5(9.807) = 179.90 \text{ kPa}$$

$$u = (14.5 + 2)9.807 = 161.81$$

$$\text{Effective pressure } \sigma' = 18.09 \text{ kPa}$$

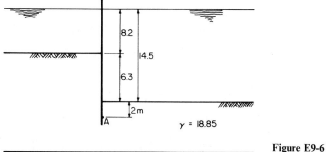

8.2

14.5

6.3

2m

A

$\gamma = 18.85$

Figure E9-6

Example 9-7

GIVEN The problem conditions and the flow net of Ex. 9-4.

REQUIRED What surcharge W, in kilopascals, will provide a safety factor $F = 2.0$ at point A for the unbalanced water head of 13.5 m?

SOLUTION
When $F = 1$, Eq. (8-28) applies, giving $\sum \sigma_v = 0$. To obtain this condition, refer to Fig. E9-7 and obtain the following:

$$W + 2(18.85) + 1.0(9.807) - (2 + 1.0 + 1.23)9.807 = 0$$

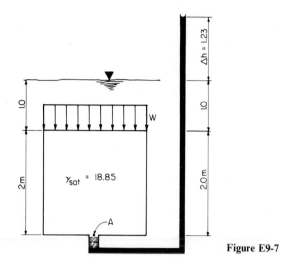

Figure E9-7

Solving, we obtain

$$W = -6.02 \text{ kPa}$$

for $F = 1$. For $F = 2$ we must have

$$\sigma_{\text{total(down)}} = 2\sigma_{\text{up(pore pressure)}}$$

The reader should verify that this is correct, not the alternative, $2\sigma_{\text{total}} = \sigma_{\text{up}}$.

$$\sigma_{\text{total(down)}} = W + 2(18.85) + 1(9.807)$$

$$= W + 47.51$$

$$2\sigma_{\text{up}} = 2(4.23)9.807 = 82.97$$

Equating,

$$W + 47.51 = 82.97$$

$$W = 35.46 \text{ kPa}$$

for $F = 2$.

9-11 FLOW NET CONSTRUCTION AT BOUNDARY OF SOILS OF DIFFERENT PERMEABILITIES

When a flow net is constructed across the boundary of two soils such that $k_2 \neq k_1$, the flow net deflects at the boundary due to the change in permeability as qualitatively shown in Fig. 9-15a. Note in this figure that we will always take k_1 as the larger of the permeabilities and construct the base flow net with squares based on this value. The flow net for k_2 will, therefore, be rectangular. From Fig. 9-15a it is

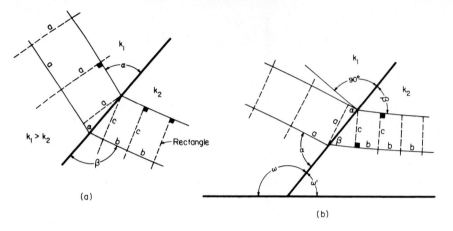

Figure 9-15 Transition of a flow net across the boundary between two soils with different permeabilities as shown.

evident that the flow quantity in any flow path will be the same on either side of the boundary between the two soils; therefore,

$$q = k_1(a) \frac{\Delta h}{a} = k_2(c) \frac{\Delta h}{b} \qquad (a)$$

or

$$\frac{k_1}{k_2} = \frac{c}{b}$$

Also the following ratios are evident:

$$\frac{a}{\sin \alpha} = \frac{c}{\sin \beta} \qquad (b)$$

and

$$\frac{a}{\cos \alpha} = \frac{b}{\cos \beta} \qquad (c)$$

Combining Eqs. (a), (b), and (c), obtain

$$\frac{a}{b} = \frac{\tan \beta}{\tan \alpha} = \frac{k_1}{k_2} \qquad (9\text{-}10)$$

From flow continuity we have

$$v_1 = k_1 i = k_1 \sin (\alpha - \omega') \qquad (d)$$

$$v_2 = k_2 i = k_2 \sin (\beta - \omega') \qquad (e)$$

$$q = a v_1 = c v_2 \qquad (f)$$

and the distance AB is

$$AB = \frac{a}{\sin \alpha} = \frac{c}{\sin \beta} \qquad (g)$$

From Eq. (b) and Eq. (9-10), obtain

$$a = c \, \frac{\sin \alpha}{\sin \beta} \quad \frac{k_1}{k_2} = \frac{\tan \beta}{\tan \alpha}$$

Substituting the appropriate terms into Eq. (f), obtain

$$\frac{\sin \alpha}{\sin \beta} \frac{\tan \beta}{\tan \alpha} \sin (\alpha - \omega') = \sin (\beta - \omega') \qquad (h)$$

Simplifying, we obtain

$$\frac{\cos \alpha}{\cos \beta} = \frac{\sin (\beta - \omega')}{\sin (\alpha - \omega')}$$

The only possible solution of this latter equation is

$$\beta = 270 - \alpha - \omega = 90 + \omega' - \alpha \qquad (9\text{-}11)$$

Solving this equation simultaneously with Eq. (9-10), which contains the coefficients of permeability, with ω given from the problem geometry, gives the flow line intersection values of α and β. This may be done graphically as in Fig. 9-16 or by using a trial procedure as illustrated in Ex. 9-8 following.

Example 9-8

GIVEN The conditions of Fig. 9-15 and the following:

$$\omega = 120° \qquad k_1 = 10 \times 10^{-3} \text{ cm/s} \qquad k_2 = 2.5 \times 10^{-3} \text{ cm/s}$$

REQUIRED What are the angles of incidence α and refraction β for flow paths at the interface of the two-soil system?

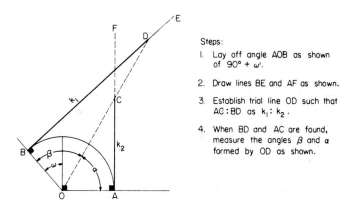

Steps:

1. Lay off angle AOB as shown of 90° + ω'.

2. Draw lines BE and AF as shown.

3. Establish trial line OD such that AC : BD as $k_1 : k_2$.

4. When BD and AC are found, measure the angles β and α formed by OD as shown.

Figure 9-16 Graphical procedure for obtaining α and β of Eq. (9-11).

SOLUTION $\beta = 270° - \alpha - \omega$

$$= 150° - \alpha$$

$$\frac{k_1}{k_2} = \frac{\tan \beta}{\tan \alpha} = \frac{10}{2.5} = 4.0$$

from which

$$\tan (150° - \alpha) = 4 \tan \alpha$$

Note that for $150 - \alpha > 90°$ the tangent function is $(-)$.
It will be convenient to set up a table of values as follows:

α	$\tan (150 - \alpha)$	$4 \tan \alpha$
65	11.43	8.58
70	5.67	10.98 (therefore solution is between 65 and 70)
66.5	8.77	9.19
66.25	9.13	9.09 (close enough)

Final angles: $\alpha = 66.25°$; $\beta = 83.75°$. (The student should verify these values using Fig. 9-16.)

Figure 9-17 gives two cases of flow nets where the soil mass has different coefficients of permeability. Figure 9-17a would be applicable to a rockfill dam with an impervious clay core and a gravel and/or rock shell. The large shell coefficient means that nearly all the head drop is across the core and the flow net can be constructed as shown. Figure 9-17b shows a situation where an earth dam is placed on a foundation with a coefficient of 0.1 of the dam material. Other cases are illustrated in the literature for the interested reader (Sherard et al., 1963; Cedergren, 1948).

9-12 CONTROL OF SEEPAGE

It is readily evident that the seepage quantity depends directly on the coefficient of permeability. Not so evident is the fact that we can increase the length of the flow path to accomplish significant reductions in seepage quantity. The effectiveness of this latter method depends upon the means used to increase the flow path distance L. Methods which have been used include using clay cores, clay blankets, grouting, and cutoff walls. Figure 9-18 illustrates several of these methods.

Where sheet piling is used as a cutoff wall to increase the flow path length, its actual effect may be overestimated if the interlocks are not continuous or if the piling becomes excessively warped so that the wall is pervious. The location of the

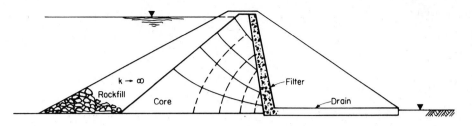

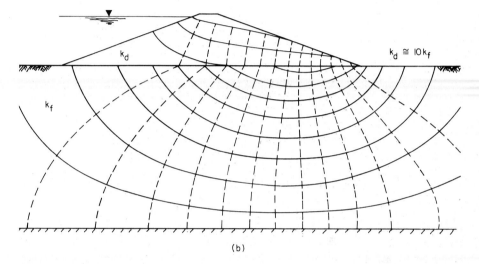

Figure 9-17 Flow nets resulting when adjacent soils have different coefficients of permeability.

cutoff wall should be investigated by placing the wall at several possible locations, drawing the flow net, and computing the seepage quantity. The location that gives the minimum seepage quantity will be the most efficient cutoff wall location.

9-13 SUDDEN DRAWDOWN AND SEEPAGE FORCES

Figure 9-19 illustrates a condition where rapid drawdown occurs behind a dam or a water reservoir. Note that the loss of the stabilizing effect of water on the upstream face may create a temporary " quick " condition. The seepage forces are analyzed for stability as in Sec. 8-8.

   Figure 9-20 illustrates the seepage force concept used to compute the uplift force acting on a concrete dam using pressures obtained from the equipotential lines of a flow net.

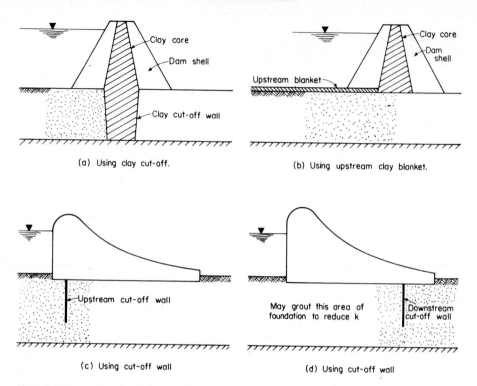

(a) Using clay cut-off.

(b) Using upstream clay blanket.

(c) Using cut-off wall

(d) Using cut-off wall

Figure 9-18 Several methods of controlling seepage. Note that foundation grouting may be used with (*b*) to (*d*). Foundation grouting may be used in lieu of the cutoff walls of (*c*) and (*d*).

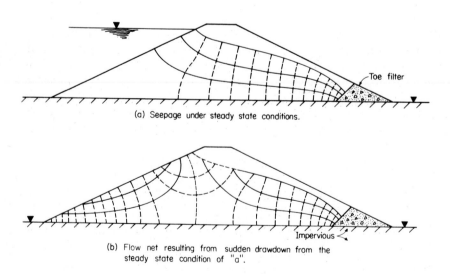

(a) Seepage under steady state conditions.

(b) Flow net resulting from sudden drawdown from the steady state condition of "a".

Figure 9-19 Flow net resulting from a sudden drawdown.

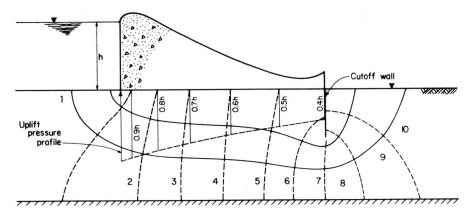

Figure 9-20 Flow net used to obtain seepage uplift forces beneath a concrete dam with a sheet-pile cutoff wall on the downstream end.

9-14 PIPING AND CONTROL OF PIPING

Soil grains will be dislodged and eroded away when the hydraulic gradient at the exit face of a percolating soil mass is large enough to overcome the interparticle bonding forces and/or cementing. This is a progressive phenomenon, with the smaller particles eroding first. As erosion of the smaller particles takes place, the resistance to flow decreases, with a corresponding increase in the hydraulic gradient. With the increase in hydraulic gradient, larger particles can now be eroded, and the process accelerates and tends to the formation of an underground flow channel or "pipe" which moves upstream, as in Fig. 9-21. Flow increases a tremendous amount when the pipe reaches the impounded water, rapidly enlarging the pipe cavity, and can cause a dam failure. The failure usually occurs through collapse of the pipe roof and overtopping in the channel thus formed, which, being of soil, rapidly erodes as a result of the large water velocity. This mode of failure may conceal the fact that piping was the real cause of the disaster.

Piping is a *progressive* failure, and monitoring of a site to observe excessive seepage flow will usually detect it in time for preventive measures. Piping can occur at a considerable time after a water retention structure has been built if an

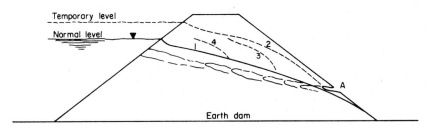

Figure 9-21 Formation of a "piping" condition.

event occurs which increases the exit gradient sufficiently to cause soil erosion at a localized point. This may happen, for example, from

1. Cavities toward upstream formed when roots decay
2. Animal burrows (those of muskrats, gophers, etc., including snake burrows)
3. Excavations downstream, either pits or in some cases simply skinning the top surface to some critical depth

Piping often occurs during floods on the land side of protective levees. Here the direction of flow is initially vertical, as the increased flood level raises the hydraulic grade line and water in the underlying pervious soil attempts to seek piezometric level. Root holes, animal burrows, etc., can provide a less resistant localized flow path, and water seeps out—slowly and carrying fine sand and silt and clay particles at first, then more rapidly until it "boils." These local pipes can be halted at relatively early stages by placing a sandbag ring around the "sand boil" to develop sufficient water head that the back pressure equalizes the hydraulic gradient. Untreated "sand boils" can result in the levee collapsing into the underground pipe cavity and being overtopped with flood water. Figure 1-2b illustrates a case where sheet piling was used in a futile attempt to halt the piping. The Teton dam (Fig. 1-2d) failure was probably caused by piping in the discontinuities between the dam and the rock abutment, which were apparently improperly grouted.

Piping can be controlled and/or eliminated in earth dams by using a *filter* or *graded filter* on the exit face. The filter should be graded to ensure that the protected soil cannot wash through the filter material (see Fig. 9-22).

Experience, supported by tests made by Bertram (1940), indicates that if the following filter criterion is met, piping will be adequately controlled:

$$\frac{D_{15(\text{filter})}}{D_{85(\text{protected soil})}} < 4 \text{ to } 5$$

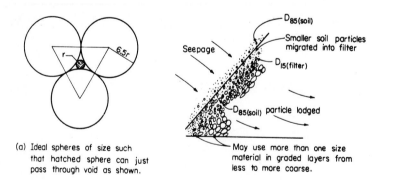

(a) Ideal spheres of size such that hatched sphere can just pass through void as shown.

(b) May use more than one size material in graded layers from less to more coarse.

Figure 9-22 Soil-filter concept. (*a*) Relative diameters of ideal spheres so that a particle can just pass through filter spheres. (*b*) Interface of exit face and filter.

This criterion states that the *piping ratio* of the D_{15} size of the filter soil is not more than four or five times the D_{85} size of the protected soil. A further criterion is:

$$\frac{D_{15(\text{filter})}}{D_{15(\text{protected soil})}} > 4 \text{ to } 5$$

This criteria states that the D_{15} size of the filter soil should be more than four or five times the D_{15} size of the protected soil. The U.S. Corps of Engineers also recommends that the ratio of the D_{50} filter and protected soil sizes should be as follows:

$$\frac{D_{50(\text{filter})}}{D_{50(\text{protected soil})}} \leq 25$$

It follows, also, from an earlier statement concerning interparticle forces and cementing that the protected soil should be cohesive.

9-15 PLAN OR RADIAL FLOW NETS

Plan flow nets may be used to estimate seepage quantities such as those into excavations or caissons. Figure 9-23 illustrates a plan flow net. Note that one needs both the plan and at least part of the profile to interpret the geometry and compute the seepage quantity. The plan flow net differs from the previously considered profile flow net in that:

1. The segments are not squares but tend to maintain a constant ratio of $rb/l =$ constant (refer to Fig. 9-23, plan and profile).
2. The segments converge both in plan and in profile at the exit perimeter.
3. If the exit sink does not fully penetrate the aquifer, the exit perimeter equipotential line does not correctly display seepage conditions; there is vertical seepage into the sink area also, which must somehow be taken into account.

Assuming that the conditions are satisfied approximately (say, 80 to 100 percent penetration of the aquifer), then:
For *artesian* flow,

$$Q = k(H - h_e)D\frac{n_f}{n_d} \tag{9-12}$$

For *gravity* flow,

$$Q = \frac{k}{2}(H^2 - h_e^2)\frac{n_f}{n_d} \tag{9-13}$$

where terms are as identified on Fig. 9-23 except for n_f and n_d, which are the numbers of flow paths and equipotential drops, respectively, as used previously.

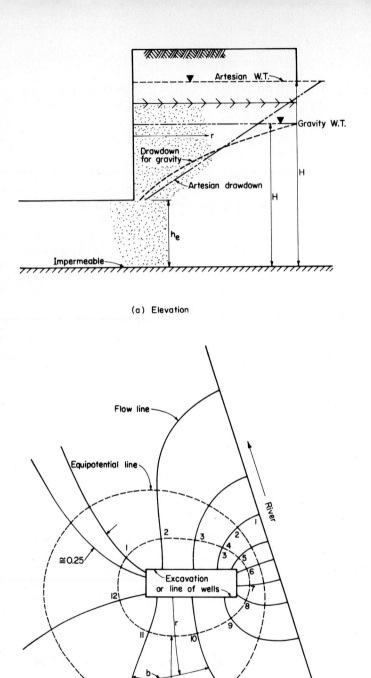

(a) Elevation

(b) Plan

Figure 9-23 Radial or plan flow net geometry.

Example 9-9 Given are the conditions shown in Fig. E9-9 and

$$H = 25.9 \text{ m} \qquad n_f = 8$$

$$h_e = 13.7 \text{ m}$$

$$k = 0.1 \text{ cm/s} \qquad n_d = 2$$

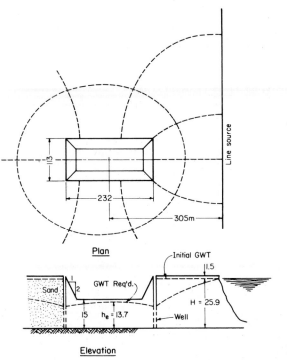

Plan

Elevation

Figure E9-9

REQUIRED Compute the flow quantity per meter of excavation perimeter.

SOLUTION Assume from the elevation view that gravity flow is developed. Substituting into Eq. (9-13), we have

$$Q = \frac{0.001}{2} (25.9^2 - 13.7^2) \frac{8}{2}$$

$$= 0.0005(483.12)(4)$$

$$= 0.97 \text{ m}^3/(\text{s} \cdot \text{m})$$

9-16 OTHER METHODS OF OBTAINING FLOW NETS

Several methods of obtaining flow nets are available, including the building of sand models with dye injected at select points to trace the flow paths. The most popular and rapid method is to use an electrical analogy. This method requires the use of a controlled voltage electrical supply, silver paint or other highly conducting material for electrodes, and a probe. The shape of the percolating soil mass is cut out of some electrically conducting material, and electrodes are placed to appropriately simulate the differential of water head. The probe is used to find a locus of points corresponding to a constant voltage loss (value), which is by definition an equipotential line. The exact details and equipment needed are outlined in Bowles (1978) for the interested reader.

9-17 ELEMENTS OF WELL HYDRAULICS

Well flow can be approximated by flow nets; however, it may be more convenient to directly compute the quantity. Several cases exist, of which those shown in Fig. 9-24 will be analyzed in additional detail to obtain flow equations.

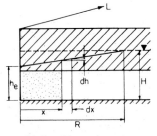

(a) Slot with artesian water

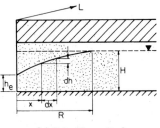

(b) Slot with gravity flow

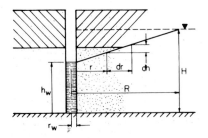

(c) Artesian well fully penetrating aquifer

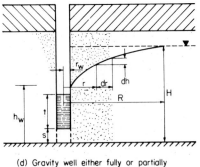

(d) Gravity well either fully or partially penetrating aquifer

Figure 9-24 Several cases of well flow.

For the slot well (Fig. 9-24a) under *artesian* conditions with a single line source (approximately doubled for source on both sides),

$$Q = kDL \frac{dh}{dx}$$

Separating variables and integrating, we obtain

$$h = \frac{Qx}{kDL} + C$$

The limits are $h = H$ at $x = R$; $h = h_e$ at $x = 0$ (at slot face). This gives

$$Q = \frac{kDL}{R}(H - h_e) \tag{9-14}$$

For the slot under *gravity* flow and a single line source (see Fig. 9-24b),

$$Q = khL \frac{dh}{dx}$$

Integrating and inserting limits of $h = H$ at $x = R$ and $h = h_e$ at the slot, we obtain

$$Q = \frac{khL}{2R}(H^2 - h_e^2) \tag{9-15}$$

For an *artesian well* fully penetrating the aquifer (see Fig. 9-24c for identification of terms), we have

$$Q = kiA \quad \text{and} \quad i = \frac{dh}{dr} \quad A = 2\pi r D$$

Substituting,

$$Q = k(2\pi r D) \frac{dh}{dr}$$

and integrating and inserting limits of $h = H$ at $r = R$ and $h = h_w$ at $r = r_w$, we obtain

$$Q = \frac{2\pi k D(H - h_w)}{\ln (R/r_w)} \tag{9-16}$$

For the *fully penetrating gravity well*,

$$i = \frac{dh}{dr} \quad \text{and} \quad A = 2\pi r h$$

and

$$Q = k(2\pi r h) \frac{dh}{dr}$$

Integrating and inserting limits, we obtain

$$Q = \frac{\pi k(H^2 - h_w^2)}{\ln (R/r_w)}$$ (9-17)

Research and observations show that the flow from partially penetrating gravity wells (the most common case encountered) is better described (Mansur and Kaufman, 1962) as

$$Q = \pi k \frac{[(H - s)^2 - t^2]}{\ln (R/r_w)} \left[1 + \left(0.3 + \frac{10r_w}{H}\right) \sin \frac{1.8s}{H}\right]$$ (9-18)

where terms are as identified in Fig. 9-24d. Note that Eq. (9-18) becomes Eq. (9-17) when the well fully penetrates the aquifer, since $s = 0$ and $t = h_w$.

Example 9-10 Given are the gravity well conditions shown in Fig. E9-10.

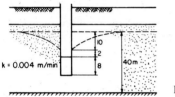

Figure E9-10

REQUIRED Estimate the flow of the well when the drawdown is 10 m as shown.

SOLUTION
With 10 m drawdown,

$$h_w = 30 \text{ m} \qquad s = 20 \text{ m}$$

$$t = 20 - 10 - 2 = 8 \text{ m}$$

$$r_w = 0.3 \text{ m} \qquad H = 40 \text{ m}$$

Using Eq. (9-18),

$$Q = 3.14(0.004)\frac{(40 - 20)^2 - 8^2}{\ln (R/0.3)} \left[1 + \left(0.3 + \frac{10(0.3)}{40}\right) \sin \frac{1.8(20)}{40}\right]$$

$$Q = 3.14(0.004)\frac{400 - 64}{\ln (R/0.3)} (1.29) = \frac{5.45}{\ln (R/0.3)}$$

Since R is unknown, we will assume several values of R and compute the corresponding flow quantity Q.

R, m	ln $R/0.3$	Q, m³/min
20	4.20	1.30
40	4.89	1.11
50	5.12	1.07
100	5.81	0.94

from which it appears that $Q \cong 1$ m³/min, and it is not necessary to have an exact value of the radius of drawdown influence R.

9-18 SUMMARY

This chapter has presented a means of obtaining seepage quantities and seepage forces for various geometrical configurations for which closed form solutions are not readily obtainable. The general procedure is to construct a flow net. The flow net consists of families of curves intersecting at 90° and forming squares, except at singular points, where judgment must be exercised. We note that if $k_x \neq k_y$, the intersection angle is still 90° but the quadrilateral is not square.

We may transform either the x or y dimension using a transformation variable

$$u = \sqrt{\frac{k_x}{k_y}} \quad \text{or} \quad v = \sqrt{\frac{k_y}{k_x}}$$

to obtain either $x' = x/u$ or $y' = y/v$, but not both simultaneously. With transformed dimensions we can use squares. The seepage quantity is computed as

$$q = kh\frac{n_f}{n_d}$$

with k = coefficient of permeability or a transformed value of $k = \sqrt{k_x k_y}$.

It was shown that uplift pressures can be obtained from a flow net. The flow net can be used in the control of piping to obtain the hydraulic gradient. The flow net is also useful in determining the optimum location of cutoff walls.

Select elements of well hydraulics have been introduced. We also note that excavation dewatering is similar to well hydraulics and that a plan flow net can be used to determine flow quantities.

HOMEWORK PROBLEMS

For Probs. 9-1 through 9-4, refer to Fig. P9-1.

9-1 Find the seepage in cubic meters per day per meter of dam width. Use a scale of 1 cm = 5 m.
  Approx. ans.: 0.71 m³/day/m.

9-2 Find the uplift force per meter of width and plot the pressure profile along the base of the dam. Use a scale of 1 cm = 5 m for dimensions and any convenient pressure scale.
  Approx. ans.: 1300 kN/m.

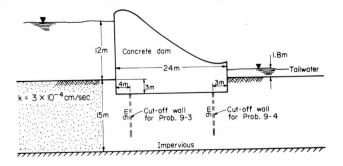

Figure P9-1

9-3 Install a cutoff wall 3.6 m downstream from the front of the dam as shown by the dashed lines in Fig. P9-1. The wall will be 9 m in length (the tip will be 10 m below the ground surface). Compute the seepage as in Prob. 9-1.

9-4 Install a cutoff wall 3 m from the downstream end of the dam as shown by the dashed lines and remove the cutoff wall from the upper end. The cutoff wall is 9 m in length. Compute the seepage as in Prob. 9-1.

   Ans.: 0.5 m³/day·m.

For Probs. 9-5 through 9-8, refer to Fig. P9-2.

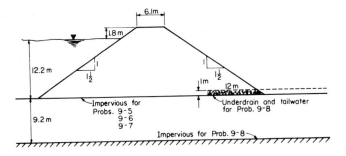

Figure P9-2

9-5 Find the seepage if the dam rests on an impervious stratum and there is no underdrain. Take $k = 3.5 \times 10^{-3}$ cm/s and any convenient scale.

   Approx. ans.: 6.5 m³/day·m.

9-6 Redo Prob. 9-5 if there is an underdrain with the dimensions shown.

9-7 Redo Prob. 9-6 if $k_x = 4 \times 10^{-4}$ and $k_y = 1 \times 10^{-4}$ cm/s.

9-8 Redo Prob. 9-5 if there is an underdrain and the soil on which the dam rests is also pervious.

   Approx. ans.: 16.9 m³/day·m.

9-9 Construct a plan flow net for the conditions shown in Fig. P9-9 and estimate the seepage quantity in cubic meters per minute.

9-10 What would be the expected well flow of a well at point A of Fig. P9-9 (center of excavation) if r_w will compute the area of the excavation and assuming a fully penetrating well with water elevation in the well the same as it would be for the excavation (i.e., 0.5 m below the bottom of the excavation)?

9-11 Same as Prob. 9-10 except $r_w = 0.5$ m and fully penetrating.

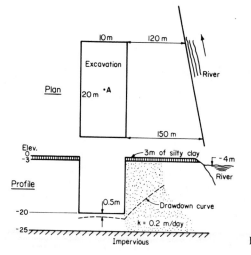

Figure P9-9

Chapter 10

Stresses, Strains, and Rheological Concepts

10-1 GENERAL CONSIDERATIONS

One of the most important functions of the study of soil mechanics is the prediction of the magnitude of the stresses under some loading which will produce excessive deformations, termed *failure stresses*. Any load will produce stresses and resulting strains which can be integrated over the stress zone of interest to obtain deformation. The deformation is usually termed *settlement*, and considerable effort is often expended to obtain a settlement prediction. This chapter and the several chapters following will be concerned in particular with stress and deformation "predictions" and some of the reasons for terming this activity a prediction.

Resisting stresses are developed when any material is subjected to a load. The study of strength of materials is devoted to prediction of:

1. Magnitude of the resulting stresses σ
2. Whether the stresses result in a material failure
3. Magnitude of strains ε

The problem is considerably less difficult when dealing with an isotropic, homogeneous, linearly elastic (obeying Hooke's law), isolated body with clearly

272

defined boundaries. We are concerned here with the stresses and strains developed in soil masses due to applied loads such as fills, building foundations, or negative loads such as from excavations. In soil masses the materials are generally:

Not	But may
Isotropic (elastic properties the same in all directions)	Be anisotropic (elastic properties not the same in different directions)
Homogeneous (constant material properties of e, w, γ, structure, etc., throughout)	Have lenses of different materials, be stratified, increase in density with depth, and contain varying amounts of moisture
Linearly elastic	Be nonlinear and be elastic only over a very limited stress range
An isolated body	Be a stressed zone in a semi-infinite medium

Since soil is a particulate material, failure is primarily by a rolling and slipping of grains and not via simple tension or compression. Because of this failure mode, the stresses of interest are *shear stresses*. The soil resistance or soil strength of interest is the *shear strength*. Conceptually, soil strength is some different from the ultimate strength of materials such as steel or concrete. The soil locally subjected to a load (or stressed) is always surrounded by the remainder of the semi-infinite medium. Exceptions occur, of course, such as adjacent to excavations or over old mine tunnels or other subsurface cavities. With the local stressing, we have a situation analogous to a ship floating in the ocean. The stressed zone will "fail" if the stresses are too large. Failure is defined as a considerable alteration or state change in soil structure (or remolding), accompanied by substantial deformation and enlargement of the stressed zone until deformation eventually halts. The resulting total deformation is the deformation under stresses up to failure plus the larger deformation occurring after failure. The soil strength after "failure" is termed the *residual strength*.

When any viscous fluid, usually water, is present in the soil void spaces, the particle rolls and slips will be resisted by the pore fluid. The amount of resistance will be proportional to the quantity of pore fluid present in a range of saturation from 0 to 100 percent. The duration of pore fluid resistance will depend on the effective coefficient of permeability k.

Rheology is the study of materials in a fluid state as a rate process. Soil deformation which depends on the coefficient of permeability becomes a rate process, and some effort has been directed toward using rheological methods for analysis. Several rheological models will be presented in Sec. 10-9.

Most efforts to predict soil response to applied loads have used theory of elasticity methods. A few researchers have used also plasticity theory. The major problem has been that both elasticity and plasticity theories are for an elastic continuum, whereas soil is an aggregation of particulate material. An additional major problem with soil is that it is state dependent, i.e., it

1. Shrinks and swells with changes in water content, with resulting differences in volume, void ratio, and unit weight (density)
2. Changes in volume during stress applications

Any of these state changes produces a different material from what was started with.

Soil has three additional significant factors to consider besides the change of state consideration, namely,

1. Any new stress condition starts from an initial *nonzero* stress condition. Soil is always subjected to in situ stresses due to overburden, deposition, and water change effects.
2. Initial in situ conditions depend on time-dependent factors (previous stress history). For particulate, or nonlinear, materials, superposition of stress effects as commonly done in structural analysis is not generally valid.
3. Stress-strain conditions in a soil mass are three-dimensional problems. Many structural mechanics problems are, or can be, treated with two-dimensional concepts. Much early, and even present, soil mechanics work treated it as a two-dimensional problem to simplify the computations.

Considering these factors, it should be readily apparent that a high degree of success in deformation predictions is not possible. It may be necessary in the future to discard the current methods and develop some new theory for a particulate medium.

10-2 GENERAL STRESSES AND STRAINS AT A POINT

Figure 10-1 illustrates the stresses on a six-sided differential element in a soil mass. Not shown on this element are the stresses on the three planes away from the reader, for additional clarity and because they would simply be equal and opposite in direction. The element strains can be shown in a similar manner, as also shown in Fig. 10-1.

Figure 10-2 illustrates an orientation of the element of Fig. 10-1 such that no shear stresses exist on the sides of the element. This orientation produces principal axes, and the normal stresses on the element faces are *principal stresses*; by analogy there exists a set of principal axes for *principal strains*. The principal axes for principal stresses may not be coincident with the axes for principal strains. In either case, however, there is always an axis orientation which produces principal stresses or principal strains.

Shown in Fig. 10-2a is the general compressive stress state existing in the soil prior to application of some soil load resulting in an incremental increase in stress of $\Delta\sigma_i$. Figure 10-2b is the stress state after the application of the stress increment

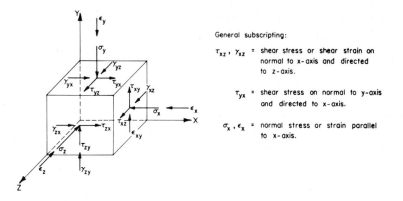

General subscripting:

τ_{xz}, γ_{xz} = shear stress or shear strain on normal to x-axis and directed to z-axis.

τ_{yx} = shear stress on normal to y-axis and directed to x-axis.

σ_x, ϵ_x = normal stress or strain parallel to x-axis.

Figure 10-1 Stresses and strains on a soil element with coordinate axes arbitrarily shown.

$\Delta\sigma_i$. We are generally interested in the deformations resulting from application of these stress increments. Deformation is related to strain by

$$\text{Deformation } \delta = \int_0^M \epsilon \, dM$$

where M may be either a length or a volume with ϵ the corresponding linear or volumetric strain.

10-3 THEORY OF ELASTICITY CONCEPTS USED IN SOIL MECHANICS PROBLEMS

A *tensor* is a vector quantity describing a physical state (such as a stress) or a physical phenomenon (such as strain), and requires three or more components for a complete description. In general, any stress or strain requires three direction

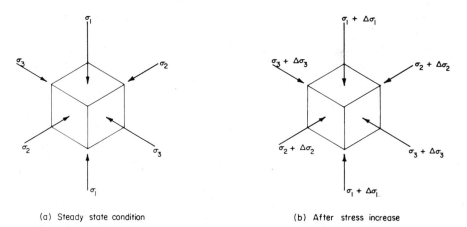

(a) Steady state condition

(b) After stress increase

Figure 10-2 A soil element subjected to incremental principal stress increases.

cosines for a complete description of magnitude and direction. In plane stress (or strain), one of the direction cosines is 0. Referring to Fig. 10-1, the stress and strain tensors (or matrices) can be developed by interchanging τ_{ij} with σ_{ij} (i.e., $\tau_{xy} = \sigma_{xy}$; $\tau_{zy} = \sigma_{zy}$, etc.) for stresses and similarly interchanging γ_{ij} with ε_{ij} for strains ($\gamma_{xy} = \varepsilon_{xy}$; $\gamma_{zx} = \varepsilon_{zx}$, etc.). The resulting stress and strain tensors are:

Stress			Strain		
σ_{xx}	σ_{xy}	σ_{xz}	ε_{xx}	ε_{xy}	ε_{xz}
σ_{yx}	σ_{yy}	σ_{yz}	ε_{yx}	ε_{yy}	ε_{yz}
σ_{zx}	σ_{zy}	σ_{zz}	ε_{zx}	ε_{zy}	ε_{zz}

The reader can readily observe that a tensor is shorthand notation for summing the stress or strain vectors along the three cartesian axes orienting the element of interest. The subscript convention is standard (refer again to Fig. 10-1) in that

1. The first subscript is the axis normal to the plane of interest.
2. The second subscript is an orthogonal axis parallel to the vector.

This subscripting produces *ii* subscripts for normal stresses or strains; *ij* subscripts are parallel to the plane and are either shear stresses or shear strains.

From moment equilibrium, and neglecting second- and higher-order differentials, we obtain three of the shear and strain quantities as complementary values:

$$\sigma_{xy} = \sigma_{yx} \qquad \sigma_{xz} = \sigma_{zx} \qquad \sigma_{yz} = \sigma_{zy}$$

$$\varepsilon_{xy} = \varepsilon_{yx} \qquad \varepsilon_{xz} = \varepsilon_{zx} \qquad \varepsilon_{yz} = \varepsilon_{zy}$$

which gives a symmetrical tensor, and only six quantities are needed for a complete description. For most soil mechanics work we are concerned with initial conditions of principal stresses. Later loading generally produces shearing stresses which must be evaluated. Figure 10-3 illustrates the boundary areas of an elementary equilateral tetrahedron (one-eighth of an octahedron) using an orthogonal coordinate system with the principal axes 1, 2, and 3 as shown. Any kind of tetrahedron can be used, but in this text only an equilateral element will be considered. The direction of the normal to the *octahedral plane* (plane *ABC* of Fig. 10-3) can be obtained by computation or by graphical means. Since we are using a differential volume, the body forces can be neglected and only surface forces (the normal stresses as shown, since these are principal planes) need be considered.

The resultant stress vector σ_r on the octahedral plane can be obtained for the general axis orientation by summing forces in the three orthogonal directions and using the appropriate direction cosines. Since force equilibrium is necessary, one cannot obtain the octahedral stresses as a vector summation of the normal and shear stress components; instead both the stresses and the stressed areas must be

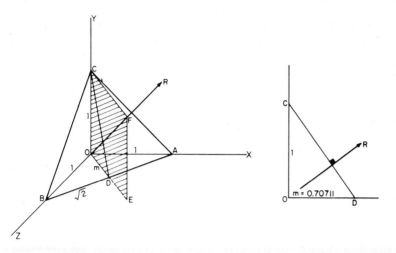

Figure 10-3 Equilateral tetrahedron oriented with respect to the principal axes shown and subjected to three principal stresses.

considered. For example (refer to Fig. 10-3), for an equilateral tetrahedron we have $OA = OB = OC = 1$; thus areas $AOC = BOC = AOB = (1 \times 1)/2 = 0.5$ unit. The area ABC is computed from the relationships $m = 0.707\,11$, $AB = \sqrt{2}$, and $OC = 1$ to obtain $ABC = 0.866\,02$ units.

The direction cosines are established as the ratio of areas as follows:

$$m_1 = \frac{AOB}{ABC} = \frac{0.5}{0.866} = \frac{1}{\sqrt{3}} \qquad m_2 = \frac{BOC}{ABC} = \frac{1}{\sqrt{3}} \qquad m_3 = \frac{AOC}{ABC} = \frac{1}{\sqrt{3}}$$

These will always be the direction cosines for an equilateral tetrahedron.

From the relationship $\sum F = 0$ along any axis, and defining the *normal stress* on plane ABC as $R = \sigma_{\text{oct}}$, we have, in general,

$$R(A_{ABC})m_i - \sigma_i A_i = 0$$

For the actual area values of this example, this produces

$$R(0.866)\left(\frac{1}{\sqrt{3}}\right) - \sigma_1(0.5) = 0$$

$$R(0.866)\left(\frac{1}{\sqrt{3}}\right) - \sigma_2(0.5) = 0$$

$$R(0.866)\left(\frac{1}{\sqrt{3}}\right) - \sigma_3(0.5) = 0$$

and adding, $\quad 3R(0.866)\left(\frac{1}{\sqrt{3}}\right) - 0.5(\sigma_1 + \sigma_2 + \sigma_3) = 0$

The octahedral normal stress is then obtained as

$$\sigma_{oct} = R = \frac{\sigma_1 + \sigma_2 + \sigma_3}{3} \tag{10-1}$$

This equation for the octahedral normal stress does not depend on equal values of principal stresses (σ_1 does not have to equal σ_2; σ_3 does not have to equal σ_2, etc.), but we are concerned here with an equilateral tetrahedron. The octahedral shear stress can be obtained (see any textbook on theory of elasticity) as

$$\tau_{oct} = \tfrac{1}{3}\sqrt{(\sigma_1 - \sigma_2)^2 + (\sigma_2 - \sigma_3)^2 + (\sigma_3 - \sigma_1)^2} \tag{10-2}$$

Equation (10-2) shows that:

1. No shear stresses are produced with $\sigma_1 = \sigma_2 = \sigma_3$. This is a hydrostatic stress condition. This condition can be produced in a soil test as an *isotropic* (equal all-around stresses) *consolidation*, considered in more detail in Chap. 13.
2. Shear stresses are produced when $\sigma_1 > \sigma_2$ or $\sigma_1 > \sigma_3$, and in a triaxial test when $\sigma_2 = \sigma_3$ and $\sigma_1 > \sigma_3$. The stress difference $\sigma_1 - \sigma_3$ is the *deviator* stress.

Two additional observations readily made at this point are:

1. Considerable simplification is obtained when two-dimensional stresses are used ($\sigma_2 = 0$).
2. In situ soil has initial shear and normal stresses on the octahedral plane, since the vertical (σ_1) and lateral (σ_2 and/or σ_3) stresses are different. Note that in steady state conditions the vertical and lateral stresses are principal stresses.

If we use matrix notation for the normal stress,

$$|\sigma_i|\{m_i\} = \{\sigma\} \tag{10-3}$$

Rearranging, dividing through by the direction cosine matrix $\{m_i\}$, and expanding, we have

$$\begin{bmatrix} \sigma_1 - \sigma & 0 & 0 \\ 0 & \sigma_2 - \sigma & 0 \\ 0 & 0 & \sigma_3 - \sigma \end{bmatrix} = 0$$

Expanding the determinant (the matrix) to 0, we obtain

$$\sigma^3 - J_1\sigma^2 + J_2\sigma - J_3 = 0$$

where $J_1 = \sigma_1 + \sigma_2 + \sigma_3$

$$J_2 = \sigma_1\sigma_2 + \sigma_2\sigma_3 + \sigma_3\sigma_1 \tag{10-4}$$

$$J_3 = \sigma_1\sigma_2\sigma_3$$

Factor $J_1 = 3\sigma_{oct}$. The J factors are termed *invariants* since the principal stresses are independent (nonvarying) of arbitrary axis orientation.

When pore pressures exist, Eq. (10-3) is modified using Eq. (2-21) to obtain

$$|\sigma_i - u|\{m_i\} = \sigma \tag{10-5}$$

Example 10-1 What are the octahedral normal and shear stresses when $\sigma_1 = 50$ kPa, $\sigma_2 = \sigma_3 = 25$ kPa, and the pore pressure $u = 10$ kPa?

SOLUTION Using Eq. (10-1), obtain

$$\sigma_{oct} = \frac{50 + 25 + 25}{3} = 33.3 \text{ kPa}$$

Using Eq. (10-2), obtain

$$\tau_{oct} = \tfrac{1}{3}\sqrt{(50 - 25)^2 + (25 - 25)^2 + (25 - 50)^2} = 11.8 \text{ kPa}$$

These values as well as several intermediate values of σ_1 have been plotted on Fig. 10-4.

σ_1	$\Delta\sigma_1$	σ_{oct}	τ_{oct} (Total stresses)
25	0	25.0	0.00
30	5	26.7	2.40
35	10	28.3	4.70
40	15	30.0	7.07
45	20	31.7	9.42
50	25	33.3	11.8 (Failure)

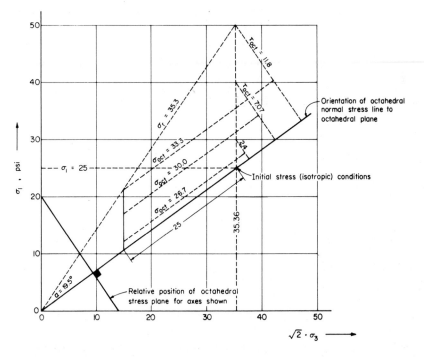

Figure 10-4 Stress path plot of octahedral normal and shear stresses. Part of the data shown are computed in Ex. 10-1.

The *effective octahedral stresses* are computed as follows:

$$\sigma'_{oct} = \frac{(50 - 10) + (25 - 10) + (25 - 10)}{3}$$

$$= 23.3 \text{ kPa} \qquad (\text{same as } 33.3 - 10)$$

$$\tau_{oct} = \tfrac{1}{3}\sqrt{(40 - 15)^2 + 0^2 + (15 - 40)^2}$$

$$= 11.8 \text{ kPa} \qquad (\text{same as before})$$

Note that shear stresses are not affected by pore pressure. We can compute the effective stress resultant σ'_r on the octahedral plane as

$$\sigma'_r = \sqrt{\sigma'^2_{oct} + \tau^2_{oct}} = 26.11 \text{ kPa}$$

which acts at an angle above the normal vector on the octahedral plane of

$$\alpha = \tan^{-1} \frac{11.8}{23.3} = 26.85°$$

10-4 STRESS-STRAIN MODULUS AND HOOKE'S LAW

The stress-strain modulus (commonly termed modulus of elasticity) E_s is defined in any mechanics of materials textbook as

$$E_s = \frac{\text{stress}}{\text{strain}} = \frac{\Delta\sigma_y}{\Delta\varepsilon_y} \tag{10-6}$$

which is simply the slope of a stress-strain curve. When an element of any material is compressed by a stress σ_y, lateral strains are introduced of

$$\varepsilon_x = \mu\sigma_y \qquad \varepsilon_z = \mu\sigma_y \tag{a}$$

where μ = Poisson's ratio. Poisson's ratio will be considered in more detail in Chap. 14, as will Eq. (10-6).

We will use $\Delta\sigma_i$ to identify the increase in stresses in a soil mass due to any loading condition (e.g., building loads; near surface loads such as footings, basement slabs, etc.; and negative loads such as excavations). The soil elements interior to the mass are always subjected to some initial vertical stresses σ_y and to lateral or radial stresses, of which σ_x and σ_z are particular coordinate values, due to overburden effects, as shown in Fig. 10-5. If these stresses have been applied for a geological time period, deformation equilibrium has been reached ($\varepsilon_i = 0$) and we have a steady state condition. The newly (recently) applied loads result in both a stress increase of $\Delta\sigma_i$ and un-steady state strain conditions. Chapters 11 and 12 consider the un-steady state strain condition when $d\varepsilon/dt$ is very small. Here and in Chap. 13 we will be concerned with $d\varepsilon/dt$ being large (time intervals on the order of hours to not more than several days).

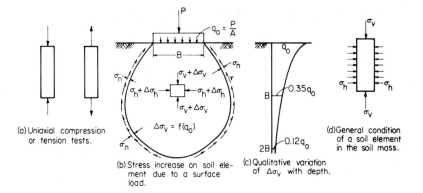

(a) Uniaxial compression or tension tests.

(b) Stress increase on soil element due to a surface load.

(c) Qualitative variation of $\Delta\sigma_v$ with depth.

(d) General condition of a soil element in the soil mass.

Figure 10-5 Soil testing and the general condition of a soil element beneath a loaded area.

When a soil element is compressed under a stress increment of $\Delta\sigma_y$, it will tend to expand laterally, producing strains of ε_x and $\varepsilon_z > 0$. This expansion is resisted by the soil surrounding the element, and stresses of $\mu\,\Delta\sigma_x$ and $\mu\,\Delta\sigma_z$ are developed, tending to restore the shape of the soil element. The combination of $\Delta\sigma_y$ and the restoring stresses $\mu\,\Delta\sigma_x$ and $\mu\,\Delta\sigma_z$ results, by combining Eq. (10-6) and Eqs. (a), in the following expression for strain:

$$\varepsilon_y = \frac{\Delta\sigma_y}{E_s} - \frac{\mu\,\Delta\sigma_z}{E_s} - \frac{\mu\,\Delta\sigma_x}{E_s} \qquad (b)$$

Simplifying Eq. (b) and extrapolating, one can obtain for the three coordinate strains

$$\varepsilon_y = \frac{1}{E_s}[\Delta\sigma_y - \mu(\Delta\sigma_z + \Delta\sigma_x)]$$

$$\varepsilon_x = \frac{1}{E_s}[\Delta\sigma_x - \mu(\Delta\sigma_z + \Delta\sigma_y)] \qquad (10\text{-}7)$$

$$\varepsilon_z = \frac{1}{E_s}[\Delta\sigma_z - \mu(\Delta\sigma_x + \Delta\sigma_y)]$$

These three strain equations are called Hooke's generalized stress-strain law.[†] In mechanics of materials courses, the Poisson's ratio effects are commonly neglected, but in soil mechanics work the Poisson's ratio effects are generally too large to neglect (although it has been commonly done).

Uniaxial compression and tension tests are widely used, as in Fig. 10-5, for steel, concrete, wood, and other common engineering materials. Since these materials form strucutural members which are isolated in space or surrounded by atmo-

---

[†] When ε_y, ε_x, and ε_z are not principal strains, there are also three shearing-strain equations as part of this stress-strain law.

spheric (0 gage) pressure, the tests produce satisfactory design data without inclusion of Poisson's ratio effects. On the other hand, isolated columns of soil are never used to carry compressive loads; the loaded soil area is always surrounded, or confined, by other soil (Fig. 10-5b) of the mass or by a wall or other structural element, as adjacent to an excavation. The single exception is the rock or soil pillars (columns) supporting the roofs in mining operations, but here the pillar is large and only the interior portion carries the roof load through a combination of "confining" effect and with the compressive forces from the mine roof.

From Eq. (10-7) it can be shown that if a uniaxial compression test is performed on a soil where $\Delta\sigma_z = \Delta\sigma_x = 0$, the strain will be larger than in situ, and since no Poisson's ratio effect is produced, the stress-strain modulus using Eq. (10-6) will be too small. Likewise, if a soil test is performed on a fully saturated sample and a vertical stress increment of $\Delta\sigma_y$ is applied, we will have a condition of $\Delta\sigma_y = \Delta\sigma_x = \Delta\sigma_z$ until pore water can drain from the soil voids. Since the strains are all zero for some time interval, this requires that Poisson's ratio be 0.5 [from Eqs. (10-7)]. Note, however, that Poisson's ratio does not remain 0.5 during this entire stress history but only for the time duration until some drainage can take place. The condition of no deformation taking place until drainage occurs is consolidation-type deformation, considered in Chaps. 11 and 12.

Since tension stresses are not possible in cohesionless soils, and since cohesive soils can carry only very limited amounts of tension stresses, the only soil stresses of practical engineering significance are compressive stresses and the shear stresses developed as the particles roll about as a result of the compressive stresses.

Example 10-2

GIVEN It is known that a soil has a Poisson's ratio of $\mu = 0.30$. An unconfined compression test is performed with lateral stresses $\sigma_x = \sigma_z = 0$. A confined compression (triaxial) test is also performed on the same soil, at the same initial water content and density, and with lateral stresses of $\sigma_x = \sigma_z = \sigma_r = 30$ kPa.

REQUIRED Find the stress-strain modulus E_s for both tests for comparison and for the conditions obtained in the unconfined compression test of $\sigma_y = \sigma_f = 70$ kPa (approximately 10 psi) and the corresponding strain $\varepsilon_y = 0.020$ cm/cm.

SOLUTION The stress-strain modulus for the unconfined compression test is simply

$$E_s = \frac{70.0}{0.020} = 3500 \text{ kPa} \qquad \text{using Eq. (10-6)}$$

For the triaxial test, we have from Eqs. (10-7)

$$E_s \varepsilon_y = \sigma_y - \mu(\sigma_x + \sigma_z)$$

from which

$$E_s \varepsilon_y = 70.0 - 0.30(30.0 + 30.0) = 52.0 \text{ kPa}$$

For the unconfined compression test, $E_s \varepsilon_y = 70.0$.

It will be necessary to assume a linear variation for the strain ε, which could produce a serious error but may not in this case, as the extrapolation is only from 52 to 70 kPa. With the strain as 0.020 for 70.0 kPa, the strain at 52 kPa is approximately

$$\varepsilon_y = \tfrac{52}{70}(0.020) = 0.0149 \text{ cm/cm}$$

The approximate value of stress-strain modulus for the triaxial compression test is

$$E_s = \frac{70.0}{0.0149} = 4698 \text{ kPa}$$

This value of stress-strain modulus is about 34 percent larger than the unconfined compression test value. The reader should observe that this computation indicates an in situ increase in E_s with depth. Why?

10-5 TWO-DIMENSIONAL STRESSES AT A POINT

Figure 10-1 showed the general case of stresses at a point within any homogeneous, isotropic, elastic body, and a soil mass in particular. In soil mechanics, it has been convenient to assume either that the principal stresses σ_2 and σ_3 are equal (and implicitly assume that E_s and μ are constants) or that $\sigma_2 = 0$, producing a plane stress condition (and often simultaneously taking $\varepsilon_2 = 0$ and terming the total stress-strain condition a "plane strain" state). Note carefully from Eqs. (10-7) that a plane strain condition can have three orthogonal nonzero stresses and a plane stress condition can have three orthogonal nonzero strains.

Let us consider the case of $\sigma_2 = \sigma_3$ and adopt the conventional practice of taking the largest of the three orthogonal principal stresses as σ_1. Field results indicate that the assumption that $\sigma_2 = \sigma_3$, while not strictly correct, does not introduce any large error and does greatly simplify the elastic analysis. In many retaining structure and excavation problems the assumption that σ_2 or $\varepsilon_2 = 0$ (plane strain) is reasonably realistic.

Figure 10-6 illustrates the conditions for a two-dimensional stress condition with the principal stresses σ_1 and σ_3 acting on the principal planes shown (element faces). A similar analysis could be made for any two of the three principal stresses, but the major and minor values are generally chosen as here.

Two of the principal planes (AC and CB of Fig. 10-6) are selected for detailed inspection, as in Fig. 10-6b. We will assume plane AB is 1 unit × 1 unit perpendic-

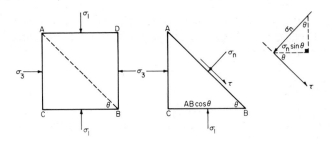

Figure 10-6 Two-dimensional principal stresses to obtain the normal and shear stresses on any plane *AB* as shown.

ular to the plane of the page, giving an area of 1 unit. From this assumption it follows that

$$BC = AB \cos \theta$$

$$AC = AB \sin \theta$$

Summing forces parallel to the X axis, we obtain $(\sum F_x = 0)$

$$\sigma_3(AB \sin \theta) + \tau(AB \cos \theta) - \sigma_n(AB \sin \theta) = 0 \qquad (a)$$

Summing forces parallel to the Y axis, we obtain

$$\sigma_1(AB \cos \theta) - \tau(AB \sin \theta) - \sigma_n(AB \cos \theta) = 0 \qquad (b)$$

Eliminating AB in both Eqs. (a) and (b), we obtain

$$\sigma_3 \sin \theta + \tau \cos \theta - \sigma_n \sin \theta = 0$$

$$\sigma_1 \cos \theta - \tau \sin \theta - \sigma_n \cos \theta = 0 \qquad (c)$$

This gives two equations in σ_1 and σ_3 and two unknown values, shear stress τ and normal stress σ_n. By the process of elimination and making use of the trigonometric relationships

$$\cos^2 \theta = 1 - \sin^2 \theta \qquad \sin^2 \theta = \tfrac{1}{2}(1 - \cos 2\theta) \qquad \sin \theta \cos \theta = \tfrac{1}{2} \sin 2\theta$$

we obtain the following equations (found in any mechanics of materials textbook but presented here for completeness and review):

$$\sigma_n = \frac{\sigma_1 + \sigma_3}{2} + \frac{\sigma_1 - \sigma_3}{2} \cos 2\theta \qquad (10\text{-}8)$$

$$\tau = \frac{\sigma_1 + \sigma_3}{2} \sin 2\theta \qquad (10\text{-}9)$$

Observe that these two equations are based on principles of mechanics and have nothing to do with material properties. Theory of elasticity deals with stresses and material properties of E_s and μ. It should be noted in passing that a similar set of equations could have been derived for the general case where planes AC and CB of Fig. 10-6 are not principal planes. The principal difference would be that shearing stresses on planes AC and CB would also have to be included. Since soil tests where Eqs. (10-8) and (10-9) are used involve starting with known principal stresses, the presentation here is preferred.

10-6 MOHR'S STRESS CIRCLE

Equations (10-8) and (10-9) are the parametric equations of a circle of stress in the XY plane with

$$\text{Radius } R = \frac{\sigma_1 - \sigma_3}{2}$$

$$\text{Origin of circle: } X = \frac{\sigma_1 + \sigma_3}{2} \qquad Y = 0$$

as shown in Fig. 10-7. It is much easier to develop Eqs. (10-8) and (10-9) from Mohr's circle† than through the procedure used in the preceding section (the reader should verify this).

---

† This circle of stress is called Mohr's circle since it is believed that Otto Mohr first proposed its use about 1871.

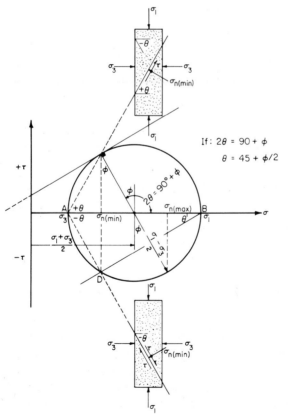

If: $2\theta = 90 + \phi$

$\theta = 45 + \phi/2$

Figure 10-7 Mohr's circle of stress, with orientation of stress planes and identification of terms used in Eqs. (10-8) and (10-9). Note that the ray from σ_3 locates $\sigma_{n(\min)}$; the angle $ADB = 90°$, as it should.

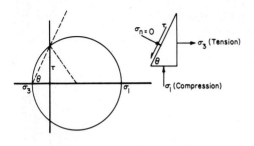

Figure 10-8 Principal stress conditions necessary to develop a condition of pure shear on some plane.

In soil mechanics, the principal stresses σ_1 and σ_3 are always compressive stresses, since soil cannot carry any significant tension stresses, and it is customary to plot Mohr's circle in the first quadrant rather than in the second quadrant as would be done in mechanics of materials textbooks.

At this point the reader should also be aware that $\Delta\sigma_1$ and σ_1 may be used interchangeably in this and the later text material. It can be understood from the context of the discussion whether initial conditions, stress increments, or stresses producing strains are being considered. Much of the soil mechanics literature also uses $\Delta\sigma_1$ and σ_1 interchangeably, with the context of usage necessary to interpret the response. In soil we always have initial stresses, which are no longer producing strains if we have steady state conditions.

A state of pure shear exists on a plane in the absence of normal stresses. Figure 10-8 illustrates that the conditions necessary to obtain pure shear on some plane require that the material be subjected to tension stresses on the minor principal plane. This situation will be very difficult to develop with a soil sample; in fact, special procedures are necessary to obtain soil tensile strengths (Fang and Chen, 1971).

10-7 BOUSSINESQ STRESSES IN AN ELASTIC HALF-SPACE

Boussinesq (ca. 1885) applied some complicated mathematical concepts to some of the equations of elasticity given earlier, together with the following boundary conditions (refer to Fig. 10-9):

1. Stresses vanish at $r \to \infty$.
2. Strains vanish at $r \to \infty$.
3. Shear stresses are zero at $y = 0$ (surface).
4. Normal stresses are zero at $y = 0$ except at the point of load application.
5. $\sum F_y = 0$.

For the point load at the surface of a semi-infinite, elastic half-space, and using the symbols on Fig. 10-9, Boussinesq obtained the following equations:

$$\Delta y = \frac{P}{4\pi R}\frac{1}{G}[2(1 - \mu) + \cos^2 \theta] \qquad \text{(settlement of point)} \qquad (10\text{-}10)$$

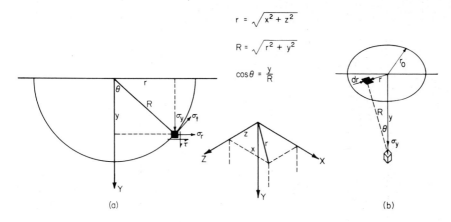

Figure 10-9 (*a*) Point load on the surface of a semi-infinite, elastic, half space for Boussinesq equations. (*b*) Application of the Boussinesq equation in reverse to obtain subsurface stresses due to application of an area load.

where G = shear modulus of elasticity (defined in Chap. 13)

$$\sigma_y = -\frac{3P}{2\pi R^2} \cos^3 \theta \qquad \text{(vertical stress)} \qquad (10\text{-}11)$$

$$\sigma_r = \frac{P}{2\pi R^2}\left(-3 \cos \theta \sin^2 \theta + \frac{1 - 2\mu}{1 + \cos \theta}\right) \qquad \text{(radial stress)} \qquad (10\text{-}12)$$

$$\sigma_t = \frac{P}{2\pi R^2}(1 - 2\mu)\left(\cos \theta - \frac{1}{1 + \cos \theta}\right) \qquad (10\text{-}13)$$

$$\tau = -\frac{3P}{2\pi R^2}(\cos^2 \theta \sin \theta) \qquad \text{(shear stress)} \qquad (10\text{-}14)$$

The Boussinesq equations have been widely used to obtain the stresses in a stratum for a surface load. Numerous modifications have been made to account for the stratum of finite depth, stratified deposits, and the load being applied at a depth below the surface of the soil mass. The Boussinesq solutions have been presented in the form of charts (Newmarks, 1942), curves (Fadum, 1948), and pressure bulbs (Bowles, 1977). With the advent of programmable electronic calculators (and the widespread availability of digital computers), it appears unnecessary to provide these types of data in this text. To obtain the vertical stress (of particular interest to geotechnical engineers) due to a footing load, which is not a point load, it is necessary to do the following:

1. Draw the footing to scale and subdivide it into n unit squares (use 0.3 × 0.3 or 0.5 × 0.5 m, or perhaps 1 × 1 ft).
2. Obtain the unit load for each unit square as $P = qA$, where q = soil pressure due to footing load.

3. Obtain the x and z coordinates for the center of each unit area with respect to the point beneath the footing where pressure is desired—usually the center, but sometimes a corner or side.

4. Compute the values of r, R, and $\cos \theta$. Since $\cos \theta = y/R$, Eq. (10-11) can be rewritten as

$$\sigma_y = \sum_1^n \frac{3P}{2\pi} \frac{y^3}{R^5} = C_1 \frac{y^3}{R^5} = C_2 \frac{1}{R^5}$$

This computation assumes the unit area is equivalent to a point load; thus, a summation of the point loads comprised in the footing area should be the correct stress increase at the point due to the footing load. It is usually not necessary to use a very small grid to obtain stresses of adequate engineering precision. Advantage may be taken of symmetry, so that $\frac{1}{4}$ of a footing may be gridded for the stress at the center (but multiply the accumulated value by 4).

5. For a given depth y, the constant C_2 can be computed and stored. The x and z coordinates are input, R is computed, and the unit contribution of stress is obtained and summed into storage. When all the unit areas have been accumulated into storage, the value is recalled as the stress value of interest.

An alternative method of obtaining the vertical stress beneath the *center* of a circular (exact), square (nearly exact), or rectangular footing is (see Fig. 10-9*b*) to rewrite Eq. (10-11) for R and depth y, neglecting $(-)$, to obtain

$$\sigma_y = \frac{3P}{2\pi y^2 [1 + (r/y)^2]^{1.5}}$$

and for $P =$ unit contact pressure q_o acting on a differential area dA, we have

$$\sigma_y = \frac{3q_o}{2\pi y} \frac{1}{[1 + (r/y)^2]^{1.5}} \, dA$$

Integrating from 0 to r_o and with $dA = 2\pi r \, dr$, we obtain

$$\sigma_y = q_o \left\{ 1 - \frac{1}{[1 + (r_o/y)^2]^{1.5}} \right\} \tag{10-15}$$

Example 10-3 What is the increase in stress at 6 m beneath the center of a circular footing of diameter $= 3$ m for a load $P = 1200$ kN?

SOLUTION

$$q_o = \frac{P}{A} = \frac{1200}{0.7854 \times 3^2} = 170 \text{ kPa}$$

From Eq. (10-15),

$$\sigma = 170 \left\{ 1 - \frac{1}{[1 + (1.5/6)^2]^{1.5}} \right\} = 170(0.09) = 14.8 \text{ kPa (exact)}$$

Example 10-4 What is the increase in stress at 1.5 and 3 m beneath the center of a square footing of $B = 3$ m? The footing load is 1800 kN.

SOLUTION

$$q_o = \frac{1800}{3^2} = 200 \text{ kPa}$$

$$\pi R^2 = A \qquad R = \sqrt{\frac{A}{\pi}} = \sqrt{\frac{9}{\pi}} = 1.69 \text{ m}$$

For $y = 1.5$ m,

$$\sigma = 200\left\{1 - \frac{1}{[1 + (1.69/1.5)^2]^{1.5}}\right\} = 200(0.71) = 141.5 \text{ kPa}$$

This value compares almost exactly with the values in Bowles (1977) and elsewhere.

For $y = 3$ m, the stress is $\sigma = 200(0.34) = 67.7$ kPa (almost exact also).

The average stress increase in the stratum is often required in settlement computations, such as the consolidation settlements of Chap. 11 or to compute the settlement as

$$\Delta H = \frac{\sigma_{av}(L)}{E_s}$$

The average increase is obtained using a numerical integration of the stresses from the several y coordinates in the stratum. One method is to use the trapezoidal formula when the y coordinate has been incremented using a constant increment Δy and for the stresses σ_{yi} to obtain

$$A = H\sigma_{y,\,av} = \Delta y\left(\frac{\sigma_{y1} + \sigma_{yn}}{2} + \sigma_{y2} + \sigma_{y3} + \cdots + \sigma_{y(n-1)}\right) \qquad (10\text{-}16)$$

Example 10-5 What is the stress at 2, 4, and 6 m beneath a rectangular footing 3×4 m in plan carrying 2000 kN? What is the average stress increase in this 6 m of depth?

SOLUTION Grid the footing into 48 units of 0.5 m, as shown in Fig. E10-5a.

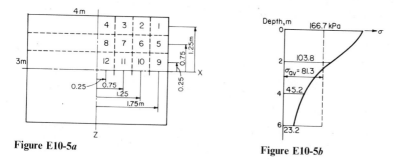

Figure E10-5a

Figure E10-5b

Step 1 Find the stresses at 2, 4, and 6 m. Due to symmetry only $\frac{1}{4}$ footing needs to be used. Coordinates are (partial list):

Element	x	z	Element	x	z
1	1.75	1.25	5	1.75	0.75
2	1.25	1.25	6	1.25	0.75
3	0.75	1.25	7	0.75	0.75
4	0.25	1.25	8	0.25	0.75

The y coordinate is successively 2, 4, then 6 m.

After inputting the 12 values, summing, recalling, and multiplying by 4, we obtain

$$\sigma_{y(0\,m)} = 166.7 \text{ kPa} \ (= \tfrac{2000}{12} \text{ as contact pressure; not computed,}$$
$$\text{as the equation is discontinuous at this point})$$

$$\sigma_{y(2\,m)} = 103.8 \text{ kPa}$$

$$\sigma_{y(4\,m)} = 45.2 \text{ kPa}$$

$$\sigma_{y(6\,m)} = 23.2 \text{ kPa}$$

Step 2 Find the average stress increase in the 6-m stratum. Refer to Fig. E10-5b and note that $\Delta y = 2$ m.

$$H\sigma_{y,\,av} = 2\left(\frac{166.7 + 23.2}{2} + 103.8 + 45.2\right) = 487.9$$

$$\sigma_{y,\,av} = \frac{487.9}{6} = 81.3 \text{ kPa}$$

10-8 SOIL DEFORMATIONS AND SETTLEMENTS

All materials are deformed when subjected to stress. This deformation is *elastic* when the material regains its original shape upon removal of the stress. *Plastic* deformations occur for stresses exceeding the elastic limit of a material, and elastoplastic deformations occur in materials which have no clearly defined elastic limit (or properties). When the stress is removed from these materials, very little deformation is recovered by elastic rebound, as illustrated in Fig. 10-10b.

A soil mass deposited over geological periods will be essentially in a steady state condition as far as having adjusted to the stresses caused by the weight of the soil above any given plane. New fill or excavations and new construction, such as buildings, dams, levees, and other embankments, will produce a temporary unsteady state deformation condition in the zones of stress influence.

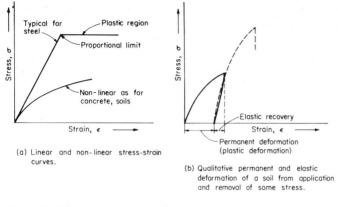

(a) Linear and non-linear stress-strain curves.

(b) Qualitative permanent and elastic deformation of a soil from application and removal of some stress.

Figure 10-10 Stress-strain characteristics.

Settlement, or deformation, under load is the total vertical movement caused by the application of load. It may be down with an increase in load and upward (swell) for decreasing loads such as excavations, rising water table, or other such load conditions. The deformations are always caused by changes in *effective stress*. In soil mechanics analyses, settlements are either

1. *Immediate.* Settlement occurs within hours to several days after load application. These types of settlements can be computed using $\Delta H = \sigma L/E_s$ type equations, with the principal difficulties being obtaining E_s and the depth of stress influence L (see Sec. 15-9).
2. *Consolidation.* Time-dependent settlements which occur within masses of saturated or partially saturated, fine-grained soils which have low coefficients of permeability. These settlements are time-dependent since pore drainage must accompany settlement.

The end result of either immediate or consolidation settlements is the same, but the method of computing the settlements is considerably different. Consolidation settlements will be considered in Chaps. 11 and 12. Chapter 12 will be concerned principally with how long it takes for consolidation settlements to occur.

10-9 RHEOLOGICAL MODELS

Rheology is a study of the behavior of a material in a fluid state. The soil skeleton under stress tends to change (flow) to a new shape to better resist the stress. If water is present in the soil pores, the change in soil structure may (or will) be accompanied by a flow of water from the pores due to the reduced pore sizes of the altered soil skeleton. This flow is due to an increase in water energy resulting from the push of the soil grains against the pore water during deformation. As water flow takes place, the pressure dissipates, analogous to a leak in a hydraulic jack.

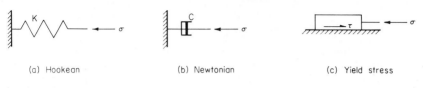

    (a) Hookean   (b) Newtonian   (c) Yield stress

Figure 10-11 Simple rheological models.

The flow of water is always a time function, as can be readily observed in water pipes, streams, surface runoff, etc.; thus, rheological models will be time-deformation-dependent stress models.

The simplest rheological model is the spring element or *hookean* model (obeying Hooke's law), as shown in Fig. 10-11*a*. If we have a dashpot (analogous to the shock absorber in an automobile), we have a *newtonian* model (Fig. 10-11*b*). If we have a condition of a certain minimum stress level causing strain (or slip), analogous to the block sliding down a plane, we have a *yield stress* model (see Fig. 10-11*c*).

The *hookean* model or spring element taken alone represents the elastic characteristics of the soil grain matrix. To have an instantaneous elastic deformation, the soil grain matrix must be continuous.

The application of stress to a fully saturated soil mass will impose the stresses initially on the pore water, since no soil skeleton deformation can occur until free space is available in the pores. Free space will be available in the soil pores only after pore drainage has occurred. Figure 10-12 displays this situation, including initial, intermediate, and final pore water and soil skeleton stresses. The dissipation of the pore pressures depends on the rate of pore flow—the size of the valve in Fig. 10-12—and may be simulated by dashpot action. In this case, however, as the pore pressure dissipates, the soil structure (spring) assumes the load, thus requiring the spring element and the dashpot to be coupled in some manner. The time

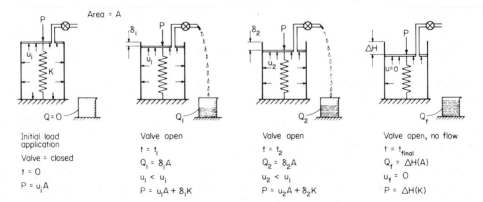

Figure 10-12 Load transfer from pore water to soil spring as a rate process.

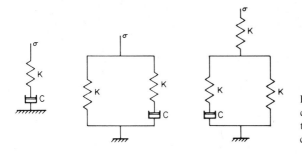

Figure 10-13 Several dashpot-spring couplings. These illustrate a few of the many possible couplings which can be made.

rate of strain is proportional to both the applied stress and how fast the pore fluid pressure can dissipate, i.e., it is a function of both the coefficient of permeability and the viscosity of the pore fluid.

Principal problems arising in the use of spring and dashpot elements are:

1. Proper couplings of elements (see Fig. 10-13), since many combinations are possible in addition to those shown
2. Proper evaluation of the spring and dashpot coefficients, which will depend on:
   (*a*) Soil structure (void ratio, gradation, grain size and shape)
   (*b*) The clay mineral
   (*c*) Water content
   (*d*) Temperature
   (*e*) Loading constraints (rate of loading, size of loaded area, magnitude of total load, etc.)

Note that in this and the next section, we are considering a *unit length of soil and a stress* rather than force; thus, the spring constants and dashpot coefficients must have dimensional units consistent with this concept.

10-10 COMPOSITE RHEOLOGICAL MODELS

Several composite rheological models will be presented to illustrate the complexity of the problem.

A St. Venant Model

One of the simpler models is the St. Venant model, illustrated in Fig. 10-14. Note that the method of loading is such the spring can only carry compression stress (or load). This is consistent with the general elastic properties of the soil mass, since soil can carry little if any tension.

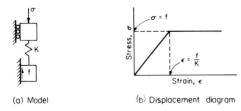

(a) Model

(b) Displacement diagram

Figure 10-14 The St. Venant model with both linear (spring) and plastic displacements.

B Kelvin Model

A parallel coupling of a hookean and a newtonian model results in the Kelvin model of Fig. 10-15. Note that the displacement of a dashpot (newtonian model) is time-dependent.

Mathematically, if we pass a plane at AA and sum stresses,

$$\sigma = K\varepsilon + C\,\frac{d\varepsilon}{dt} \tag{10-16}$$

where $d\varepsilon/dt$ = the velocity of deformation
$\quad\varepsilon$ = unit strain
$\quad K$ = spring constant
$\quad\sigma$ = total applied stress (force)
$\quad C$ = dashpot coefficient

If Eq. (10-16) is solved using standard solutions for differential equations of the form $dx/dy + Mx = N$ and boundary conditions of $\varepsilon = 0$ at $t = 0$ for σ = constant stress, we obtain

$$\varepsilon = \frac{\sigma}{K}\left(1 - \exp{-\frac{Kt}{C}}\right) \tag{10-17}$$

An analysis of units indicates that

$$K = \text{force/area} = \text{stress}$$

$$C = \text{force-time/area} = \text{viscosity}$$

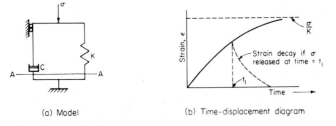

(a) Model

(b) Time-displacement diagram

Figure 10-15 Kelvin model. With parallel coupling all displacements are time-dependent.

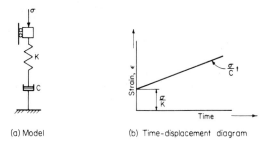

(a) Model   (b) Time-displacement diagram

Figure 10-16 The Maxwell model. Spring compression occurs at $t = 0$.

C Maxwell Model

The Maxwell model consists of a hookean and a newtonian model in series as in Fig. 10-16.

Since σ is common for both elements, the strain is

$$\varepsilon = \varepsilon_{spring} + \varepsilon_{dashpot}$$

and substituting values for strain,

$$\varepsilon = \frac{\sigma}{K} + \frac{\sigma t}{C} \tag{10-18}$$

one obtains an instantaneous deformation due to the applied load (stress) followed by a time-dependent deformation as given by the dashpot.

D Bingham Model

Figure 10-17 illustrates the Bingham model. In this model, as long as $\sigma \leq f$, the strain is simply

$$\varepsilon = \frac{\sigma}{K}$$

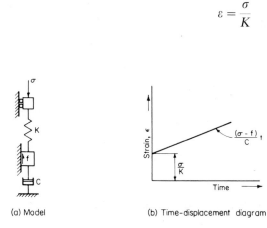

(a) Model   (b) Time-displacement diagram

Figure 10-17 The Bingham model. The friction model and the dashpot both restrain displacement. The spring displacement is independent of time.

When $\sigma > f$, the friction block begins to move. As the friction block moves, the dashpot influences the deformation so that the strain is now

$$\varepsilon = \varepsilon_{spring} + \varepsilon_{dashpot}$$

or
$$\varepsilon = \frac{\sigma}{K} + \frac{(\sigma - f)t}{C} \qquad (10\text{-}19)$$

E Burgers Model

Figure 10-18 illustrates the Burgers model. The solution of this rheological model is the sum of the solutions of the Maxwell model (Eq. 10-18) and the Kelvin model (Eq. 10-17), or

$$\varepsilon = \frac{\sigma}{K_1} + \frac{\sigma t}{C_1} + \frac{\sigma}{K_2}\left(1 - \exp - \frac{K_2}{C_2}t\right) \qquad (10\text{-}20)$$

These few simple rheological models illustrate the complexity of the problem of providing an adequate model to describe soil behavior for any given soil deposit. Rheological models for soil behavior are not used at present because of the problem of evaluating the spring and dashpot constants and the problem of selecting the appropriate model or coupling mode. Some researchers are of the opinion, however, that the Burgers model holds the most future promise for describing soil behavior. The Bingham model may be more appropriate for analysis in critical states, where the soil is in a continuous yield state.

Example 10-6

GIVEN The Burgers model of Fig. 10-18 with the following:

$$K_1 = 100 \text{ kN/cm}^3 \qquad C_1 = 0$$
$$K_2 = 150 \text{ kN/cm}^3 \qquad C_2 = 100 \text{ kN} - \text{month/cm}^3$$
$$\sigma = 10 \text{ kN/cm}^2$$

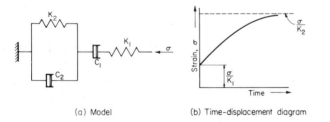

(a) Model   (b) Time-displacement diagram

Figure 10-18 The Burgers model as a coupling of the Maxwell and Kelvin models.

REQUIRED

(*a*) Plot strain unit ε versus time for $t = 0 - 4$ months.

(*b*) What is the total deformation in 10 cm of soil if $\sigma = 10 \ \text{kN/cm}^2$ at the end of four months?

SOLUTION

(*a*) With $C_1 = 0$, the reader must realize that this is a case of *no dashpot* term and not a case of $\sigma t/0$.

$$\varepsilon_i = \frac{\sigma}{K_1} + \frac{\sigma t}{C_1} + \frac{\sigma}{K_2}\left(1 - \exp - \frac{K_2}{C_2} t\right) \qquad (10\text{-}20)$$

Simplifying, one can make a table as follows:

$$\varepsilon_i = \tfrac{10}{100} + \tfrac{10}{150}(1 - \exp - 15t)$$

t, months	ε_i, cm/cm
0	0.10
0.5	0.135
1.0	0.152
1.5	0.160
2.0	0.163
2.5	0.165
3.0	0.166
4.0	0.166

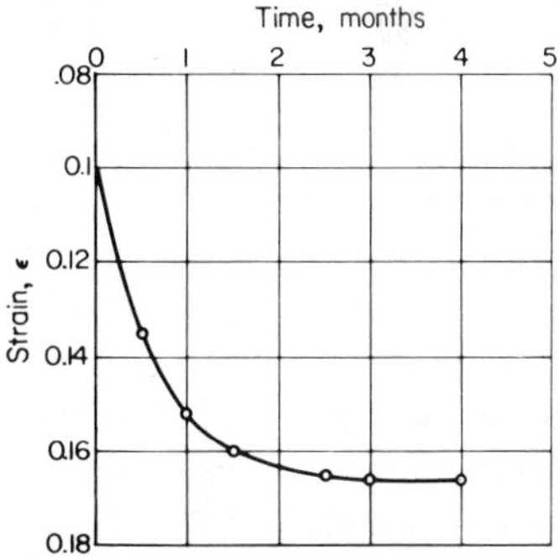

Figure E10-6

(*b*) The deformation in a 10-cm length of soil at the end of four months is

$$e = \varepsilon L = 0.166(10) = 1.67 \ \text{cm}$$

Note that this is not a highly realistic problem for a real soil, as C_2 is a viscosity term for fluid. Thus, one can question how C_1 can be zero or different from C_2, i.e., how can we both have and not have a fluid?

10-11 SUMMARY

This chapter has presented an introduction of the following: (1) Selected theory of elasticity concepts; (2) the concept of octahedral stresses; (3) principal stresses; (4) Mohr's circle for plane stress conditions; (5) elements of rheology and simple rheological models; and (6) selected coupling of simple rheological models.

In particular, the reader should put in perspective these several attempts to describe a particulate system using continuum models. The problem is formidable due to the general limitations associated with the particulate aggregation of lenses, stratification, variations in moisture content, void ratio, and other factors. Later chapters will illustrate that reasonable predictions can be made provided that the appropriate laboratory (or in situ) tests are carefully performed and interpreted on good-quality soil samples.

HOMEWORK PROBLEMS

10-1 Plot the octahedral effective stress path for Ex. 10-1.

10-2 For an initial triaxial stress condition of $\sigma_1 = \sigma_2 = \sigma_3 = 40$ kPa, what are the octahedral stresses? When the deviator stress $(\sigma_1 - \sigma_3)$ is 70 kPa, what are the new octahedral stresses?

 Ans.: For deviator stress $= 70$, $\tau_{oct} = 33$ kPa.

10-3 A two-dimensional stress condition has $\sigma_1 = 30$ kPa and a deviator stress $= 30$ kPa. If $\phi = 36°$, find the shear stress at failure and graphically display the orientation of the failure surface.

 Ans.: 12.1 kPa.

10-4 Verify the σ_y stresses of Ex. 10-5.

10-5 Redo Ex. 10-5 using unit areas of 1 m × 1 m and compare the σ_y stresses with those of the example.

10-6 If the depth of interest in Ex. 10-5 increases to 12 m, what is the average stress increase due to the footing load?

 Ans.: 46.8 kPa.

10-7 Redo Ex. 10-6 if $C_2 = 100$ kN $-$ year/cm^3.

10-8 Plot a curve of ε versus time for the Kelvin model of Fig. 10-15 if $K = 30$ kN/cm^3 and $C = 10$ kN $-$ month/cm^3.

10-9 Redo Ex. 10-6 if $\sigma = 1.0$ kN/cm^2.

Chapter 11

Consolidation
and Consolidation
Settlements

11-1 SOIL CONSOLIDATION PROBLEMS

When a saturated, fine-grained soil is subjected to an increase in compressive stress from some loading, the soil skeleton undergoes deformation or strain. Recall from Chap. 10 that strain is a cumulative effect of grain distortion (minor) and particle rolling and slipping. This strain results in a reduction in void ratio or voids volume, which can only take place as pore fluid is displaced. Since fine-grained soils have low coefficients of permeability (Chap. 8), the pore fluid displacement is a rate process, or time-dependent. The strains are, however, caused by *effective* stresses.

In the past many building problems have been caused by failure to recognize that settlements could continue for many years, with the total accumulated settlement being very large. It is easy to find cases in the literature where settlements of 0.3 m or more have occurred over periods of 5 to 20 years.

At present the potential for long-term settlements can be determined by soil exploration. Methods of predicting the magnitude of the time settlement and the length of time for the major portion of the settlement to occur will be considered in this and the following chapter. We will find that the best method of prediction will involve laboratory testing of carefully recovered ("undisturbed") samples, but that estimates can be made using less refined means.

Soil *creep* or *secondary compression* is a phenomenon in which soil strains continue for some time after excess pore pressures have essentially dissipated and result in additional void ratio reductions. Soil creep is a major portion of the total settlement for many organic or peaty soils and can be substantial for other soils. Soil creep prediction will also be considered in this chapter.

11-2 SOIL CONSOLIDATION

When the compression of a soil mass is time-dependent, it is termed *consolidation*. As for all soil settlements, consolidation is the elastoplastic deformation resulting in a permanent reduction in void ratio due to an increase in stress. The essential difference between ordinary compression and consolidation settlement is that *consolidation is time-dependent*.

In consolidation theory the following assumptions are made:

1. The soil is, and remains, saturated ($S = 100$ percent). Consolidation settlements can be obtained for nonsaturated soil, but the predicted time for settlement to occur is extremely unreliable.
2. Water and soil grains are incompressible.
3. There is a linear relationship between applied pressure and volume change [$a_v = \Delta e/\Delta p$, as defined in Eq. (11-6)].
4. The coefficient of permeability k is a constant. This is essentially true in situ, but in the laboratory there may be large errors associated with this assumption which will tend to produce error in the time for settlement to occur.
5. Darcy's law is valid ($v = ki$).
6. There is a constant temperature. A change in temperature from about 10 to 20°C (typical field and laboratory temperature, respectively) results in about a 30 percent change in the viscosity of water. It is important that the laboratory test be performed at a known temperature, or preferably at the in situ temperature.
7. Consolidation is one-dimensional (vertical), that is, there is no lateral flow of water or soil movements. This is exactly true in the laboratory test and is generally nearly so in situ.
8. Samples are undisturbed. This is a major problem in that no matter how carefully the sample is taken, it is unloaded of the in situ overburden. Additionally, the static water table pore pressure is usually lost. In sensitive soils serious errors may result; in other soils the effects may be much less. Careful interpretation of data can reduce the effect of sampling errors somewhat.

The consolidation characteristics (or parameters) of a soil are the *compression index C_c* and the *coefficient of consolidation c_v*. The compression index relates to

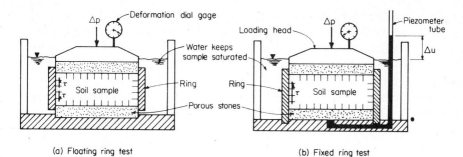

Figure 11-1 Schematics of the consolidation test. The load is applied to a sample through the load head onto porous stones to the sample. In the *floating-ring test*, the base porous stone provides lateral drainage from beneath the confining ring, and compression takes place from both faces of the soil sample. Friction effects for both tests are qualitatively shown. In the *fixed-ring test*, the sample can drain only through the top porous stone, as the ring fits tightly to the base, so that water at the sample base (qualitatively) rises in the piezometer tube immediately on application of Δp, then slowly drains back through the base and up through the soil sample. The piezometer tube (say, 100-mL burette) can be used to perform a falling-head permeability test to obtain the coefficient of permeability at the end of each load increment to obtain k versus e.

how much consolidation or settlement will take place. The coefficient of consolidation relates to how long it will take for an amount of consolidation to take place.

The consolidation parameters can be obtained (or at least estimated) from a laboratory consolidation test, schematically shown in Fig. 11-1. The carefully trimmed soil specimen [usual diameter from 6.3 to 11.3 cm (2.5 to $4\frac{7}{16}$ in)] is placed inside a metal confining ring. Uniform soil pressure is applied through the loading block, and the porous stones allow the excess pore pressure due to the load increment to freely escape as the soil voids are compressed. A dial gage or LVDT† is used to measure the amount of compression at varying time intervals; thus, volume changes can be computed.

A new increment of load is periodically applied to the soil. Research (Leonards, 1962) has found that best results are obtained when the load is doubled producing a ratio of $\Delta p/p = 1$; thus, a typical sequence would be 25, 50, 100, 200, 400, 800, 1600, 3200 kPa,‡ or 0.25, 0.50, 1, 2, 4, 8, 16 tsf in the fps system. There is also evidence (Leonards, 1962) that if the initial load increment is too low, the excess pore pressure gradient may not be sufficient to initiate pore water flow in some clay soils. This should only affect the initial part of the curve of void ratio vs. pressure, as the later load increments will be large enough to avoid this problem; also, initial loads on the order of 25 kPa appear adequate to avoid it.

† Linear voltage displacement transducer.
‡ The reader may consult Bowles (1978) for a method of converting existing fps consolidation test equipment to use this load sequence directly.

The test loads are changed on the sample when consolidation under the current load increment is complete. This may be taken as the time when the dial reading has remained relatively unchanged for three successive readings, where the elapsed time of each reading is approximately double that of the previous reading. One may arbitrarily change loads every 24 h (commonly done in commercial testing laboratories), which is generally satisfactory for samples of the usual 2 to 3 cm thickness and using the floating ring test equipment.

The results of a consolidation test are presented in the form of curves of settlement (or dial reading) vs. time, as shown in Fig. 12-4a and b or 12-5, and void ratio e versus log pressure or strain ε versus log pressure, as shown in Fig. 11-2. Sometimes a plot of e versus p is used instead of the semilogarithmic plots, as in Fig. 11-6. Note that the *effective* pressure is used in these plots.

The semilogarithmic plot of either e versus log p or ε versus log p for "undisturbed"[†] cohesive soils has the following characteristics:

1. The initial branch of the curve has a relatively flat slope (due primarily to the initial loading being small pressure increments that are less than the in situ overburden, and due also to soil expansion from loss of overburden pressure, which always occurs during sample recovery).
2. At a pressure close to the in situ overburden pressure (p_o), the curve becomes much steeper and a curved portion exists. For relatively insensitive clays the curve is rather flat (Fig. 11-3); for sensitive[‡] clays it is much sharper (Fig. 11-4).
3. Beyond the p_o point the curve is nearly linear for insensitive clays. For sensitive clays, the curve may exhibit some concavity, as shown in Fig. 11-4. This relatively steep and characteristic concave shape may be in part due to a structure collapse at a pressure greater than p_o.
4. If an undisturbed soil sample is loaded to some pressure, such as p_1 of Fig. 11-2, then unloaded and reloaded, the curves form a hysteresis loop. The reload curve has been found to consistently have the general shape of the initial branch of the load curve. When the new reload cycle exceeds p_1, the reload curve will become an approximate extension of the original load curve.
5. (a) It has been found that a remolded soil always produces a "virgin" type compression curve as qualitatively shown in Figs. 11-3 through 11-5.
   (b) This curve always has a regular slope but less than the "undisturbed" samples, probably because the structure is more oriented.
   (c) When the remolded samples are unloaded, hysteresis loops are formed as in Fig. 11-2.

[†] There is no such thing as an undisturbed sample, but if the sample recovery effects are kept small, the sample is described as being "undisturbed." Hand carved block samples from a test pit are likely to be of the highest quality, but these can seldom be economically justified. Thin-walled tube samples are most common and should be at least 6 mm larger than the consolidation ring. "Undisturbed samples" is standard terminology for the samples described here and will be used as such in this text.

[‡] Sensitivity and the concept of sensitive vs. insensitive clays are defined and considered in Sec. 13-8.

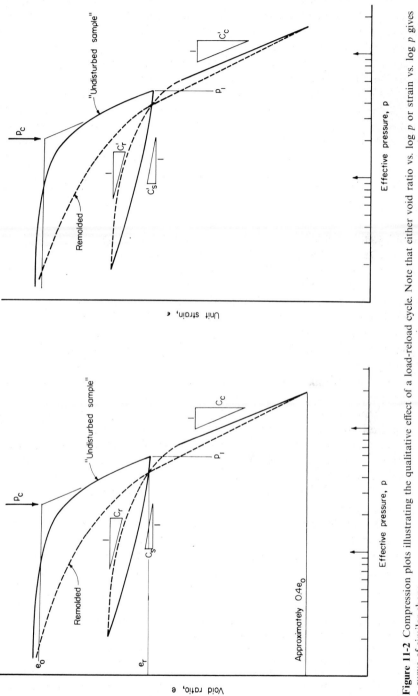

Figure 11-2 Compression plots illustrating the qualitative effect of a load-reload cycle. Note that either void ratio vs. log p or strain vs. log p gives a curve of similar shape.

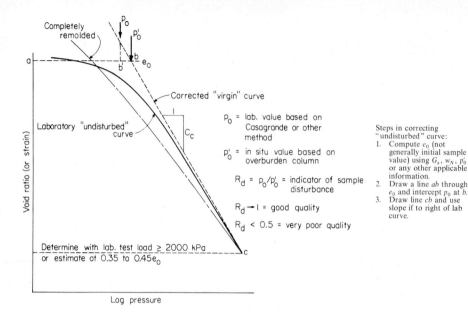

P_o = lab. value based on Casagrande or other method

P_o' = in situ value based on overburden column

$R_d = P_o/P_o'$ = indicator of sample disturbance

$R_d \rightarrow 1$ = good quality

$R_d < 0.5$ = very poor quality

Steps in correcting "undisturbed" curve:
1. Compute e_0 (not generally initial sample value) using G_s, w_N, P_0' or any other applicable information.
2. Draw a line ab through e_0 and intercept p_0 at b.
3. Draw line cb and use slope if to right of lab curve.

Figure 11-3 Qualitative void ratio (or strain) vs. log p curve for a normally consolidated clay with a sensitivity $S_t < 4$.

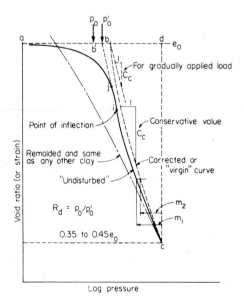

Notes:
1. Point b' is located as extension of line ij from point of inflection i of "undisturbed" curve.
2. Extend line ab to d.
3. Draw perpendicular from c to d.
4. Draw field curve such that

$$m_2 = m_1 \frac{bd}{b'd}$$

and using several values of m_1.

Figure 11-4 Qualitative void ratio (or strain) vs. log p curve for a sensitive soil and the method of correcting C_c. (*After Terzaghi and Peck, 1967.*)

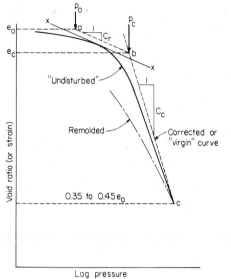

Steps:
1. Compute p_0 and e_0; initial void ratio of sample may be used if rebound is minimal.
2. Draw a line $e_0 a$ and locate p_0.
3. Obtain p_c using valid method.
4. Obtain slope of rebound by eye as line x-x and slope C_r.
5. Through a and parallel to x-x draw line ab.
6. Locate b as the intercept of the "virgin" curve; verify e_c using in situ water content.
7. Obtain line cb and the slope as C_c.

Figure 11-5 Correcting the laboratory void ratio (or strain) vs. log p curve for preconsolidated soils. (*After Schmertmann, 1955.*)

(*d*) From (*b*) and (*c*) it is concluded that:

(*i*) The initial branch of an undisturbed sample is a recompression branch of the field virgin curve, since the sample has been unloaded of the overburden pressure.

(*ii*) The virgin curve would be expected to be somewhat to the right of the undisturbed curve.

(*iii*) The laboratory undisturbed curve should be corrected to the virgin curve to obtain the compression parameters (as shown in Figs. 11-3 through 11-5). Note that corrections should not be used unless e_o and e_c are reliably known, since "just making corrections" may produce a nonconservative solution.

The pressure p_c is equal to the *effective* in situ overburden pressure for *normally consolidated* clays. The effective overburden pressure is computed from the column of soil from the point of interest to the ground surface. If p_c [as obtained from the e versus log p or ε versus log p curves (Fig. 11-2)] is larger than the in situ overburden pressure, the soil has been subjected to a pressure at some time in the geologic past larger than the present pressure p_o, and this past pressure may have been due to:

1. A greater amount of overburden which has since been eroded away
2. Drying and resulting shrinkage stresses
3. A change in the water table
4. A combination of drying and wetting in the presence of certain sodium, calcium, or magnesium salts (particularly in uplifted marine deposits)

In this case the soil is said to be *preconsolidated* (or overconsolidated) and the symbol p_c is used. The overconsolidation ratio is

$$\text{OCR} = \frac{p_c}{p_o}$$

For *normally* consolidated clay, as the in situ pressure is increased from p_o to $p_o + \Delta p = p_2$, the void ratio should decrease by an amount Δe, as shown in Fig. 11-2a. The slope of the straight line portion of the virgin compression curve can be denoted as the *compression index* C_c, computed as

$$C_c = \frac{\Delta e}{\log (p_o + \Delta p) - \log p_o}$$

And rearranging,

$$\Delta e = C_c \log \frac{p_o + \Delta p}{p_o} \tag{11-1}$$

For the ε versus $\log p$ plot of Fig. 11-2b,

$$\Delta \varepsilon = C'_c \log \frac{p_o + \Delta p}{p_o} \tag{11-2}$$

Similarly, and letting $p_2 = p_i + \Delta p$, obtain the following:

e versus $\log p$	ε versus $\log p$
For swell:	
$\Delta e = C_s \log \dfrac{p_2}{p_i}$	$\Delta \varepsilon = C'_s \log \dfrac{p_2}{p_i}$
For recompression:	
$\Delta e = C_r \log \dfrac{p_2}{p_i}$	$\Delta \varepsilon = C'_r \log \dfrac{p_2}{p_i}$
where $\Delta e, \Delta \varepsilon$ = change in void ratio or strain between p_i and p_2	
p_i = any pressure along the appropriate curve	

A relationship exists between C'_c and C_c as follows:

1. From the definition of void ratio e, obtain

$$\Delta e = \frac{\Delta V_v}{V_s} = \frac{\Delta H}{H_s} = \Delta H \qquad \text{when } H_s = V_s = 1.0$$

from which

$$\Delta H = C_c \log \frac{p_2}{p_1} \tag{a}$$

2. Introducing the initial (or in situ) void ratio e_o, we obtain for strain

$$\Delta \varepsilon = \frac{\Delta H}{H_i} = \frac{\Delta H}{H_s + H_v} = \frac{\Delta H}{1 + e_o}$$

During compression only the voids compress; therefore,

$$\Delta \varepsilon = \frac{\Delta H}{1 + e_o} = \frac{\Delta H_y}{1 + e_o}$$

from which

$$\frac{\Delta H_v}{1 + e_o} = C_c' \log \frac{p_2}{p_1} \qquad (b)$$

Equating Eqs. (a) and (b),

$$C_c \log \frac{p_2}{p_1} = C_c'(1 + e_o) \log \frac{p_2}{p_1}$$

and cancelling, obtain

$$C_c' = \frac{C_c}{1 + e_o} \qquad (11\text{-}3)$$

Equations (11-1) and (11-2) cannot be used to compute Δe or $\Delta \varepsilon$ for *preconsolidated* soils when the value of p_o is less than p_c. For these conditions either C_r, C_s, C_r', or C_s' must be used, depending on whether swell or compression (recompression) is desired. For settlements in preconsolidated soil, the change in void ratio is computed in two stages as follows:

Stage 1 For $\Delta p + p_o \le p_c$:

$$\Delta e_1 = C_r \log \frac{p_o + \Delta p}{p_o}$$

$$\qquad (11\text{-}4)$$

$$\Delta \varepsilon_1 = C_c' \log \frac{p_o + \Delta p}{p_o}$$

Stage 2 For $p_2 = p_o + \Delta p > p_c$, i.e., when the increase in pressure extends beyond the preconsolidation pressure into the linear portion of the pressure plot, compute this extension as $\Delta p' = p_o + \Delta p - p_c$. The additional void (or strain) change is computed as

$$\Delta e_2 = C_c \log \frac{p_c + \Delta p'}{p_c}$$

$$\qquad (11\text{-}5)$$

$$\Delta \varepsilon_2 = C_c' \log \frac{p_c + \Delta p'}{p_c}$$

The total change is the sum of the values,

$$\Delta e = \Delta e_1 + \Delta e_2$$

$$\Delta \varepsilon = \Delta \varepsilon_1 + \Delta \varepsilon_2$$

The principal advantages in using a plot of ε versus log p instead of e versus log p are:

1. It is possible to use electronic data acquisition equipment which makes the plot as the load increment changes are made.
2. Most of the plot can be made prior to completion of the test, since H_i is known. Often in a consolidation test, H_s (required to compute the void ratio) is obtained from the final sample height and oven-drying of the soil cake to obtain the saturated water content, since the soil should be in a two-phase state at this point.
3. The plot is simpler to make, with less chance of computation errors.
4. The settlement computation becomes simply $H\varepsilon$.

11-3 ARITHMETIC PLOT OF CONSOLIDATION SETTLEMENT RELATIONSHIPS

Occasionally e versus p plots are made as in Fig. 11-6. From a tangent to the curve as illustrated in Fig. 11-6, obtain the *coefficient of compressibility* a_v as

$$a_v = \frac{\Delta e}{\Delta p} \qquad \text{units of } L^2 F^{-1} \tag{11-6}$$

Neglect the negative sign of Δe versus Δp. The *coefficient of volume compressibility*

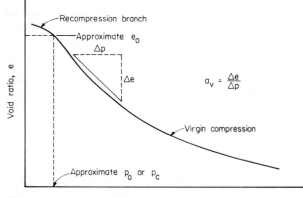

Figure 11-6 Arithmetic plot of e versus pressure (effective) to obtain the coefficient of compressibility a_v. Note that a_v is heavily dependent on tangent location and is not a constant.

m_v, is defined as

$$m_v = \frac{a_v}{1 + e_o} = \frac{\Delta e}{\Delta p(1 + e_o)} \qquad \text{units of } L^2 F^{-1} \qquad (11\text{-}7)$$

The tangent location for a_v should approximate the in situ p_o and Δp from anticipated soil loads.

11-4 APPROXIMATE METHODS TO DETERMINE C_c AND C'_c

Due to the time involved in performing a consolidation test to obtain C_c, it is desirable to relate the compression index to some other index property that is more easily determined. Azzouz et al. (1976) lists several equations as given in Table 11-1.

Table 11-1 Empirical and regression equations for C_c and C'_c as indicated†

(a) Equations from several literature sources for C_c and C'_c

Equation	Regions of applicability.
$C_c = 0.007(w_L - 7)$	Remolded clays
$C_c = 0.01w_N$	Chicago clays
$C_c = 1.15(e_o - 0.35)$	All clays
$C_c = 0.30(e_o - 0.27)$	Inorganic cohesive soil; silt, silty clay, clay
$C_c = 0.0115w_N$	Organic soils, peats, organic silt and clay
$C_c = 0.0046(w_L - 9)$	Brazilian clays
$C_c = 1.21 + 1.055(e_o - 1.87)$	Motley clays from São Paulo city
$C_c = 0.009(w_L - 10)$	Normally consolidated clays
$C_c = 0.75(e_o - 0.50)$	Soils with low plasticity
$C'_c = 0.208e_o + 0.0083$	Chicago clays
$C'_c = 0.156e_o + 0.0107$	All clays

(b) Regression equations used, with reliability R and coefficient of variation S indicated

R	S‡	Regression equation
0.85	0.077	$C_c = 0.40(e_o - 0.25)$
0.86	0.074	$C_c = 0.37(e_o + 0.003w_L - 0.34)$
0.85	0.007	$C_c = 0.40(e_o + 0.001w_N - 0.25)$
0.81	0.085	$C_c = 0.009w_N = 0.002w_L - 0.10$
0.86	0.074	$C_c = 0.37(e_o + 0.003w_L + 0.0004w_N - 0.34)$
0.74	0.038	$C'_c = 0.14(e_o + 0.007)$
0.76	0.039	$C'_c = 0.126(e_o + 0.003w_L - 0.06)$
0.74	0.038	$C'_c = 0.142(e_o - 0.0009w_N - 0.006)$
0.71	0.040	$C'_c = 0.003w_N + 0.0006w_L - 0.004$
0.76	0.037	$C'_c = 0.135(e_o + 0.01w_L - 0.002w_N - 0.06)$

† From Azzouz et al. (1976) and edited.
‡ A small value of S is preferred, indicating not too much interdependence between variables.
  Symbols: e_o = in situ void ratio; w_N = in situ water content; w_L = liquid limit.

Terzaghi and Peck (1967), based on research on undisturbed clays of low to medium sensitivity, proposed

$$C_c = 0.009(w_L - 10) \tag{11-8}$$

which has a reliability range of about ± 30 percent. This equation is widely used, in spite of the reliability range given above, to make initial consolidation settlement estimates. This equation should not be used where the sensitivity of the clay is greater than 4 and under no circumstances where $S_t > 8$.

11-5 DETERMINATION OF THE PRECONSOLIDATION PRESSURE

The preconsolidation pressure can be estimated with sufficient precision by using judgment and extending the straight line portion of the e versus $\log p$ or ε versus $\log p$ plot to a point at approximately the breaking of the two branches of the curve. Alternatively, one may use a method proposed by Casagrande (1936) to obtain the approximate p_c as follows (referring to Fig. 11-7):

1. By eye estimate the sharpest point of curvature and draw a tangent line.
2. Through the point of tangency draw a horizontal line to form an angle α.
3. Bisect the α angle.
4. Project the straight line part of the curve to intersect with the α angle bisector of step 3.

The intersection obtained in step 4 above provides both p_c (projected vertically to the pressure axis) and e_o (projected to the void ratio axis). If $p_c \cong p_o$, the soil is probably normally consolidated, with disturbance accounting for the discrepancies. If $p_c > p_o$, the soil is preconsolidated.

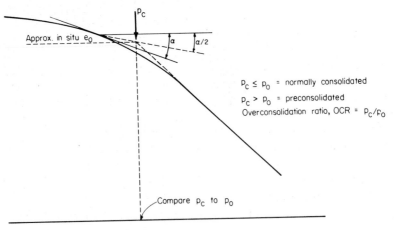

Figure 11-7 Casagrande method for obtaining preconsolidation pressure.

11-6 SOIL STRUCTURE AND CONSOLIDATION

Consolidation is the gradual breaking down of the soil skeleton. The shape of the void ratio vs. pressure curve is an indicator of the changes taking place within the sample. Considering Fig. 11-3, one may note that little structure change takes place under increasing load until the pressure p_o (or p_c) is reached. The reason is that the soil has been accustomed to a pressure up to the overburden or preconsolidation pressure and has reached equilibrium under this stress. Only when applied pressures exceed this equilibrium condition does the soil skeleton again undergo significant changes.

If, however, the soil is sensitive to disturbances which can cause a structure collapse, a curve like the one shown in Fig. 11-4 can be obtained. Again, up to p_o the soil skeleton undergoes little change. When the applied pressure exceeds p_o, the structure essentially fails (or collapses) and a very steep plot (or break) is obtained on the void ratio vs. pressure curve.

A completely remolded soil loses all its natural soil structure characteristics. When it is loaded in a consolidation test, the particles, which are more likely to be in a dispersed structure state, are forced ever closer together, resulting in a smooth curve of void ratio vs. pressure. A remolded soil cannot be used to obtain the preconsolidation pressure. It follows that the accuracy of the measured preconsolidation pressure is very dependent on the degree of disturbance of the soil sample. An indication of sample disturbance can be made for normally consolidated soils using the disturbance ratio R_d shown on Figs. 11-3 and 11-4.

11-7 SECONDARY COMPRESSION

Strictly, consolidation is the void ratio reduction taking place under conditions of excess pore pressure. When the excess pore pressure has dissipated under an imposed load condition, consolidation (often termed *primary* consolidation) is complete (see Fig. 12-4). Actually the soil continues to compress under the load, although at a much slower rate, for some considerable period of time after consolidation is complete. The compression taking place after consolidation is termed *secondary compression* or *creep* (also called *secondary consolidation*).

Evaluation of the amount of secondary compression is difficult, and it is often ignored, especially for inorganic soil; however, creep may be the principal settlement for highly organic soils (Weber, 1969). Lo (1961) made an extensive study of the problem and concluded that the secondary compression may

1. Gradually decrease with time
2. Continue at a rate proportional to the logarithm of time
3. Undergo a sudden increase in rate of compression

In all cases, however, secondary compression eventually halts.

From observations of settlement vs. log time curves, such as Fig. 12-4, taken over times sufficient to plot well into the secondary compression range, it appears that the general shape of the secondary compression curve is constant for a particular soil. If the slope of the secondary branch is constant, it is only necessary to continue a single load increment for a time sufficient to determine the basic slope on a plot of settlement vs. log time; then the secondary compression branch for other load increments is approximately parallel *when plotted to the same settlement scale*. This could result in some time savings in a consolidation test by allowing the changing of loads earlier (e.g., as soon as it is estimated that a curve point has been obtained in the secondary compression region).

A *coefficient of secondary compression* C_α can be obtained from the settlement vs. log time plot (Fig. 12-4) as

$$C_\alpha = \frac{\Delta H/H_i}{\log\left[(t_i + \Delta t)/t_i\right]} \qquad (11\text{-}9)$$

or, since $\Delta H_s/H_i = \Delta\varepsilon$,

$$C_\alpha = \frac{\Delta\varepsilon}{\log\left[(t_i + \Delta t)/t_i\right]} \qquad (11\text{-}9a)$$

and the secondary settlement ΔH_s is

$$\Delta H_s = HC_\alpha \log \frac{t_i + \Delta t}{t_i} \qquad (11\text{-}10)$$

11-8 COMPUTATION OF CONSOLIDATION SETTLEMENTS

The time-dependent settlement of a layer of soil can be computed using consolidation parameters (refer to Fig. 11-8) as follows:

By proportion,

$$\frac{\Delta H}{H} = \frac{\Delta e}{1 + e}$$

or, the settlement

$$\Delta H = \frac{\Delta e}{1 + e} H \qquad (a)$$

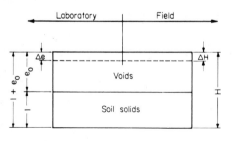

Figure 11-8 Settlement of a soil sample or layer of soil of thickness H in situ.

Substituting Δe from Eq. (11-1), obtain

$$\Delta H = \frac{C_c H}{1 + e} \log \frac{p_o + \Delta p}{p_o} \tag{11-11}$$

Alternatively, since $C_c' = C_c/(1 + e)$, obtain from Eq. (a)

$$\Delta H = C_c' H \log \frac{p_o + \Delta p}{p_o} \tag{11-12}$$

which is equivalent to

$$\Delta H = H\varepsilon \tag{11-12a}$$

Note that by analogy one may also use C_s, C_s', C_r, and C_r' in Eq. (11-9), depending on the parameter and the in situ conditions of p_o, p_c, and Δp.

The in situ effective overburden pressure p_o can be somewhat in error, since usual field exploration techniques often obtain only estimates of unit weights. The unit weight of cohesionless soil is frequently estimated. The unit weight of cohesive soil can be obtained with reasonable reliability using a method in Bowles (1978), but again the values are often estimated. Estimates generally tend to be low, which increases the computed settlement and is conservative. Care should be taken, however, not to be too conservative.

The arithmetic plot value of the coefficient of compressibility a_v can be used in Eq. (a) to obtain

$$\Delta H = \frac{a_v}{1 + e} \Delta p H = m_v \Delta p H \tag{11-13}$$

Note that with units of $m_v = 1/E_s$, Eq. (11-13) is simply using the mechanics of materials equation $\delta = \sigma L/E$, where here we have

$$\Delta H = \frac{\Delta p H}{E_s} = H\varepsilon$$

11-9 EXAMPLES

Example 11-1
GIVEN The e versus $\log p$ plot of Fig. E11-1.

REQUIRED Compute C_c, C_s, C_c', and C' if $H = 2$ cm.

SOLUTION For C_c: $\Delta e = 0.80 - 0.51 = 0.29$

$$p_2 = 2000 \text{ kPa} \qquad p_1 = 800 \text{ kPa}$$

$$C_c = \frac{0.29}{\log \frac{20}{8}} = \frac{0.29}{0.398} = 0.73$$

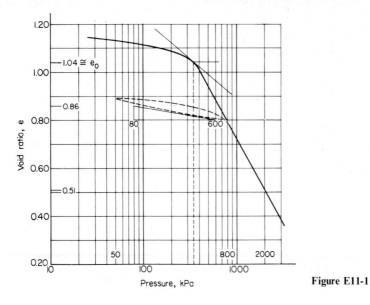

Figure E11-1

$$\text{For } C_s: \Delta e = 0.86 - 0.80 = 0.06$$

$$p_2 = 80 \qquad p_1 = 600 \text{ kPa}$$

$$C_s = \frac{0.06}{\log \frac{8}{60}} = \frac{0.06}{0.875} = 0.069$$

Note that strictly, $C_s = -0.069$
Computing C'_c and C'_s:

$$C'_c = \frac{C_c}{1 + e_o} = \frac{0.73}{1 + 1.04} = 0.36$$

$$C'_s = \frac{C_s}{1 + e_o} = \frac{0.069}{1 + 1.04} = 0.033$$

Note: The author did not attempt to "correct" C_c or C'_c, since e_o is not precisely known (see Prob. 11-1).

Example 11-2
GIVEN The ε versus $\log p$ plot and soil profile shown in Fig. E11-2.

REQUIRED Compute the expected settlement for the load causing the *net pressure* profile shown in the clay stratum.

SOLUTION Settlement computations are based on *average* values of p_o, C'_c, and e_o in some stratum thickness H. Assume C'_c is constant in the entire thickness of 9 m of clay.

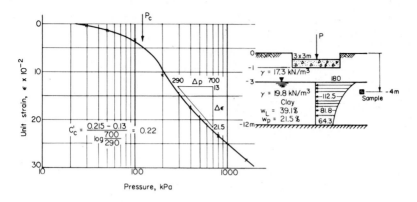

Figure E11-2

Step 1   Compute average p_o. Since p_o varies linearly, the middepth value is also the average value in the stratum.

$$3(17.3) = 51.9 \text{ kPa}$$

$$+ 4.5(19.8 - 9.807) = 44.97 \text{ kPa}$$

$$p_o = 96.87 \text{ kPa}$$

Since the estimated p_c from the ε versus log p curve shows 120 kPa (and also, w_N is slightly closer to w_P than to w_L), assume that the stratum is moderately preconsolidated.

Step 2   Compute average Δp using the pressure profile shown (which could be obtained using the Boussinesq method of Sec. 10-7). Use numerical integration to find the area of the diagram, and note that the area is $H_c\,\Delta p_{average}$. Use the trapezoidal rule [Eq. (10-15)] to obtain the area.

$$A = H\,\Delta p = 3\left(\frac{180 + 64.3}{2} + 112.5 + 81.8\right) = 949.35$$

$$\Delta p = \frac{949.35}{9} = 105.5 \text{ kPa}$$

Step 3   Compute the expected settlement using C_c', shown on Fig. E11-2. Note that the value shown did not use any virgin curve correction.

$$\Delta H = C_c'H \log \frac{p_o + \Delta p}{p_o}$$

$$= 0.22(9) \log \frac{96.9 + 105.5}{96.9}$$

$$= 0.22(9)(0.3199) = 0.633 \text{ m}$$

This is a large settlement which few buildings could tolerate. This large settlement might be expected considering the shape of the ε versus log p curve and the rather high w_N. Also the thickness of the clay stratum is substantial. The reader should ask whether the author should have refined the computations to include preconsolidation effects based on a C'_r of approximately 0.15.

11-10 CONTROLLING CONSOLIDATION SETTLEMENT

From careful analysis of the material and equations presented in this chapter, it is evident that several methods of controlling consolidation settlements can be used.

1. Since the ratio of $(p_o + \Delta p)/p_o$ is a significant factor, it is evident that control of $p_o + \Delta p$ can be accomplished by excavation of soil so that

$$p_o - p_{\text{exc}} + \Delta p \rightarrow p_o$$

If this is done we have

$$\frac{p_o + \Delta p}{p_o} = \frac{p_o}{p_o} = 1 \quad \text{and} \quad \log 1 = 0$$

i.e., if enough soil is excavated, no settlement will occur. This is commonly done and produces a "floating" foundation.
2. Since C_r is less than C_c, if we have sufficient time we may preload (preconsolidate) the soil. This is also frequently done by:
    (a) Placing excess fill
    (b) Where impervious surface soil is present, building a dike and filling the enclosure with water
    (c) Enclosing the zone of interest with an airtight membrane and producing a vacuum in the soil beneath the membrane so that the differential air pressure becomes a surcharge
    (d) Lowering the water table to increase the effective pressure

11-11 SUMMARY

Several methods of using consolidation test data to obtain settlement parameters have been presented and are summarized as follows:

e versus log p	ε versus log p	e versus p
Compression index C_c	Compression ratio C'_c	Coefficient of compressibility a_v
Recompression index C_r	Recompression ratio C'_r	Coefficient of volume compressibility m_v
Swell index C_s	Swell ratio C'_s	

Several empirical relationships (Table 11-1) for C_c and C'_c have also been presented.

The use of the settlement parameters has been illustrated in computation of consolidation settlements. Concepts of previous stress history and the effect of preconsolidation p_c have been examined. From observations of the load-unload-reload curves of e versus log p for remolded samples, a method of obtaining p_c has been developed. The reader should carefully note that all the consolidation settlement equations are

$$\Delta H = \int_{H_1}^{H_2} \varepsilon \, dh \rightarrow \sum_{i=1}^{n} H_i \varepsilon$$

from which:

1. Δp is the average increase in pressure in the increment of stratum H_i or of the total stratum thickness, which is preferred since its use reduces computations.
2. e_o is the average value of the *in situ* void ratio and not the sample value. Equation (11-1) shows that e_o varies with depth logarithmically. For, say, $H_i = 1$ to 2 m one may use the average of top and bottom or simply the midheight value. For thicker strata several values should be computed and numerical integration using Eq. (10-15) performed to obtain the "best" value of e_o.
3. p_o is the average existing *effective overburden pressure* existing in a stratum of thickness H_i. The value at midheight is commonly used, since the variation is linear with depth if we assume a constant unit weight. This will produce a small error, since there is some increase in unit weight with depth along with the corresponding decrease in e_o.

Secondary compression, or creep, has also been considered. Careful analysis of secondary settlement computations shows that this is simply computed as

$$\Delta H = H\varepsilon$$

since the C_α term is simply a time-dependent strain.

HOMEWORK PROBLEMS

11-1 In Ex. 11-1, if it is known that $e_o = 1.15$, what is the corrected value of C_c? If $G_s = 2.72$, what is the in situ value of w_N?

　　Ans.: $C_c \cong 0.84$; $w_N = 42.3$ percent

11-2 Recompute the settlement of Ex. 11-2 using both C_r and C_c. If the footing is placed on the clay (2 m farther down), what is the approximate settlement?

　　Ans.: $\Delta H \cong 0.47$ m.

11-3 Given are the following void ratio vs. pressure data from laboratory tests. The initial sample height is 20.00 mm and diameter = 62.3 mm.

	Test 1	Test 2
p, kPa	void ratio e	void ratio e
0	1.02	1.11
25	0.98	1.09
50	0.975	1.085
100	0.954	1.079
200	0.880	0.942
400	0.781	0.831
800	0.688	0.762
1600	0.575	0.705
3200	—	0.638

REQUIRED: Plot e versus p and e versus log p of the assigned test and obtain:
(a) Coefficient of compressibility a_v
(b) Coefficient of volume compressibility m_v
(c) Compression index C_c
Partial ans.: Test 1 $C_c = 0.35$.

11-4 For the assigned soil test of Prob. 11-3, back compute the void ratio data using the given height and diameter to obtain corresponding strains, and plot ε versus log p and obtain C_c'.
Partial ans.: Test 1 $\varepsilon_{25} = 0.0198$; $\varepsilon_{1600} = 0.220$; $C_c' = 0.18$. and using $e_0 = 1.02$

11-5 If the sample used in Prob. 11-3 to plot and obtain C_c' or C_c came from the strata as shown in Fig. P11-5,
(a) Obtain the preconsolidation pressure p_c and determine if the soil is preconsolidated.
Ans.: Test 1 $R_d \cong 1.06$.
(b) Obtain the expected settlement ΔH if the clay stratum is 3 m thick and the average increase in pressure is $\Delta p = 25$ kPa.
Ans.: Test 1 $\Delta H \cong 4$ cm.

11-6 Given the following consolidation test data, plot e versus log p, compute C_c and C_r, and find p_c.

	Void ratio e		
p, kN/m²	First load	Unload	Reload
0	1.20	1.04	—
5	1.18	1.00	1.01
12.5	1.17	0.981	0.990
25	1.15	0.949	0.968
50	1.12	0.923	0.945
100	1.07	0.880	0.917
200	0.99	0.865	0.888
400	0.855		0.843
800			0.734
1600			0.589

Partial ans.: $p_c \cong 110$ kPa.

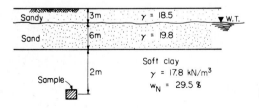

Figure P11-5

11-7 Back compute the void ratio data of Prob. 11-6 to obtain the strain ε if the sample was 20 mm $\times$ 63.5 mm diameter, and obtain C'_c, C'_r, and p_c.

11-8 If the sample used in Prob. 11-6 (or 11-7) came from in situ as shown in Fig. P11-8, is the soil preconsolidated? If the pressure profile due to the surface load is as shown, what is the expected consolidation settlement?

Partial ans.: Yes; $p_c \cong 110$ versus 41.6 kPa., $\Delta H \cong 1.7$ cm.

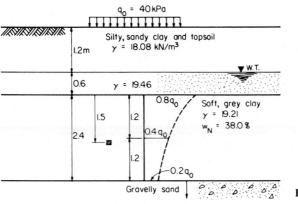

Figure P11-8

Chapter 12

Rate of Consolidation

12-1 THE COEFFICIENT OF CONSOLIDATION

The second consolidation parameter of interest from a consolidation test is the *coefficient of consolidation* c_v. This parameter is obtained from an equation to be developed in the following text material and a time value obtained from a plot of settlement (or strain) vs. logarithm time (or $\sqrt{\text{time}}$).

The equation for c_v is developed based on *one-dimensional flow* and *saturated soil conditions* (assumptions 1 and 7 of Sec. 11-2). Since consolidation under these conditions is directly dependent on the extrusion of pore water from the soil voids, one may develop the needed equations by considering continuity of flow (referring to Fig. 12-1) as

$$Q_{\text{in}} = v_y \, dx \, dz \, dt \qquad Q_{\text{out}} = \left(v_y + \frac{\partial v_y}{\partial y} \, dy \right) dx \, dz \, dt$$

and the volume change as a rate process is

$$\text{Volume change} = \frac{\partial V}{\partial t} \, dt = Q_{\text{out}} - Q_{\text{in}}$$

Equating as indicated, one obtains

$$\frac{\partial v_y}{\partial y} \, dy \, dx \, dz \, dt = \frac{\partial V}{\partial t} \, dt \qquad\qquad (a)$$

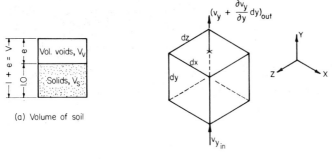

(a) Volume of soil

(b) One-dimensional flow

Figure 12-1 Idealization of soil and fluid flow for development of the one-dimensional theory of consolidation.

Since the soil grains and water are incompressible for all practical purposes and for the pressures involved, the change in soil volume must be due to a change in the volume of soil voids, or

$$\frac{\partial V}{\partial t} = \frac{\partial V_v}{\partial t} \qquad (b)$$

The definition of void ratio gives

$$V_v = eV_s$$

and differentiating with respect to time and using the product rule, obtain

$$\frac{\partial V_v}{\partial t} = e\,\frac{\partial V_s}{\partial t} + V_s\,\frac{\partial e}{\partial t}$$

Observing that $e(\partial V_s/\partial t) = 0$, since there is no change in the volume of soil grains, we have

$$\frac{\partial V_v}{\partial t} = V_s\,\frac{\partial e}{\partial t} \qquad (c)$$

From Fig. 12-1 note that $V_s/V = 1/(1 + e)$, and solving for V_s, obtain

$$V_s = \frac{V}{1 + e} = \frac{dx\,dy\,dz}{1 + e} \qquad (d)$$

Substituting Eq. (d) into Eq. (c) and multiplying by dt,

$$\frac{\partial V}{\partial t}\,dt = \frac{dx\,dy\,dz}{1 + e}\,\frac{\partial e}{\partial t}\,dt \qquad (e)$$

Substituting Eq. (e) into Eq. (a) and cancelling the product of the four differentials, obtain

$$\frac{\partial v_y}{\partial y} = \frac{1}{1 + e}\,\frac{\partial e}{\partial t} \qquad (f)$$

From assumption 3 of Sec. 11-2, that there is a linear relationship between applied pressure and volume change, and noting that volume change depends on pore water extrusion from the soil voids when the soil is saturated, obtain

$$\frac{\partial e}{\partial t} = \frac{de}{dp}\frac{\partial u}{\partial t} = a_v \frac{\partial u}{\partial t} \tag{g}$$

where $a_v = de/dp$ (or $\Delta e/\Delta p$), the coefficient of compressibility
$\quad\quad u =$ pore water pressure

From Darcy's law, the velocity of water is $v = ki$, and with the total head $h = u/\gamma_w$, we have

$$\frac{\partial v_y}{\partial y} = \frac{k}{\gamma_w}\frac{\partial^2 u}{\partial_y^2} \tag{h}$$

Substituting Eqs. (h) and (g) into Eq. (f), obtain

$$\frac{a_v}{1+e}\frac{\partial u}{\partial t} = \frac{k}{\gamma_w}\frac{\partial^2 u}{\partial y^2}$$

and rearranging,

$$\frac{\partial u}{\partial t} = \frac{k(1+e)}{a_v \gamma_w}\frac{\partial^2 u}{\partial y^2}$$

which can be written as

$$\frac{\partial u}{\partial t} = c_v \frac{\partial^2 u}{\partial y^2} \tag{i}$$

where

$$c_v = \frac{k(1+e)}{a_v \gamma_w} \tag{12-1}$$

The solution of Eq. (i) takes the form (Taylor, 1948) of a series solution to give the instantaneous value of the excess pore water pressure u at a specified point in the soil mass as

$$u = \sum_{n=1}^{\infty} \left(\frac{1}{H}\int_0^H u_i \sin\frac{n\pi y}{H}\,dy\right)\left(\sin\frac{n\pi y}{H}\right)\exp\left(-\frac{1}{4}n^2\pi^2 T\right) \tag{12-2}$$

where $n =$ any integer (generally 0, 2, 3, and 4 are sufficient)
$\quad\quad y =$ depth into a stratum of length of drainage path H
$\quad\quad H =$ length of longest drainage path in soil sample or mass
$\quad\quad T =$ dimensionless number termed a *time factor*, or

$$T = \frac{c_v t_i}{H^2} \tag{12-3}$$

$\quad\quad t_i =$ time of interest
$\quad\quad u_i =$ initial pore pressure distribution; use constant, linear variation, sine wave, or other shape

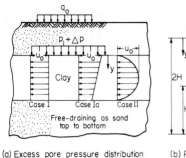
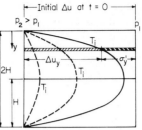
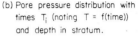

(a) Excess pore pressure distribution assumptions in consolidating clay layer with foundation load q_o causing a stress increase from p_1 to $p_1 + \Delta q$.

(b) Pore pressure distribution with times T_i (noting $T = f(\text{time})$) and depth in stratum.

Figure 12-2 Excess pore pressure distribution assumptions for an increased effective stress within a stratum and qualitative pore pressure distribution as a function of elapsed time.

For the case of constant (actually any *linear* variation) initial hydrostatic pressure, Eq. (12-2) simplifies to

$$u = \sum_{m=0}^{\infty} \frac{2u_i}{M} \sin \frac{My}{H} \exp -M^2 T \qquad (12\text{-}4)$$

where $u_i =$ constant or $u_i = u_o + u_1(H - y/H)$ as in Fig. 12-2
$\quad M = \frac{1}{2}\pi(2m + 1)$, where m is any integer from 0 to ∞

Referring to Fig. 12-2, if we apply a pressure increment Δp to a fully saturated soil which has fully consolidated ($u = 0$) under the existing pressure p_1, the new total pressure is $p_2 = p_1 + \Delta p$. The pore water carries the load at $t = 0$ (Fig. 12-2a), since drainage is not instantaneous, and $u_i = \Delta p$ for $S = 100$ percent. We also assume a thin stratum such that Δp is essentially constant with depth. At $t = 0$ the consolidation has just begun, or the percent consolidation $U = 0$ percent. At some times t_i (as in Fig. 12-2b), the pore pressure patterns (isochrones) due to drainage being more rapid at or near the free surfaces are at, or approaching, zero as shown. At any point y from the free surface one may compute the status of the pore fluid using the percent consolidation concept of the following section.

12-2 PERCENT CONSOLIDATION

At the free surfaces, $u = 0$ and consolidation is complete ($U = 100$ percent). At interior points the consolidation U_y would, by inspection of Fig. 12-2b, be

$$U_y = \frac{u_o - u_i}{u_o} = 1 - \frac{u_i}{u_o} \qquad (12\text{-}5)$$

To obtain U_y in percent, multiply the ratio by 100. One can now use Eq. (12-4) to plot a graphic display of the variation of pore pressure (or percent consolidation) with depth in a stratum of thickness $2H$ (half thickness $= H$) as

$$U_y = 1 - \sum_{m=0}^{\infty} \frac{2}{M} \sin \frac{My}{H} \exp -M^2 T \qquad (12\text{-}6)$$

Equation (12-6) is obtained by dividing Eq. (12-4) by u_o and subtracting 1, as indicated in Eq. (12-5). A solution can be easily obtained by programming Eq. (12-6) on a computer for the following variables:

$M = \frac{1}{2}\pi(2m + 1)$, which depends only on incrementing the integer m

$T = $ constant, say, 0.05, 0.1, 0.15, ..., 0.90

$y/H = 0, 0.1, 0.2, ..., 1.0$. Values larger than 1 are not needed, since the resulting curve is symmetrical about the half depth of $y = H$

Sufficiently precise results are obtained by varying m from 0 to about 4 or 5 for computing M. A plot of Eq. (12-6) for selected values of T is shown in Fig. 12-3.

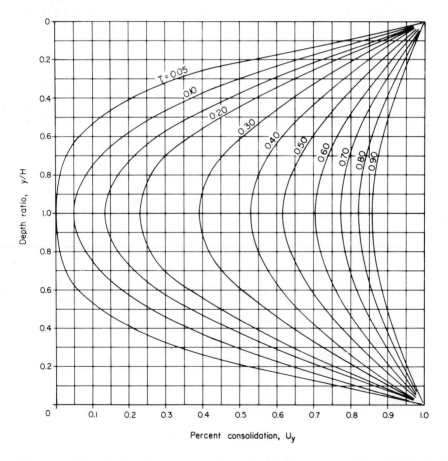

Figure 12-3 Plot of pore pressure isochrones from Eq. (12-6) for selected values of time factor T.

While plots of the type shown in Fig. 12-3 give an indication of pore pressure variations within the stratum, it is often of more immediate interest to obtain the *average* consolidation U for the entire stratum, for example, when an amount of consolidation of, say, 10, 50, or 80 percent is complete. For this estimate it is necessary to integrate U_y for the entire stratum thickness as

$$U = 1 - \frac{\int_0^{2H} u_i \, dy}{\int_0^{2H} u_o \, dy} \tag{12-7}$$

Substituting Eq. (12-3) in the numerator and u_o = constant in the denominator of Eq. (12-7) gives

$$U = 1 - \sum_{m=0}^{\infty} \frac{y}{M^2} \exp{-M^2 T} \tag{12-8}$$

Often the first term (with $m = 0$) provides a solution of sufficient precision for practical purposes, and tabular values can be made once and for all in terms of percent consolidation U versus T by rearranging and solving Eq. (12-8) for T, to obtain

$$T \cong \frac{\ln{(2/M^2)} - \ln{(1-U)}}{M^2} \tag{12-9}$$

Note that the time factor T is not defined when $U = 1.00$ (100 percent consolidation, which theoretically never occurs), since the logarithm of 0 is ∞. Selected solutions of Eq. (12-8) are given in Table 12-1 (with $m > 0$) for several assumed

Table 12-1 Time factors for percent consolidation and indicated pore pressure distributions shown

U, percent	Cases I and Ia	Case II
0	0.000	0.000
10	0.008	0.048
20	0.031	0.090
30	0.071	0.115
40	0.126	0.207
50	0.197	0.281
60	0.287	0.371
70	0.403	0.488
80	0.567	0.652
90	0.848	0.933
100	∞	∞

Case I   Case Ia   Case II

initial pore pressure distributions caused by a stress increase Δp as shown. Superposition of Cases I and II or Cases Ia and II can provide solutions for other pore pressure distributions.

Of some interest, the basis of the $\sqrt{\text{time}}$ method used in Sec. 12-4, the percent consolidation U for one-way drainage of a layer of soil extending to an infinite depth ($2H \rightarrow \infty$) is

$$U = \sqrt{\frac{T}{\pi}} \qquad (12\text{-}10)$$

12-3 METHODS OF OBTAINING THE TIME OF INTEREST FOR COMPUTING c_v

A Semilog Plot

Usually one plots the settlement (or alternatively, dial gage readings or strain) vs. logarithm time, as illustrated in Fig. 12-4, for the purpose of obtaining the time at a given percent of consolidation. The same shaped curve will be obtained whether one plots dial readings vs. log time or strain vs. log time.

The use of settlement or strain vs. log time curves requires finding the initial dial reading D_o (dial reading at $t = 0$, which cannot be plotted since $\log 0 = \infty$). The initial dial reading D_o may be taken as the actual dial reading at $t = 0$ for that load increment; however, the remainder of the initial branch of the time settlement curve may indicate that the readings started from some other apparent dial

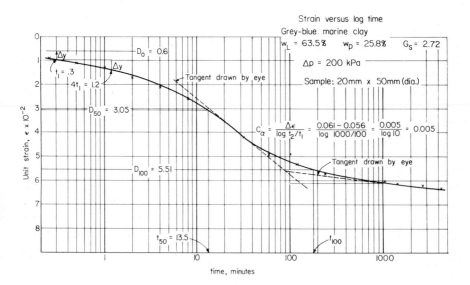

Figure 12-4 Plot of strain vs. log time and data obtained from plot. Note that a plot of dial reading (or settlement) would give a curve of identical shape.

reading due to dial gage slack, seating, and other factors. If the initial branch of the curve is parabolic or nearly so as observed by eye, one may obtain the apparent initial dial reading D_o as follows:

1. Find some time t_i of, say, 0.1 or 0.2 min (illustrated in Fig. 12-4).
2. Take a time $4t_i$ (i.e., $4 \times 0.1 = 0.4$ or $4 \times 0.2 = 0.8$ min).
3. Obtain the vertical offset from t_i and $4t_i$ as Δy.
4. Lay off the Δy distance above t_i to represent D_o. This may be repeated one or more times, and the average of the several points is taken as D_o.

To find the dial reading at $U = 100$ percent consolidation (D_{100}), draw a tangent to the middle branch of the settlement curve by eye and another tangent to the end branch of the curve. The intersection of these two tangents is arbitrarily taken as D_{100}, and a projection to the curve and then down to the time axis is taken as t_{100}. The dial reading for D_{50} is usually used to obtain t_{50}, however, and is obtained as

$$D_{50} = \frac{D_0 + D_{100}}{2}$$

and t_{50} can be found by entering at D_{50} and projecting to the curve and down to the time axis as shown in Fig. 12-4.

B $\sqrt{\text{Time}}$ Plot

The plot of $U = 2\sqrt{T/\pi}$ is essentially a straight line in the initial stage of consolidation, before it curves and becomes asymptotic at $U = 1$. Taylor (1948) proposed that one could use this observation as an alternative method of presenting time-settlement curves to obtain the time at various percents of consolidation. That is, we can plot settlement versus $\sqrt{t}$ as in Fig. 12-5 and *draw as a best fit* a straight line to the settlement ordinate of the first several data points, locating point A and continuing this straight line to the time axis to locate a point b. Since this neglects the constant $2/\sqrt{\pi} \simeq 1.15$ (actually 1.13), we draw a second line $\overline{Ac}$ from A which is 15 percent larger than $\overline{Ob}$. The point d where the experimental curve intercepts Ac is at $U \cong 0.9$ (approximately 90 percent consolidation). Now that we know the settlement or strain for 90 percent consolidation (the distance Ae of Fig. 12-5), we can obtain the time at 90 percent consolidation by projecting to the time axis, and $t_{90} = N_{90}^2$. We can find the settlement or strain for 50 percent consolidation by assuming that D_o occurs at point A (this is simpler than the semilog plot) and that $\frac{5}{9}Ae$ is the settlement corresponding to t_{50}; then rearranging Eq. (12-3) and using T factors from Table 12-1,

$$c_v = \frac{0.848H^2}{t_{90}} \quad \text{or} \quad c_v = \frac{0.197H^2}{t_{50}}$$

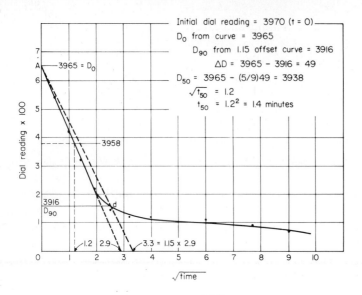

Initial dial reading = 3970 (t = 0)
D_0 from curve = 3965
D_{90} from 1.15 offset curve = 3916
ΔD = 3965 − 3916 = 49
D_{50} = 3965 − (5/9)49 = 3938
$\sqrt{t_{50}}$ = 1.2
t_{50} = 1.2^2 = 1.4 minutes

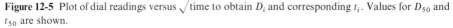

Figure 12-5 Plot of dial readings versus $\sqrt{\text{time}}$ to obtain D_i and corresponding t_i. Values for D_{50} and t_{50} are shown.

Both values computed above should, of course, be equal—in actual practice they will be close. These values of c_v should compare reasonably well with those from the semilog plot.

With c_v computed, one can estimate the field consolidation time as

$$t_{i(\text{field})} = \frac{T_i H_{\text{field}}^2}{c_v} \tag{12-11}$$

12-4 RATE OF CONSOLIDATION BASED ON STRAIN

Experimental evidence indicates that there is no constant of proportionality between Δp and u_i. That is, the value of u_i at $t = 0$ for any pressure increment Δp is not

$$u_i = \Delta p$$

as commonly assumed, but depends on the stress-strain relationships of the soil.† Since these are, in general, nonlinear, and since Δp is not constant with depth,

---

† This can be observed with consolidometers equipped with pore pressure equipment. It usually takes some time after applying a load increment before a large change in u is noted.

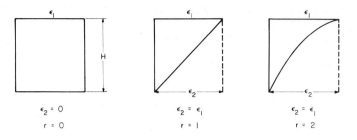

Figure 12-6 Strain distribution assumption in any increment of stratum thickness H as defined by the integer r.

except possibly for very thin laboratory samples (2 to 3 cm thick), a procedure should be used which incorporates the stress (and strain) into the solution. Janbu (1965) proposed using the soil strain at various depths to obtain strain profiles, as in Fig. 12-6. The percent consolidation U could be computed using

$$U = \frac{U_o - f_s F_r(T)}{1 - f_s}$$

where $f_s = \dfrac{r\varepsilon_2}{(1+r)\varepsilon_1}$

r = integer = 0, 1, and 2
ε_1 = strain at top of layer
ε_2 = strain at base or midheight of layer, depending on drainage
U_o = value of percent classical consolidation, Eq. (12-5)

$$F_r = 1 - 2(r+1) \sum_{N=0}^{\infty} \frac{\sin^{2+r} N}{N^{2+r}} \exp -N^2 T$$

Several values of F_r are as follows:

T	$r = 0$	$r = 1$	$r = 2$
0.01	0.1128	0.0199	0.0276
0.10	0.3568	0.1977	0.2285
1.00	0.9313	0.9125	0.9164

Figure 12-6 gives the significance of r, ε_1, and ε_2.

Using superposition the six cases shown on Fig. 12-9 can be obtained, and from a complete table of F_r the six curves of Figs. 12-7 and 12-8 can be drawn.

The most convenient method of computing the rate of consolidation based on strain is that used in the table shown as Fig. 12-9. The steps are as follows:

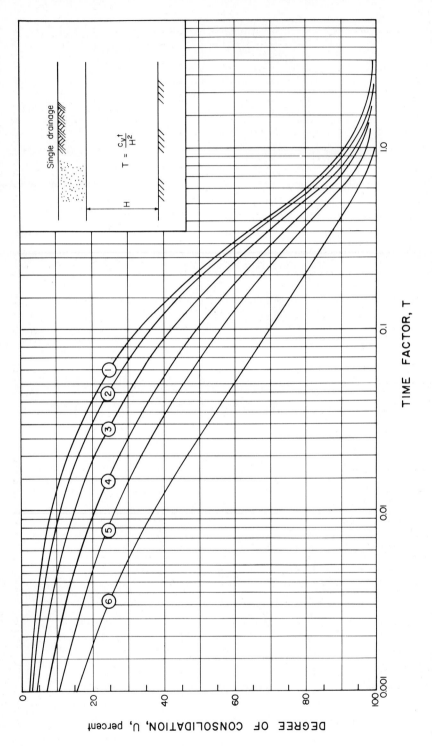

Figure 12-7 Time factor vs. degree of consolidation curves for single drainage. Note that curve 2 is a plot of Case I of Table 12-1. Refer to Fig. 12-6 for pore pressure distributions for curve numbers. (*After Janbu, 1965.*)

Within the figure:

Single drainage

$$T = \frac{c_v t}{H^2}$$

H

TIME FACTOR, T

DEGREE OF CONSOLIDATION, U, percent

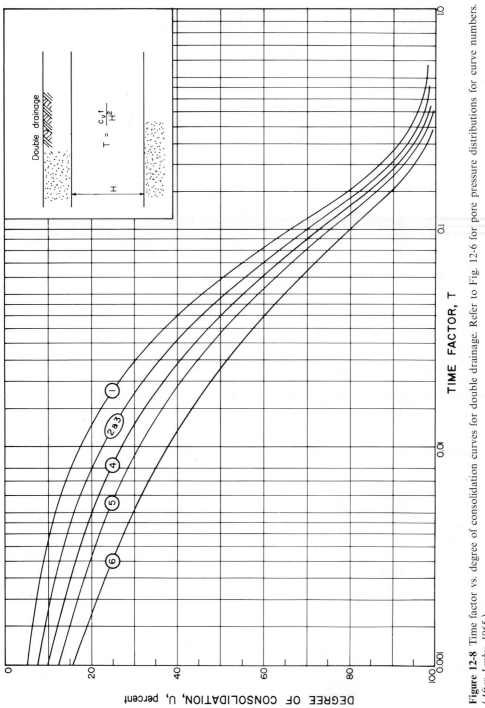

Figure 12-8 Time factor vs. degree of consolidation curves for double drainage. Refer to Fig. 12-6 for pore pressure distributions for curve numbers. (*After Janbu, 1965.*)

331

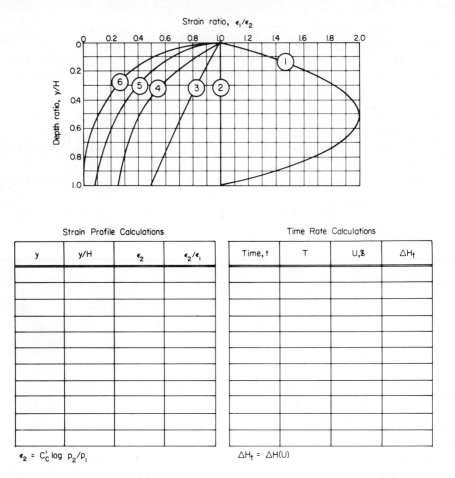

Strain Profile Calculations

y	y/H	ϵ_2	ϵ_2/ϵ_1

$\epsilon_2 = C'_c \log p_2/p_1$

Time Rate Calculations

Time, t	T	U,%	ΔH_t

$\Delta H_t = \Delta H(U)$

Figure 12-9 Table for calculating rate of settlement. Plot strain ratio on the figure to find the correct T versus U curve. Use c_v and arbitrary selected times t to compute T. Enter the T versus U curve and find U. Using the total settlement, compute ΔH_t.

1. Fill in the table on the left side using

y = depth below top of consolidating layer

y/H = ratio of y to total layer thickness

H = thickness of consolidating layer

ε_2 = strain in element of thickness y computed using the following:

$$\varepsilon_2 = C'_c \log \frac{p_o + \Delta p}{p_o}$$

and p_o = average effective overburden pressure in increment of y

Δp = average increase in effective pressure in increment

C'_c = compression ratio as obtained from Eq. (11-8) or a plot of ε versus log p

Note that the general considerations of preconsolidation pressure may apply to adjust the above equation

ε_1 = strain at *top* of consolidation layer, computed in the same manner as for ε_2 above, using p_o and Δp at the top of the consolidating soil layer

Values of $\varepsilon_2/\varepsilon_1$ are plotted on the graph above the table, and that numbered curve which most nearly coincides with the plot is used to identify the curve of either Fig. 12-7 or Fig. 12-8, depending on stratum drainage. If the plot falls between curves, it will be necessary to interpolate.

2. Using Eq. (12-3) and c_v from the laboratory test, for arbitrary selected values of time, compute T.
3. Using the appropriate curve of Fig. 12-7 or 12-8 from step 1 and the T values from step 2, find the corresponding percent(s) of consolidation U.
4. Compute the settlement for this time interval t as

$$\Delta H_t = \Delta H(U)$$

where ΔH = total stratum settlement as computed in Chap. 11.
5. Make a plot of t versus ΔH_t as illustrated in Ex. 12-6.

This method of obtaining the rate of settlement is recommended for thicker consolidating strata, say, greater than 5 m. For thinner layers the average method with values obtained from Table 12-1 or Fig. 12-3 will be sufficiently correct. Several reasons for this include:

1. The difficulty of obtaining a reliable value (or values) of c_v. In the laboratory, when thin samples are consolidated an amount ΔH of 2 mm in a H of 20 mm, this is a 10 percent change in height and a large change in void ratio and the resulting coefficient of permeability k. In the field a 20-mm change in a thickness of, say, 2 or 3 m is a negligible change in void ratio and in k; thus, c_v may be considerably in error for the field rate of consolidation.
2. The coefficient of consolidation c_v is dependent on the viscosity of the pore fluid, which, in turn, depends on the temperature of the test. Laboratory tests are commonly run at the temperature of the laboratory (20 to 25°C). In the field the groundwater temperature is much less—generally on the order of 8 to 12°C.
3. The difficulty of obtaining the increase in effective stress Δp at the various depths in the strata. The Boussinesq method (as in Chap. 11) is commonly used for obtaining the stresses. The Boussinesq method, however, idealizes the soil to a homogeneous, isotropic, elastic, semi-infinite half space. A layered soil mass does not satisfy these assumptions, and the amount of deviation from the ideal soil mass may be substantial—or at least enough to not justify elaborate or refined computations.

12-5 ILLUSTRATIVE EXAMPLES

The following examples will be presented to illustrate the material presented in this chapter.

Example 12-1 Data from a consolidation test are as follows:

$t_{50} = 12.2$ min (see Fig. 12-4 for method of obtaining)
Load increment $= 1600$ kPa (approx. 8 tons/ft^2)
Conditions are such that Case I, constant pore pressure through H, exists at $t = 0$ and drainage is on *both* faces of the sample.
The average one-half sample height is $(D_f + D_o)/2 = 0.7913$ cm.

REQUIRED Compute c_v for this load increment.

SOLUTION

$$c_v = \frac{TH^2}{t_{50}}$$

From Table 12-1, at $U = 50$ percent, which corresponds to t_{50}, obtain $T = 0.197$.

$$c_v = \frac{0.197(0.7913)^2}{12.2} = 0.0101 \text{ cm}^2/\text{min}$$

Example 12-2 Data from a consolidation test are as in Ex. 12-1 except that the sample drains from *one* side.

REQUIRED Compute the coefficient of consolidation.

SOLUTION With one-side drainage, $H = $ full sample height, but T remains 0.197 as obtained from Table 12-1 at $U = 50$ percent.

$$H = 2(0.7913) = 1.5826 \text{ cm}$$

$$c_v = \frac{0.197(1.5826)^2}{12.2} = 0.0404 \text{ cm}^2/\text{min}$$

From these two computations (Exs. 12-1 and 12-2), one may conclude that doubling the length of the drainage path will increase the time of consolidation four times.

Example 12-3 Computation of time for field consolidation.

GIVEN Laboratory data of Ex. 12-1: $H = 0.7913$ cm (one-half height), $t_{50} = 12.2$ min, $c_v = 0.0101$ cm^2/min (computed from laboratory data).

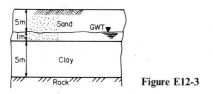

Figure E12-3

REQUIRED Length of time it will take for consolidation settlement to occur for the field conditions shown in Fig. E12-3.

SOLUTION Inspection of the figure indicates drainage on one face; therefore, $H = 5$ m. For full consolidation, $U \to 100$ percent, obtain $T = 0.848$ from Table 12-1.

$$t = \frac{TH^2}{c_v}$$

$$= \frac{0.848(5 \times 100)^2}{0.0101} = 2.099 \times 10^7 \text{ min} = 14\,576 \text{ days} = 39.9 \text{ years}$$

Example 12-4 Finding the instantaneous pore pressure at a point in a soil mass.

GIVEN A load is applied to the soil of Ex. 12-3 (see Fig. E12-3). At the instant of load application, a piezometer located 2 m below the top of the clay layer indicates an excess pore pressure Δu of 2 m, as shown in Fig. E12-4. Assume Case I conditions (Table 12-1) for the initial pore pressure distribution in the clay layer.

REQUIRED What is the pore pressure at this point when settlements corresponding to $T = 0.10$ and $T = 0.50$ have taken place?

SOLUTION Referring to Fig. 12-3 at $y/H = \frac{2}{5} = 0.4$, obtain the approximate values of U_y as

T	U_y
0.10	0.375
0.50	0.780

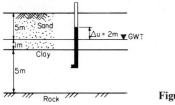

Figure E12-4

The corresponding pore pressures in meters of water are computed as follows:

$$u_{10} = 2 - 2(0.375) = 1.25 \text{ m}$$

$$u_{50} = 2 - 2(0.78) = 0.44 \text{ m}$$

Note that the values of U_y of 0.375 and 0.78 are the percent consolidations at the point $y = 2$ m below the top of the clay stratum.

Example 12-5 Computation of the average consolidation at selected values of T.

GIVEN Time factors of $T = 0.10$ and 0.50, as in Ex. 12-4.

REQUIRED The average percent consolidation U of the soil stratum when the pore pressures of $u = 1.25$ and $u = 0.44$ m of water remain as computed in Ex. 12-4.

SOLUTION Draw a curve of U versus T using the data of Table 12-1 and Case I as in Fig. E12-5.
From Fig. E12-5, obtain $U = 0.35$ for $T = 0.10$ and $U = 0.76$ when $T = 0.50$, or

$$T = 0.10 \qquad \text{Percent consolidation is 35 percent}$$

$$T = 0.50 \qquad \text{Percent consolidation is 76 percent}$$

These average values compare to 37.5 and 78 percent, respectively, in Ex. 12-4. One would expect the average values to be smaller than point values near free surfaces, larger than point values at the maximum depth of the drainage path, and approximately the same near the middle of the drainage path.

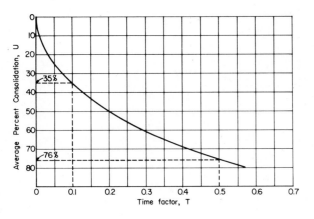

Figure E12-5

Example 12-6

GIVEN The soil profile of Fig. E12-6a and the e versus $\log p$ curve of Fig. E12-6b. Assume that the several plots of settlement versus log time give

$$c_v = 0.032 \text{ cm}^2/\text{min}$$

Unit weights on the soil profile can be obtained using displacement methods of placing a sample of known weight in a container of known volume and

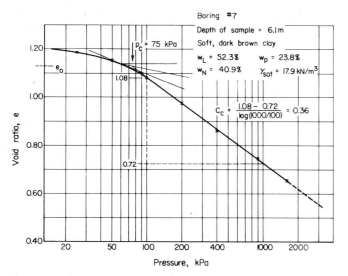

Figure E12-6a

Figure E12-6b

filling it with water to obtain the sample volume as $V = V_c - V_w$, from which the unit weight is readily computed.

REQUIRED Compute the expected total settlement for the footing load shown and make a plot of time versus settlement so that settlement rates can be predicted.

SOLUTION From the e versus log p curve of Fig. E12-6b, the value of p_c can be plotted on the effective stress profile of Fig. E12-6a, and the soil appears to be *normally* consolidated.

Spot check the data for reasonableness:

$$C_c \cong 0.009(w_L - 10) = 0.009(52.3 - 10) = 0.38 \text{ versus } 0.36 \quad \text{(O.K.)}$$

$e_o \cong 1.13$ from e versus log p curve;

$$e_o \cong w_N G_s$$

from which, solving for G_s, $G_s \cong 1.13/0.409 = 2.76$ which is not unreasonable for clay

With $G_s = 2.76$ and a block diagram as shown in Fig. E12-6c, and taking $V_s = 1.0$, we have

$$\gamma_{sat} = \frac{1.13 + 2.76}{1 + 1.13}(9.807) = 17.9 \text{ kN/m}^3 \text{ versus } 17.9 \quad \text{(O.K.)}$$

therefore, the data appear satisfactory and we shall continue. With 8 m of soil involved in consolidation, use the method of Sec. 12-4 instead of a single layer. Divide the soil into four sublayers of 2 m each as shown in Fig. E12-6a.

Step 1 Calculate the effective overburden pressure p_o at the top of the stratum and at the midheight of each sublayer:

Point	y, m		p_o	
1 (top)	2.6	$1.8(16.9) + 0.8(20.3 - 9.8)$	=	38.8 kPa
2	3.6	$38.8 + 1(17.9 - 9.8)$	=	46.9
3	5.6	$46.9 + 2(8.1)$	=	63.1
4	7.6	$63.1 + 2(8.1)$	=	79.3
5	9.6	$79.3 + 2(8.1)$	=	95.5
Bottom	10.6	$95.5 + 1(8.1)$	=	103.6

Note that the points are shown on the soil stress profile of Fig. E12-6a.

Figure E12-6c

Step 2 Obtain e_o from Fig. E12-6b as $e_o \cong 1.13$, obtain $C_c = 0.36$ from the computation shown on Fig. E12-6b, and compute C_c' as

$$C_c' = \frac{C_c}{1 + e_o} = \frac{0.36}{1 + 1.13} = 0.17$$

One might adjust e_o for each H_i, but this is generally too much refinement.

Step 3 Obtain Δp due to the footing load of 1150 kN on a footing of 3×3 m using the Boussinesq method of Sec. 10-7, assuming a homogeneous soil and using a computer program to treat the footing as a series of 36 point loads (unit area of 0.5×0.5 m) with a point load value on each unit area of $1150/36 = 31.94$ kN. Summing the contribution from the 36 unit areas gives the Δp values shown on the stress profile of Fig. E12-6a.

Step 4 Develop data to fill in the left side of the table in Fig. E12-6d. In the table, $y =$ distance below top of consolidating cohesive layer; $H =$ total thickness of consolidating layer ($= 8$ m); ε_1 and $\varepsilon_2 =$ strain values using C_c' from step 3, and computations are as follows:

Point	H_i	p_o	At midheight of any sublayer Δp	ε_2	ΔH_i
1	—	38.8 kPa	52.3 kPa	$0.063 = \varepsilon_1$	—
2	2 m	46.9	33.0	0.039	0.078
3	2	63.1	15.6	0.016	0.032
4	2	79.3	9.0	0.008	0.016
5	2	95.5	5.8	0.004	0.008
Bottom	—	103.6	4.7	0.003	—
				$\sum \Delta H_i = \Delta H = 0.134$ m	

Typical computations:

$$\varepsilon_1 = C_c' \log \frac{p_o + \Delta p}{p_o} = 0.17 \log \frac{38.8 + 52.3}{38.8} = 0.063 \text{ m/m}$$

$$\varepsilon_2 = 0.17 \log \frac{46.9 + 33.0}{46.9} = 0.039 \text{ (second line of above table)}$$

$$\Delta H_1 = H_1(\varepsilon_2) = 0.078 \text{ m (also second line of above table)}$$

The remaining table entries are computed similarly. Note that the point number identifies the point plotted on the stress profile in Fig. E12-6a.

These data can be used to fill in the left side of the table as shown. Plot y/H versus $\varepsilon_2/\varepsilon_1$ as shown on the graph above the table of Fig. E12-6d to obtain the curve number.

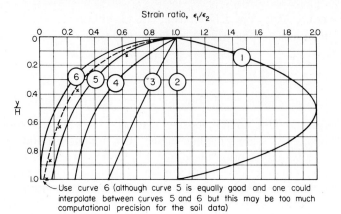

Use curve 6 (although curve 5 is equally good and one could interpolate between curves 5 and 6 but this may be too much computational precision for the soil data)

Strain Profile Calculations				Time Rate Calculations			
y, m	y/H	ϵ_2	ϵ_2/ϵ_1	Time, t	T	U,%	ΔH_t
1	0.125	0.038	0.603	0.1 yr.	0.003	25	0.033
3	0.375	0.016	0.254	0.5	0.013	40	0.054
5	0.625	0.008	0.127	1.0	0.026	52	0.070
7	0.875	0.004	0.063	2.0	0.053	66	0.088
8	1.000	0.003	0.048	4.0	0.105	85	0.114
				6.0	0.157	94	0.126
$\epsilon_1 = 0.063$							

$\epsilon_2 = C_c' \log p_2/p_1$

$y/H = 1/8 = 0.125$

$\epsilon_2/\epsilon_1 = 0.038/0.063 = 0.603$

$\Delta H_t = \Delta H(U)$ $\Delta H_t = 0.134(.25) = 0.033$ m

$T = 0.0263t = 0.0263(.1) = 0.003$

Figure E12-6d

Step 5 Fill in the right side of Fig. E12-6d as follows:
  (a) Rearrange Eq. (12-11) to give

$$T = \frac{c_v t_i}{H^2}$$

where t = time in field (assume several values sufficient to draw curve)
  H = full thickness of stratum in field (8 m in this example)
  $c_v = 0.032$ cm^2/min $= 0.032(1440 \times 365)/(100)^2 = 1.682$ m^2/year

from which

$$T = \frac{1.682 t_i}{8^2} = 0.0263 t_i$$

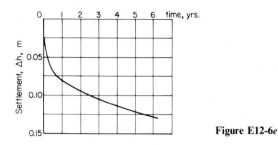

Figure E12-6e

Use the values of t_i of 0.1, 0.5, 1, etc., as shown in the table.

(b) For the computed values of T using t_i, enter Fig. 12-8 (double drainage, from Fig. E12-6a) and obtain the corresponding U values in percent. For $t_i = 0.1$,

$$T = 0.0263(0.1) = 0.003$$

Entering Fig. 12-8 at $T = 0.003$, project to curve 6 (from step 4) and obtain

$$U = 25 \text{ percent}$$

(c) Compute the amount of settlement at the various times t_i using the equation shown in Fig. E12-6d,

$$\Delta H_t = \Delta H(U)$$

In the example from step (b), with the value of ΔH obtained from step 4 of 0.126 m,

$$\Delta H_t = 0.134(0.25) = 0.034 \text{ m}$$

Step 6 Plot t_i versus ΔH_t as shown in Fig. E12-6e above.

12-6 CONSOLIDATION RATES FOR LAYERED MEDIA

When the stratum consists of several layers of soil subject to consolidation theory, the total settlement can be computed by obtaining C_c or C_c' from consolidation tests on samples from the several strata, computing the ΔH_i values for the individual strata, and summing to obtain

$$\Delta H = \sum_1^n \Delta H_i$$

The rate of consolidation will be quite complicated, however, as the settlements of the lower or interior strata depend on c_v of the particular stratum as well as on c_v of the exterior strata. It is evident that drainage (see Fig. 12-10) must travel through both the stratum under consideration and the exterior strata to the free surface.

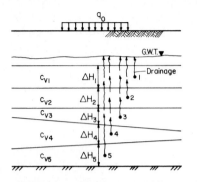

Figure 12-10 Layered soil with one-way drainage. Drainage from stratum 5 must travel through the other four strata at a rate dependent on $c_{v(i)}$ of the other strata.

Drainage, then, depends on the coefficient of permeability, void ratio, stress response, and applied stress in the strata, and from Eq. (12-1),

$$c_v = f(k, e, a_v, \gamma_w)$$

From this it is evident that to even approximate a solution, consolidation tests are required for each stratum so that values of c_v and C_c for that stratum can be obtained. Theoretical solutions for this case are beyond the scope of this text, and the reader should refer to Rowe (1964), De Leeuw (1965), Martins (1965), or Poskitt (1969) for analytical solutions. Finite difference solutions are also available and require the use of a computer. Finite difference solution references include Abbot (1960) and Gibson and Lumb (1952).

12-7 THREE-DIMENSIONAL CONSOLIDATION

Three-dimensional consolidation will occur in situations involving drainage to a central source, such as sand drains used beneath fills to accelerate drainage by reducing the drainage path and therefore accelerate consolidation. This type of analysis is also beyond the scope of this text, and the reader is referred to references such as Berry and Wilkinson (1969), Rowe (1964), and Aboshi and Monden (1961).

12-8 SUMMARY

This chapter has presented a means of evaluating the rate of one-dimensional consolidation using the coefficient of consolidation c_v. The coefficient of consolidation is obtained from a plot of settlement versus log time, strain versus log time, or settlement versus $\sqrt{\text{time}}$ by using the plot to obtain the time t_i for a percentage of consolidation to take place.

The analytical theory of consolidation was used to obtain a dimensionless time factor T, which was found to depend on an assumption of pore pressure distribution within the soil layer under some pressure change Δp and for some percent of consolidation U.

The rate of consolidation also depends on the length of the drainage path H, which is dependent on the type of laboratory consolidation test (fixed or floating ring) and, in the field, on the location of the groundwater table and the arrangement of free-draining and consolidating soil layers.

With t_i, T, and H, the coefficient of consolidation was computed as

$$c_v = \frac{TH^2}{t_i}$$

Not so directly presented is the fact that larger loads settle faster than smaller ones. This is because (1) strains are larger, and (2) the excess pore pressure, being larger, promotes faster drainage ($v = ki$). This latter aspect can be used to speed up preload consolidation by using a load that is larger than the service load.

HOMEWORK PROBLEMS

12-1 Redo Ex. 12-1 if $t_{50} = 8.1$ min.

12-2 Redo Ex. 12-3 using the value of c_v from Prob. 12-1 above ($t_{50} = 8.1$ min).

12-3 Redo Ex. 12-4 for a piezometer at a 3-m depth instead of 2 m.

12-4 Redo Ex. 12-4 for a piezometer at a 5-m depth. When will U be 10 percent at the 5-m depth using $c_v = 0.0101$ cm^2/min?

12-5 Redo Ex. 12-5 for the piezometer at 5 m into the soil instead of 2 m.

  Ans.: $U_{av} = 35$ percent for $T = 0.10$ when $U_i = 5$ percent.

For Probs. 12-6, 12-7, and 12-8, use the data in Table P12-1 as appropriate.

12-6 Using data from Table P12-1 as required,

  (a) Draw ε versus log t curves and compute c_v. Average the values from the several curves for a final value.

  (b) Draw a curve of ε versus log p and compute C'_c.

12-7 For data of Table P12-1,

  (a) Draw settlement versus log t curves and compute c_v.

  (b) Draw e versus log p curves and compute C_c and C'_c.

12-8 For the data of Table P12-1,

  (a) Draw settlement versus $\sqrt{\text{time}}$ curves and compute c_v.

  (b) Draw a curve of e versus p and obtain a_v and m_v.

12-9 Redo Ex. 12-6 if the footing load is 792 kN.

Table P12-1 Time–dial reading data for a consolidation test*

Time	25 kPa	Time	50 kPa	Time	100 kPa	Time	200 kPa	Time	400 kPa
0	0000	0	305	0	570	0	975	0	1490
0.25	102	0.25	386	0.25	663	0.25	1051	0.25	1544
0.50	121	0.50	397	0.50	681	0.50	1065	0.50	1566
1.0	133	1.0	410	1.0	701	1.0	1079	1.0	1595
2.0	154	2.0	428	2.0	726	2.0	1107	2.0	1626
4.0	184	4.0	451	4.0	757	4.0	1146	4.0	1674
8.0	208	8.0	468	8.0	803	8.0	1190	8.0	1727
16	229	16	493	16	841	32	1321	16	1800
32	244	32	510	32	879	60	1370	60	1976
60	254	231	543	60	902	120	1408	137	2045
126	269	406	553	120	925	240	1434	256	2075
250	274	600	559	285	944	1150	1465	1276	2115
883	285	1585	570	415	960	1440	1471	2510	2130
1043	296	1746	570	1440	972	2756	1481	5761	2154
1440	302			4272	975	4320	1490		
2320	305								

Time	800 kPa	Time	1600 kPa	Time	3200 kPa
0	2154	0	2652	0	3300
0.25	2164	0.25	2703	0.25	3317
0.50	2195	0.50	2721	0.50	3373
1.0	2205	1.0	2736	1.0	3395
2.0	2253	2.0	2771	2.0	3431
4.0	2296	4.0	2813	4.0	3475
8.0	2362	8.0	2876	8.0	3537
32	2487	16	2954	16	3603
60	2542	32	3039	32	3684
224	2602	60	3128	60	3755
1164	2633	105	3183	147	3828
1539	2642	360	3245	211	3838
2664	2652	1395	3278	475	3871
		1640	3281	1465	3892
		2886	3300	4725	3912
				5670	3913

* Soil sample data: $G_s = 2.73$; $w_L = 62.1$ percent; $w_p = 27.1$ percent;
$H = 25$ mm; diameter $= 50$ mm;
$H_f = 25 - 3913(0.0025) = 15.218$ mm;
dial gage: 1 div $= 0.0025$ mm

Chapter 13

Shear Strength of Soils

13-1 INTRODUCTION

Chapters 10, 11, and 12 have been concerned with means of determining stresses in a soil mass and the effect of these stresses in producing long-term (consolidation) settlements. This and the following chapter will be concerned with evaluation of soil shear strength parameters and selected stress-strain considerations.

Soil invariably fails under a combination of normal compressive and shear stresses on the failure plane, with the normal stresses providing a portion—up to all—of the friction or shear resistance. In a soil mass, as shown in Fig. 10-5b, the soil in compression beneath the loaded area tends to bulge from the Poisson's ratio effect, which in turn develops a bulb-shaped zone of stresses with both normal and shearing stresses developed along the perimeter. Summing shear forces in the vertical direction gives static equilibrium, and it is self-evident that horizontal components of the shear forces cancel.

Shear strength evaluation is necessary in most soil stability problems. These problems include:

Providing adequate slopes for embankments or excavations (including road cuts)
Determining the load a soil can safely carry, including loads from embankment
   fills and levees on the underlying soil
Determining the bearing capacity for spread footings and mat foundations
Determining the shear resistance developed between the soil and piles or caissons

We will find that shear strength is not a unique value but is heavily influenced in situ by environmental factors such as loading, unloading, and, particularly, changes in water content. In the laboratory, test methods, sample disturbance, and load (or strain) rate markedly influence the strength value obtained. These factors will be examined in some detail in the following sections.

The first hypothesis on the shear strength of a soil was presented by Coulomb (ca. 1773) as

$$s = c + \sigma_n \tan \phi \tag{13-1}$$

Terzaghi (ca. 1925) pointed out the necessity for considering the effect of pore water pressure on soil strength. Hvorslev (1937) used laboratory data to verify the use of effective stress parameters to give what is now often called the Coulomb-Hvorslev shear strength equation,

$$s = c' + \sigma' \tan \phi' \tag{13-2}$$

where s = shear strength [both Eqs. (13-1) and (13-2)]

c = soil cohesion; c' = effective value when σ' is used. Cohesion is the interparticle attraction effect; it is independent of normal stress but depends considerably on water content and strain rate

σ = normal stress on the critical plane; $\sigma' = \sigma - u$ = effective normal stress as obtained using Eq. (2-21)

ϕ = angle of internal friction; ϕ' = effective angle of internal friction as obtained using σ'. Note that tan ϕ = coefficient of friction as used in any friction problem

An inspection of either Eq. (13-1) or (13-2) shows that it is the equation of a straight line which can be plotted tangent to Mohr's circle to represent failure conditions, as shown in Fig. 13-1.

Cohesionless soils tested in the usual working range of pressures, say 7 to 150 kPa, produce an experimental plot of the failure envelope which tends to be very nearly linear, according to Eq. (13-2), with a cohesion intercept $c \cong 0$. When the normal pressures increase to values that result in crushing of the soil grains at the contact points, the experimental plot becomes nonlinear.

The failure envelope for *cohesive soils* described by Eq. (13-1) is generally not straight but nonlinear over almost the entire testing range. Effective stress tests on normally consolidated cohesive soils described by Eq. (13-2) produce an effective

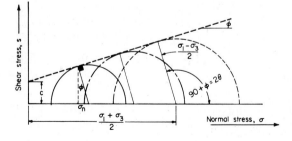

Figure 13-1 Mohr-Coulomb failure envelope for obtaining the limiting soil parameters c and ϕ.

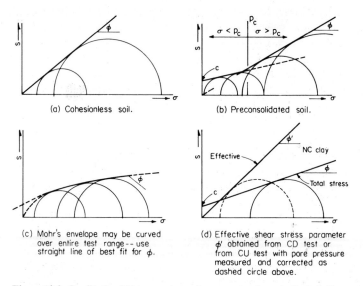

(a) Cohesionless soil.

(b) Preconsolidated soil.

(c) Mohr's envelope may be curved over entire test range -- use straight line of best fit for ϕ.

(d) Effective shear stress parameter ϕ' obtained from CD test or from CU test with pore pressure measured and corrected as dashed circle above.

Figure 13-2 Qualitative failure envelopes for several soils as indicated. Care must be taken to obtain a representative ϕ (and c) when the failure envelope is curved and when the circles require considerably more interpretation to develop the Mohr envelope than shown here. (*a*) Cohesionless soil; (*b*) preconsolidated cohesive soil. (*c*) Mohr's envelope may be curved over the entire test range— use the straight line of best fit for ϕ. (*d*) Effective shear stress parameter ϕ' obtained from CD test or from CU test with pore pressure measured and corrected as dashed circle above.

stress envelope which generally passes through the origin ($c \cong 0$), as illustrated in Fig. 13-2*d*, and is reasonably linear. The slope of the failure envelope for preconsolidated cohesive soils contains a discontinuity at approximately the preconsolidation pressure, as illustrated in Fig. 13-2*b*. The Mohr's circles qualitatively illustrated in Fig. 13-2 may be obtained from either direct shear tests or triaxial tests, described in the next section.

13-2 SOIL TESTS TO DETERMINE SHEAR STRENGTH PARAMETERS

The soil tests commonly employed to obtain the strength parameters include (in order of increasing costs):

1. Unconfined compression or q_u tests. The compressive strength obtained from this test is always identified as q_u. This test is also called an unconsolidated-undrained test, or simply an undrained test or U test. The *undrained shear strength* is usually identified as s_u.
2. Direct shear tests
3. Confined compression or triaxial tests
   (*a*) Triaxial strain (three-dimensional)
   (*b*) Plane strain—requires special equipment

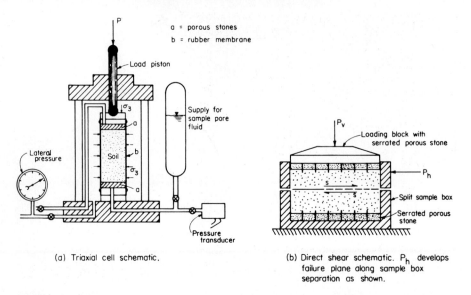

(a) Triaxial cell schematic.

(b) Direct shear schematic. P_h develops failure plane along sample box separation as shown.

Figure 13-3 Schematics of triaxial and direct shear tests. (See Fig. 13-5a for schematic of unconfined compression test.) (a) Triaxial cell schematic; (b) direct shear schematic. P_h develops a failure plane along the sample box separation as shown.

These tests are schematically illustrated in Fig. 13-3, and photographs of somewhat typical equipment are shown in Fig. 13-4. A critical evaluation of shear tests has been presented by Sowers (1963). There is a considerable body of literature on shear testing, and periodically someone puts forth a new device which purports to be an improvement over the shear testing devices schematically illustrated in Fig. 13-3 and commonly used in most soils laboratories. Most of the so-called improved shear testing devices are too complicated for practical use, at least to date. The plane strain triaxial test is also rather complicated, and only a few university and larger commercial laboratories have constructed and/or use this test.

The unconfined compression test is very widely used worldwide. It is a simple test where atmospheric pressure surrounds the soil sample. The corresponding Mohr's circle is illustrated in Fig. 13-5c. From the single circle for a series of tests we can only extrapolate the slope of the failure envelope as $\phi = 0$ and the undrained shear strength s_u is

$$s_u = c = \frac{q_u}{2}$$

Some hold the opinion that if water is present, an effective "confining" pressure exists internally in an unconfined compression test sample due to capillary effects and that these effects may effectively confine the sample just as if it were in situ. If this be correct, then it is also true that degree of saturation, grain size, stress cracks or fissuring, and laboratory humidity would be very important considerations. In

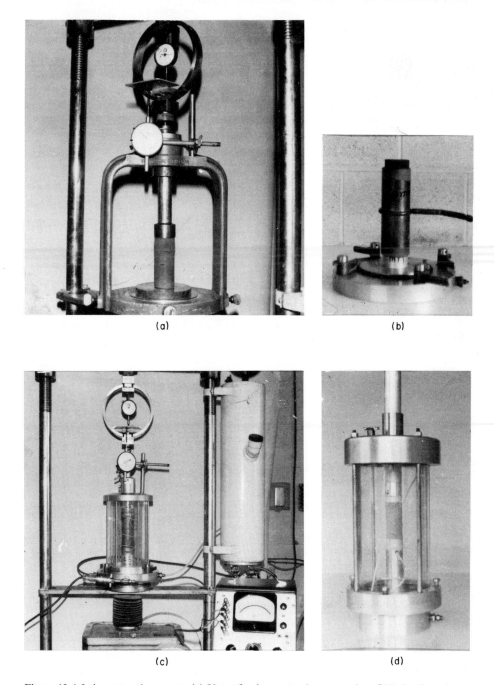

(a)

(b)

(c)

(d)

Figure 13-4 Laboratory shear tests. (*a*) Unconfined compression test using CBR loading piston; (*b*) cohesive sample partially inserted into sample membrane using oversize membrane stretcher; (*c*) triaxial test with pore pressure transducer and electronic pressure indicator; (*d*) triaxial cell modified so that the piston is the same size as the sample for anisotropic consolidation.

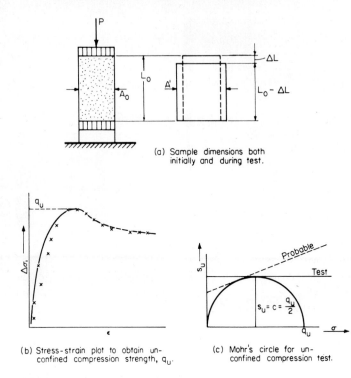

(a) Sample dimensions both
initially and during test.

(b) Stress-strain plot to obtain unconfined compression strength, q_u.

(c) Mohr's circle for un-
confined compression test.

Figure 13-5 The unconfined compression test.

any case, few (if any) laboratories attempt to allow for capillary confinement; q_u is taken as the maximum compressive stress and plotted on Mohr's circle as in Fig. 13-5c. Since each test would be plotted with $\sigma_3 = 0$, the strength parameters c (with $\phi = 0$) generally provide a conservative solution as long as the working stresses are greater than s_u.

The unconfined compression (and triaxial) test is made by obtaining a thin-walled tube or other sample as nearly "undisturbed" as possible. Sometimes q_u is gotten from the sample obtained from the standard penetration test of Sec. 3-9. The sample is divided, where the length is sufficient,† into several samples with a length/diameter ratio

$$2 < \frac{L}{d} < 3$$

to ensure that a failure plane does not intersect the loading heads. The sample is made as square as practical (no simple matter when small pieces of gravel are

---

† Some organizations use a sample on the order of 12-cm diameter so that three cylindrical samples of a diameter of approximately 5 cm can be trimmed from any vertical location to reduce sample depth variations.

present) and placed in a compression machine adjusted to a deformation rate on the order of under 1.5 mm/min, and deformations vs. corresponding sample loads are obtained. These data are used to plot a stress-strain curve of σ versus ε to obtain the maximum value of compressive stress, which is q_u for the unconfined compression test. It is evident that an unconfined compression test can be made only on cohesive soils.

The compressive stress σ (or $\Delta\sigma_1$ for a triaxial test) is computed differently than for steel, concrete, or other materials. For these materials the compressive (deviator) stress is computed as

$$\Delta\sigma_1 = \frac{P}{A}$$

with A being the original cross-sectional area, which is a conservative computation. As the soil strains vertically, it produces lateral strains which increases the effective cross-sectional area which resists the stress. Since this always occurs with the laboratory sample, and on a similarly enlarged area in the field due to the Poisson effect, it is considered more correct to base the compressive stress on this enlarged cross-sectional area A'. This area can be computed using the assumption that the volume of the sample remains constant (Fig. 13-5a); thus,

$$A_o L_o = A'(L_o - \Delta L)$$

from which we obtain, since strain $\varepsilon = \Delta L/L_o$,

$$A' = \frac{A_o}{1 - \varepsilon} \tag{13-3}$$

with all terms identified in Fig. 13-5. The instantaneous deviator stress $\Delta\sigma_1$ is

$$\Delta\sigma_1 = \frac{P}{A'} \tag{13-4}$$

where $\Delta\sigma_1$ is the change in major principal stress, as it is assumed that there is no shear stress on the ends of the soil sample. Since friction is developed as the sample attempts to expand, a small error is introduced; however, research indicates the error is essentially negligible (Barden and McDermott, 1965). The error may not be negligible in tests where the sample is in contact with the end platens for a very long time, as may happen in research projects using cohesive soils, and it is necessary to fully consolidate the sample prior to testing. Research by the author indicates that Eq. (13-3) does describe the bulging reasonably well (i.e., measure the sample diameter and compare change to corresponding strain). The use of smooth metal end platens with silicone grease may reduce end restraint effects.

The *direct shear* test is a simple, straightforward test to perform and is next in order of increasing cost. The test is made by placing a soil sample into the shear box illustrated in Fig. 13-3b. The box is split as shown, with the bottom half fixed and the top half free to float and translate. The box is available in several sizes but commonly is 6.4 cm in diameter or 5.0 × 5.0 cm square. The sample is carefully placed in the box; a loading block, which includes a serrated porous stone for rapid drainage, is placed on the sample. Next a normal load P_v is applied. The two halves are separated slightly and the loading block and top half are clamped together and either

1. The test is begun immediately and horizontal displacements obtained for the corresponding horizontal load P_h, so that a plot of P_h versus δ_h can be made to find $P_{h, \text{max}}$. This is called an undrained (U) test.
2. The test is begun after consolidation of the soil under P_v is complete as determined by observation of a dial gage in contact with the load head. If the test proceeds:
   (a) Rapidly, pore pressures will develop in wet or saturated cohesive soils due to the low coefficient of permeability and the test is called a consolidated-undrained (CU) test.
   (b) Very slowly, so that no pore pressures develop, the test is called a consolidated-drained (CD) test. This test usually gives for normally consolidated soils a value of $c' \cong 0$ or

$$s = \sigma'_n \tan \phi' \tag{13-5}$$

   Two or more additional tests at larger values of P_v are performed to make a scaled plot of

$$s = \frac{P_v}{A_o} \qquad \text{versus} \qquad \sigma_n = \frac{P_h}{A_o}$$

as in Fig. 13-6 so that a graphical solution of Eq. (13-1) can be obtained. This is necessary, as there are two unknowns (c and ϕ) in Eq. (13-5) or (13-6) and a minimum of two tests is required to obtain two values of s so that a simultaneous equation solution can be made. Qualitative results of the U, CU, and CD direct

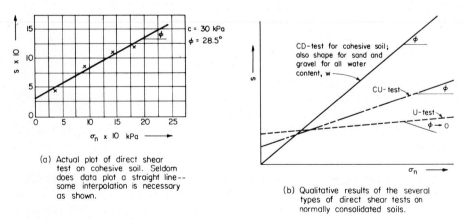

(a) Actual plot of direct shear test on cohesive soil. Seldom does data plot a straight line—some interpolation is necessary as shown.

(b) Qualitative results of the several types of direct shear tests on normally consolidated soils.

Figure 13-6 Direct shear test.

shear tests are shown in Fig. 13-6b, from which it is apparent that the shear strength parameters are not unique values but depend heavily on the test procedure.

The thickness of the direct shear test sample is on the order of 2 to 3 cm; thus, drained conditions are quite probable for cohesionless soils (and are commonly assumed). In the undrained test with cohesive soils, some drainage is likely to take place; consolidation usually does not take a very long time for CU tests, which represents a considerable economic advantage. Pore water conditions cannot be controlled or measured in this test, however. Some researchers have attempted to correlate the water content along the failure surface (say the zone 4 to 6 mm along the shear box separation line) to shear strength. The water content in this zone will be different from the average due to particle displacement, resulting in a soil void ratio decrease.

The direct shear test forces the direction and location of the failure plane, i.e., at the location of the box split and parallel to the horizontal load. Practically, this condition may not be obtained, but the results are nevertheless considered to be satisfactory. This test is considered by some persons as meeting the requirements of plane strain sufficiently to term this a plane strain test. Considering that the sample is confined so that only lateral and vertical movements (strains) can take place, which is by definition plane strain, the reader can decide whether to use the data as "plane strain" data. A second test deficiency is that the shear area A_o decreases as the test proceeds. A correction for area reduction can easily be made with square shear boxes, but it is not very practical to make this correction for round boxes due to the considerable mathematics involved in computing the instantaneous area as a function of δ_h.

The triaxial test is considered to provide the best soil parameters and stress-strain data. This is true only if "undisturbed" soil samples are obtained and great care is exercised in trimming them to test size and inserting them into the rubber membrane (Fig. 13-4b). The general sample dimensions range from 3.6- to 7.6-cm diameter with an L/d ratio of 2.2 to 3.0, as with the unconfined compression test. Larger diameter samples may be tested; however, with sample diameters larger than 7.6 cm, sample recovery may become a significant economic factor. It is necessary to test the sample using in situ pore pressure conditions if correct soil parameters are to be obtained. The particular advantages of the triaxial test are

1. Means of controlling the confining pressure.
2. Control of the pore water pressure. If the pore water pressure is manipulated, this is termed applying "back pressure."
3. When the cell is appropriately modified (Fig. 13-4), both isotropic and aniso-tropic initial conditions may be simulated.

The pore water pressure is commonly taken from the base of the sample using a pore pressure measuring device (a transducer if electronic equipment is used), as schematically illustrated in Fig. 13-3a, but can be taken from midheight if a pore pressure needle is used (which requires a large sample and a fluid-tight entry through the rubber sample membrane). Unless the pore pressure is measured exactly in the failure plane, the value may not be correct, since it is on the failure plane that maximum soil structure reorientation and resulting maximum developed excess pore pressure are occurring.

A triaxial stress-strain plot is made exactly the same way as that for an unconfined compression test, computing the deviator stress as

$$\Delta \sigma_1 = \frac{P}{A'}$$

with the stress-strain plot as in Fig. 13-7a. Figure 13-7b shows a plot of σ_1 versus ε. This plot is not used, since the stress $\Delta \sigma_1 = \sigma_3$ at $\varepsilon = 0$. For the triaxial test the major principal stress is computed as

$$\sigma_1 = \sigma_3 + \Delta \sigma_1$$

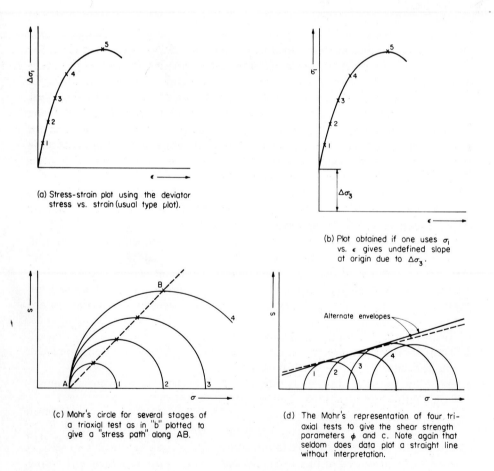

(a) Stress-strain plot using the deviator stress vs. strain (usual type plot).

(b) Plot obtained if one uses σ_1 vs. ε gives undefined slope at origin due to $\Delta \sigma_3$.

(c) Mohr's circle for several stages of a triaxial test as in "b" plotted to give a "stress path" along AB.

(d) The Mohr's representation of four triaxial tests to give the shear strength parameters ϕ and c. Note again that seldom does data plot a straight line without interpretation.

Figure 13-7 Presentation of triaxial stress-strain and Mohr's circle data.

and the instantaneous values could be plotted with Mohr's circles as illustrated in Fig. 13-7c. Usually the maximum, or failure, circle which produces a point on the Mohr failure envelope is the only one of interest for a particular confining pressure. The results of several tests are plotted, as in Fig. 13-7d, and the failure envelope draws as a best fit to obtain c and ϕ.

In general, three strength tests should be performed as an absolute minimum, as averaging the results of two tests may give a poor value if one is high and the other low. With three tests, an unreasonably high or low value can be discarded.

13-3 STRESS PATHS AND pq DIAGRAMS

Instead of plotting a Mohr's circle for each stress condition as illustrated in Fig. 13-7c, it may be more convenient to plot stress coordinates of

$$p = \frac{\sigma_1 + \sigma_3}{2} \qquad q = \frac{\sigma_1 - \sigma_3}{2} \qquad (13-6)$$

These p and q values are the coordinates of a point on the Mohr's circle at the origin of the circle (p) and at the maximum shear stress (q or the radius). The locus of the several points of a single test is illustrated in Fig. 13-7c and also in Fig. 13-8 and is termed a *stress path*. A stress path can be made from a single test, but is of more value when a plot is made from several tests as in Fig. 13-8. Stress paths can be drawn for either triaxial or consolidation tests using either effective or total stresses. The locus of the pq points for a test series on a soil forms a failure line termed the K_f line, such as line $BB'B''$ of Fig. 13-8. This method of plotting seems to have been first used by Simons (1960).

The relationship between the K_f line (of a pq plot) and the Mohr's failure envelope (called the ϕ *line*) is as shown in Fig. 13-9. Since the origin is common for both lines (Fig. 13-9b), we have

$$\frac{R}{m + x} = \tan \alpha = \sin \phi = \frac{a}{m}$$

Also

$$\frac{c}{\tan \phi} = \frac{a}{\tan \alpha}$$

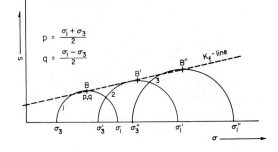

Figure 13-8 Stress path for a triaxial test series. The locus of pq points as B, B', and B'' may require a "best fit" to produce the K_f line as shown. Refer to Fig. 13-7c for the stress path for a single test to describe intermediate stress stages.

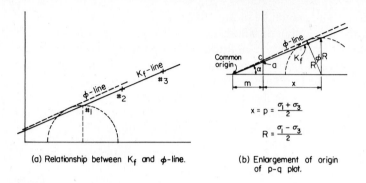

(a) Relationship between K_f and ϕ-line.

(b) Enlargement of origin of p-q plot.

$$x = p = \frac{\sigma_1 + \sigma_3}{2}$$

$$R = \frac{\sigma_1 - \sigma_3}{2}$$

Figure 13-9 Relationship between shear strength parameters and the K_f line from the stress path plot.

But with $\tan \alpha = \sin \phi$ we have for the cohesion intercept

$$c = \frac{a}{\cos \phi} \tag{13-7}$$

From this, the soil parameters c and ϕ may be obtained from a pq diagram by

1. Scaling α, a, and m
2. Computing $\alpha = \tan^{-1}(a/m)$ (or scaling it)
3. Computing $\cos \phi$ and then computing the cohesion c

The results of a triaxial test can be plotted in the stress space of plane $OCFE$ of Fig. 10-3. Referring to Fig. 13-10, we will rotate the stress plane investigated in Fig. 13-10a parallel to the plane of the paper for two-dimensional plotting of the stress paths, as shown in Fig. 13-10b. We may plot either total or effective stresses in a plot of this type. The failure envelope for compression is analogous to Mohr's failure envelope for a compression test. The failure envelope for extension tests is similar to that for compression but results in a smaller failure stress. The extension envelope is obtained by decreasing the axial load in the triaxial test while holding σ_3 constant. This requires either a special loading setup in which dead loads are used rather than a strain controlled compression machine or modification of the triaxial cell as done by the author in Fig. 13-4d, where the load piston is the same as the specimen diameter, so that the external load applies the vertical pressure regardless of the cell pressure.

The line OB of Fig. 13-10b represents hydrostatic stress conditions as described in Sec. 10-3 and has a slope, as shown, of $1/\sqrt{2}$ if we have isotropic consolidation producing the octahedral stress conditions.

For anisotropic soil tests it is generally better to consider the octahedral stresses and/or to use a plot similar to Fig. 13-10b to describe the changed stress conditions to failure. A Mohr's circle type plot will start with an initial stress circle rather than a point stress.

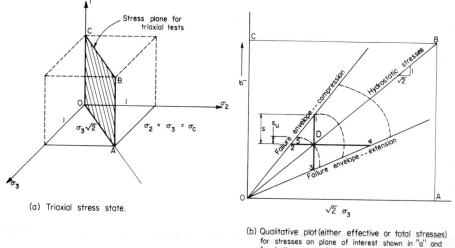

(a) Triaxial stress state.

(b) Qualitative plot (either effective or total stresses) for stresses on plane of interest shown in "a" and for both compression and extension type tests.

Figure 13-10 Plotting of triaxial stress data in stress plane $OABC$. The plot in (b) is on the stress plane after rotation, so that the plot is two-dimensional. (*After Henkel, 1960.*)

For selected axial load increments, pore pressure is measured, and effective (solid) or total (dashed) stress paths can be plotted as shown in Fig. 13-10b. In undrained tests the stress path depends on pore water pressure characteristics of the soil sample, initial consolidation conditions such as isotropic or anisotropic, the overconsolidation ratio, and the degree of saturation. The *drained shear strength* is the vertical distance from the initial conditions (such as point D) on the hydrostatic stress line to the intersection with the failure line (such as $D1$ or $D3$). The *undrained shear strength* is the vertical distance from the initial hydrostatic stress conditions to the point of intersection of the stress path and the failure line (as $D5$ for s_u shown on Fig. 13-10b). Lines $D2$ and $D4$ represent drained conditions for decreasing or increasing cell pressure σ_3, respectively. The stress path for undrained conditions is, according to Henkel (1960), a line of constant water content. From this it follows that one may perform several triaxial tests at select OCR, plot the results, and carefully determine the water content at failure (and in the failure zone) to identify the contour of constant water content. Additional lines of constant water content are interpolated, and then the plot can be used to study shear strength and changes in water content of the soil for other initial conditions, such as OCR, initial water content, or both.

Example 13-1

GIVEN A cohesive soil was tested in a direct shear test with the following values: Square shear box 5.5 × 5.5 cm; height = 2.1 cm.

			Computed stresses, kPa	
Test	P_v, kg	P_h, kg	σ_n	s
1	4	2.9	13.0	9.4
2	8	4.3	26.0	13.9
3	12	5.1	39.0	16.5

REQUIRED

(a) Obtain c and ϕ (c in kilopascals).

(b) Find the orientation of principal planes for test 2.

SOLUTION

Step 1 Compute σ_n and s as shown in the table above.

$$\sigma_n = \frac{4(9.807) \times 10^4}{(5.5)^2 \times 10^3} = 13.0 \text{ kPa}$$

The remaining stresses are computed similarly.

Step 2 Plot values and "fair" the failure envelope as shown in Fig. E13-1a, and measure

$$\phi = 17.5°$$

$$c = 5.0 \text{ kPa}$$

Step 3 To obtain the orientation of principal planes, construct a Mohr's circle for test 2 and lay off AB at θ as shown in Fig. E13-1b, noting that principal planes are oriented with θ as shown and 90° apart and that with the actual failure plane horizontal, the principal planes are as shown.

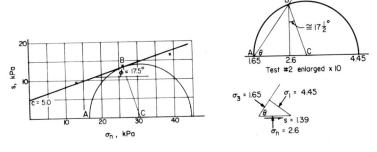

Figure E13-1a

Figure E13-1b

Example 13-2

GIVEN A CU triaxial test on an unsaturated soil with the following data:

Test	σ_3, kPa	$\Delta\sigma_1$ = deviator stress	σ_1 (computed)
1	40	63.0	103.0
2	80	86.0	166.0
3	120	102.0	222.0

REQUIRED Find the apparent soil parameters ϕ and c and orient the failure plane for test 1.

SOLUTION

Step 1 Compute

$$\sigma_1 = \sigma_3 + \Delta\sigma_1$$

$$= 40 + 63.0 = 103.0 \text{ for test 1, etc.}$$

Step 2 Plot Mohr's circle for the three tests as in Fig. E13-2, fair a failure envelope, and scale

$$\phi = 12°$$

$$c = 17.5 \text{ kPa}$$

Step 3 From σ_3 of test 1, draw line AB at $\theta = 45 + \phi/2 = 51°$ as shown, and place on the element the stresses shown. Values of σ_n and s can be computed from Eqs. (10-8) and (10-9) but can be scaled with satisfactory precision.

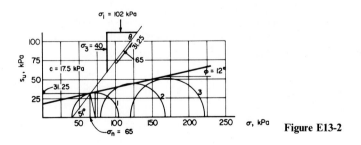

Figure E13-2

Example 13-3

GIVEN Triaxial data of Ex. 13-2.

REQUIRED Plot a pq diagram, obtain c and ϕ, and compare to Ex. 13-2.

SOLUTION

Step 1 Compute p and q for each test (note that $q = \Delta\sigma/2$).

Test	σ_3	σ_1	p	q, kPa
1	40	103.0	71.5	31.5
2	80	166.0	123.0	43.0
3	120	222.0	171.0	51.0

Step 2 Plot p versus q as in Fig. E13-3, fair the K_f line as shown, and scale

$$a = 18 \text{ kPa} \qquad m = 94 \text{ kPa} \qquad \alpha = 11°$$

Step 3 Compute

$$\sin \phi = \tan 11° = 0.1944$$

$$\phi = \sin^{-1}(0.1944) = 11.21° \text{ (versus } 12° \text{ of Ex. 13-2)}$$

then $\qquad c = \dfrac{a}{\cos \phi} = \dfrac{18}{\cos 11.21°} = 18.35 \text{ kPa (versus } 17.5)$

These discrepancies are a combination of scale effects and interpreting the location of the K_f and ϕ lines on the separate plots.

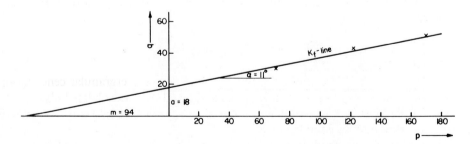

Figure E13-3

Example 13-4

GIVEN Triaxial tests on a cohesionless soil.

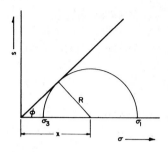

Test	σ_1	σ_3, kPa
1	78.5	20
2	186.0	50

Figure E13-4

REQUIRED The angle of internal friction ϕ.

SOLUTION Compute the angle from the relationships shown in Fig. E13-4. From Fig. E13-4:

$$\sin \phi = \frac{R}{X} = \frac{\sigma_1 - \sigma_3}{\sigma_1 + \sigma_3}$$

For test 1,

$$\phi = \sin^{-1} \frac{78.5 - 20}{78.5 + 20} = 36.4°$$

For test 2,

$$\phi = \sin^{-1} \tfrac{136}{236} = 35.2°$$

Take the average as the best value to obtain $\phi = 35.8°$.

13-4 SOIL FAILURE AND CONCEPT OF RESIDUAL STRENGTH

Soil does not fail, in general, as other materials; rather, after the peak or ultimate compressive or shear stress has been reached, the soil continues to carry a substantial load. The postpeak stress is termed the *residual* stress.

Soil stress-strain curves may exhibit a *brittle*, or relatively sudden, failure when:

1. The soil is dense, in the case of dry or wet sand.
2. The soil is cohesive and dry.
3. The soil is an "undisturbed" natural soil in which intergranular cementation exists, as in some sands and both normally and overconsolidated clays.
4. The soil has been compacted and tested at a water content on the dry side of the optimum moisture content.
5. The confining pressure is much larger than 70 kPa in triaxial tests (see Fig. 13-16b).

Loose sands and most remolded cohesive soils, and soils compacted on the wet side of optimum, tend to a more progressive failure particularly in q_u tests or triaxial tests using low cell pressures. These conditions are qualitatively illustrated in Fig. 13-11a and b. Note that the brittle failure occurs at low strains—on the order of 1 to 3 percent (0.01 to 0.03 cm/cm).

When the soil is cohesionless and loose or cohesive and wet, the failure stress may not be clearly defined, as in Fig. 13-11b. In this case the failure may be defined arbitrarily at some strain, with the strain value of $\varepsilon = 0.20$ (20 percent strain) commonly used as illustrated.

The gradual development of the ultimate stress of Fig. 13-11b can be termed *progressive* failure, as contrasted to the more sudden brittle failure of Fig. 13-11a. That is, along any failure surface portions of the soil matrix "fail"; the stresses are transferred to "unfailed" zones, where again the higher stressed zones "fail," with further transferring of load, etc. The ultimate resistance is reached as a combination of "failed," nearly failed, and other stresses of uncertain magnitude. Evidence indicates that most soil failures are progressive over a finite area rather than having a singular stress value.

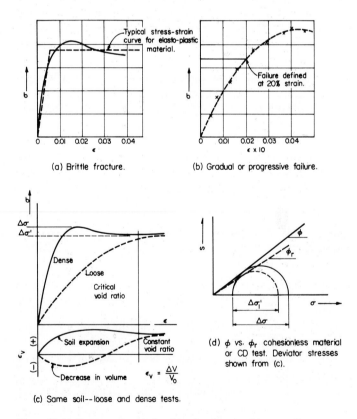

(a) Brittle fracture.

(b) Gradual or progressive failure.

(c) Same soil--loose and dense tests.

(d) ϕ vs. ϕ_r cohesionless material or CD test. Deviator stresses shown from (c).

Figure 13-11 Stress-strain relationships for dense and loose soil.

Figure 13-12 (*a*) Dilatency and (*b*) collapsing effects during shear in cohesionless soils.

For cohesionless samples, experimental evidence indicates that the brittle or dense curve of Fig. 13-11*a* and the loose curve of Fig. 13-11*b* will converge as in Fig. 13-11*c* for the same soil, at some void ratio termed the critical void ratio by some early researchers. This critical void ratio will generally be somewhere between the two test void ratios. This observation means that dense soils tend to *dilate*, or expand (property of dilatency), and loose soils tend to densify, or collapse, during shear. The tendency to dilation occurs from the interlocking effects of the dense soil and is obtained as the net effect of the large numbers of random displacements as grains are forced up, over, and around adjacent and confining grains as in Fig. 13-12*a*. When the soil is loose, the change in position results in the net effect being a decrease in volume as the grains roll about and "fall in" the voids.

The critical void ratio may be approximately determined by measuring volumetric changes for a dense and a loose soil sample in the direct shear test to obtain the type of data illustrated in Fig. 13-11*c*. The critical void ratio is less easily determined in a triaxial test.

One may use some value of the residual strength, say, at 20 to 30 percent strain, to plot a modified Mohr's circle as illustrated in Fig. 13-11*d* to obtain *the residual angle of internal friction* ϕ_r, which may provide a more realistic field parameter where large soil strains can be tolerated and/or anticipated.

13-5 PORE PRESSURE EFFECTS AND UNCONSOLIDATED-UNDRAINED TESTS (UNDRAINED OR U TESTS)

The results from unconsolidated-undrained tests will depend heavily on the type of test, soil, and degree of saturation. We will consider these several factors in the following discussions.

Cohesionless Soil

It will be nearly impossible to perform an undrained test on cohesionless soil in the direct shear apparatus due to the 20- to 25-mm sample thickness and large coefficient of permeability. In the triaxial apparatus, if the drainage tubes to the sample are open, some drainage will occur during the test, so that a ϕ angle intermediate between 0 and ϕ' will be obtained. The exact value will depend on the degree of saturation, coefficient of permeability and strain rate, i.e., how much instantaneous pore pressure develops will depend on how filled the voids are and/or how fast the pore pressure can dissipate via drainage.

It would be pointless to perform a truly undrained test on a saturated loose to medium dense cohesionless soil, since the test could not be performed for the reasons illustrated in Fig. 13-13a and b. An undrained test on a dense, saturated sample will give a small apparent cohesion due to negative pore pressures as sample dilates. This cohesion is very small and difficult to measure and should be neglected as it would dissipate with time in the field.

Cohesive Soil—Remolded and Compacted

Figure 13-13a illustrates a completely remolded, saturated sample confined in a rubber membrane as in a triaxial test, with initial isotropic stress conditions of $\sigma_1 = \sigma_3 = \sigma_c$ and pore pressure $= 0$. A confining pressure $\Delta\sigma_3$ is now applied. With no drainage allowed, no soil structure change can occur, and, as in the consolidation test, the pore water will carry the stress, producing $u = \Delta\sigma_3$. When the deviator stress $\Delta\sigma_1$ is applied, the free body shown in Fig. 13-13b can be

(a) Confined sample.

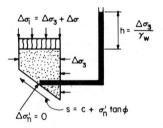

(b) Free-body of U-test on cohesive soil.

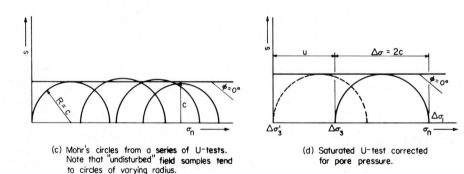

(c) Mohr's circles from a series of U-tests. Note that "undisturbed" field samples tend to circles of varying radius.

(d) Saturated U-test corrected for pore pressure.

Figure 13-13 Unconsolidated-undrained shear testing.

obtained. From this free body it is evident that the effective stress on the plane AA (neglecting the effects of unit weight of soil) is:

$$\Delta\sigma'_n = \Delta\sigma_n - u = 0$$

since the pore pressure acts equally in all directions. From this it follows that any $\Delta\sigma_1$ applied is carried by the soil cohesion, since

$$s = c + \sigma \tan \phi$$

but with $\sigma' = 0$ the $\sigma \tan \phi$ term which represents friction resistance in terms of total stresses still produces a zero contribution, and we have for the shear strength $s = c$. This value of shear strength is commonly given the symbol s_u for the undrained shear strength. Thus, the load applied to a saturated cohesive soil in undrained conditions is carried by the cohesion parameter. From this it necessarily follows that a series of triaxial tests can only plot a series of Mohr's circles of radius c displaced from left to right along the σ axis depending on the cell pressure $\Delta\sigma_3$. Any differences of circle size are due to normal variation of degree of saturation S and cohesion from sample to sample. The reader should now refer again to the last sentence in "cohesionless soil" immediately preceding.

This undrained condition produces an apparent $\phi = 0$ condition. The term "apparent" is used since it is σ' which produces the friction resistance of

$$\sigma' \tan \phi' = 0$$

regardless of whether total or effective stresses are used, and ϕ is not zero, as drained tests would indicate.

Since the pore pressure depends on $\Delta\sigma_3$, we may write

$$u = B \Delta\sigma_3$$

where $B = 1$ for saturated conditions but is generally not known for $S < 100$ percent and must be measured for any reliability (see Sec. 13-12). In any undrained test, if the pore pressure is known we may correct for it as follows:

Writing for effective stresses

Generally	When $s \to 100$ percent
$\sigma_3 = \sigma_3 - u$	$\sigma_3 = 0$
$\sigma'_1 = \sigma_1 - u$	$\sigma_1 = \sigma_3 + \Delta\sigma_1 - u = \Delta\sigma$

Subtracting the general case, obtain

$$\sigma'_1 - \sigma'_3 = \sigma_1 - \sigma_3 \tag{13-8}$$

This shows that the diameter of Mohr's circle is unchanged by the pore pressure, but the origin of the circles is shifted to the left (in the case of $S = 100$ percent, all the way to the origin) by the pore pressure u, as in Fig. 13-13d. This set of Mohr's circles is theoretically identical to those obtained from the unconfined compression test.

Cohesive Soil—Normally Consolidated

If we take a sample from in situ, the tendency for sample expansion due to loss of overburden pressure will produce negative pore pressures. If we place the sample in a triaxial cell and

Apply	We produce (qualitatively) pore pressures of
$\sigma_c < p_o$	$-u$
$\sigma_c = p_o$	$u \to 0$
$\sigma_c > p_o$	$+u$

Application of the deviator stress then tends to apply a $+\Delta u$ to that existing unless the sample structure collapses. The results of a test series usually produce $\phi \to 0$, as qualitatively shown in Fig. 13-14a. In general, in soil mechanics literature, "undrained shear" implies $\phi = 0$.

The unconfined compression test may in this situation produce reasonably good data, since with $\sigma_c = 0$ the negative pore pressure, which acts as a confining agent, is a maximum (maximum possible $\to 103$ kPa) and may be approximately p_o. These data may be adequate for the additional reason that with good quality samples the extra disturbance of sample insertion into the triaxial setup and, coupled with the uncertain pore water conditions—especially if $\sigma_c > p_o$—produce results of dubious quality.

In general, for these several reasons the unconfined compression test should be used when unconsolidated-undrained or simply "undrained" tests are used, since the data are about as good as triaxial data and the unconfined compression test (Fig. 13-14b) is considerably more economical to perform.

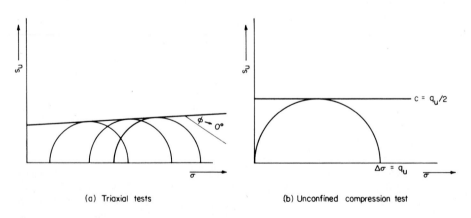

(a) Triaxial tests   (b) Unconfined compression test

Figure 13-14 Unconsolidated-undrained shear tests.

13-6 PORE PRESSURE EFFECTS IN CONSOLIDATED-UNDRAINED (CU) TESTS

Consolidated-undrained tests are obtained when a sample is placed in a triaxial cell in which the sample volume changes can be measured and either:

1. Consolidated under the cell pressure σ_c, which produces *isotropic* consolidation (Fig. 13-15) and is not generally the same stress state as in situ even if σ_c is computed based on the existing depth of overburden pressure and K_o of Sec. 15-1.
2. Consolidated under a combination of cell pressure σ_c and an additional vertical pressure of $\sigma_1 = \sigma_c + \Delta\sigma_1$ to produce *anisotropic* consolidation. This may be more realistic due to overburden effects and/or preconsolidation. Note that with considerable ingenuity and cell modification, the vertical and lateral pressure on the sample can be supplied in any desired ratio of $K = \sigma_h/\sigma_v$.

Volume changes under consolidation require allowing drainage during consolidation. The consolidated-undrained test is performed after consolidation is complete, with no further external drainage allowed. Consolidation is considered complete when the slope of the volume change vs. time curve is sufficiently small. The process is accelerated by placing strips of filter paper inside the membrane against the sample or, if the sample is sufficiently large and gravel free, by coring a very small diameter hole in the center which is filled with fine sand.

Consolidation of the sample produces increased density, a small decrease in water content, and an increase in shear strength. The increase in shear strength will be due partly to the decrease in water content, resulting in a closer spacing of clay particles so that the interparticle attraction is larger, and partly due to the interlocking effect from the denser particle arrangement. At the end of consolidation, the excess pore pressure produced by the consolidation stresses should be approximately zero (the definition of end of consolidation). Since the consolidated-undrained test results are some different for normally consolidated and overconsolidated clays, these will be considered separately.

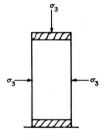

(a) Isotropic

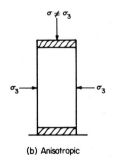

(b) Anisotropic

Figure 13-15 Two methods of sample consolidation for a consolidated-undrained test.

Normally Consolidated Clay

Figure 13-16a illustrates the field conditions for a cohesive soil. Under normal field loadings of overburden, water table, etc., the soil consolidates along path ABC. The present status of the sample is point B, corresponding to the *effective* overburden pressure p_o. No excess pore pressure exists in the sample by definition, since it is consolidated. When the sample is removed from the ground, the overburden pressure is lost and hydrostatic pressure conditions are considerably changed. This is qualitatively shown as dashed line BD, which represents sample expansion due to the loss of confining stress. If drainage is prevented, i.e., the sample is not allowed to adsorb water from any source including laboratory humidity, negative pore pressures will develop as capillary effects resist sample expansion.

If the sample is placed in the triaxial membrane and a confining cell pressure is applied in increments, as in the consolidation test, to a level which reproduces the in situ value of p_o (and without drainage), we obtain the qualitative curve branch DB'. At this point the pore pressure is zero. The sample is also some to considerably disturbed, since B' is below B. The relative positions of B and B' would, if it were possible to reproduce curve ABC, indicate the degree of disturbance (refer also to Fig. 13-18). We can only qualitatively state that:

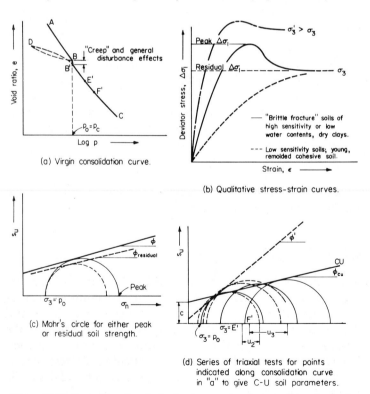

(a) Virgin consolidation curve.

(b) Qualitative stress-strain curves.

(c) Mohr's circle for either peak or residual soil strength.

(d) Series of triaxial tests for points indicated along consolidation curve in "a" to give C-U soil parameters.

Figure 13-16 Consolidated-undrained tests on normally consolidated cohesive soil.

1. If the clay is sensitive, the disturbance will be large.
2. If we use isotropic recompression stresses instead of the in situ anisotropic stress state of $\sigma'_3 = K_o \sigma'_1$, we will also introduce a disturbance.
3. If the recompression stress $\sigma_c < p_o$, the pore pressure will remain slightly negative; if $\sigma_c > p_o$, there will be a positive pore pressure, or in general,

$$u = \sigma_c - p_o$$

If we allow no drainage and now apply a deviator stress $\Delta\sigma_1$, we obtain the consolidated-undrained stress-strain curve qualitatively shown in Fig. 13-16b, and using the peak or residual value of strength, we may plot the solid or dashed Mohr's circles of Fig. 13-16c.

We may now take a second sample from the same in situ elevation and as close as practical, put it in the membrane and recompress it to p_c, then further consolidate *with drainage* along the compression curve BB'C to point E'. When the volume change is complete, the pore pressure is again zero in this sample. Closing the drainage system and applying the deviator stress, we obtain a second Mohr's circle, and with at least one additional sample at $\sigma_c = F'$, we obtain the series of circles shown in Fig. 13-16d. The failure envelope obtained as a "best fit" gives the CU shear strength parameters, which may be:

Peak shear strength values	Residual strength values
ϕ_{cu}	$\phi_{cu(residual)}$
c_{cu}	$c_{cu(residual)} \to 0$

If we measure the pore pressure which develops during application of the deviator stress, we will obtain the dashed circles shown in Fig. 13-16d, which give Mohr's rupture line as approximately

$$s = \sigma' \tan \phi'$$

where ϕ' is the drained angle of internal friction and is considered to be the absolute, and maximum, value possible for the soil.

Intact Overconsolidated Clay

An overconsolidated clay may be produced as in Fig. 13-17. Figure 13-17a represents a clay deposit, as in the bed of a lake or bay. Subsequent deposition produces the overburden of Fig. 13-17b and c. At this time the deposit may be several hundred to 1000 or more meters in depth. Subsequent erosion produces the overconsolidated condition of Fig. 13-17d. The removal of pressure is accompanied by a slight increase in water content, but far less than the decrease occurring during the consolidation process. Thus, although the clay in b and d is at the same present overburden pressure p_o, the density and shear strength are larger

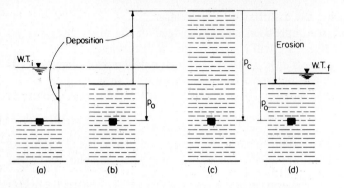

Figure 13-17 Natural formation of an overconsolidated clay deposit.

and the water content is smaller for *d*. Other factors such as desiccation, changes in salt content, and creep may produce the effect of overconsolidation. Figure 13-18 illustrates how creep may produce the effect of preconsolidation. The curve shown is interpreted as follows:

1. Soil is deposited to an in situ overburden pressure of p_o, which produces e_o at the end of primary consolidation as point *A*.
2. Subsequent creep reduces the in situ void ratio e_o to e'_o at point *B*.
3. A sample removed would expand to *C*, then in a consolidation test recompress along *CDEF*, giving an apparent preconsolidation of p_c at *E*.

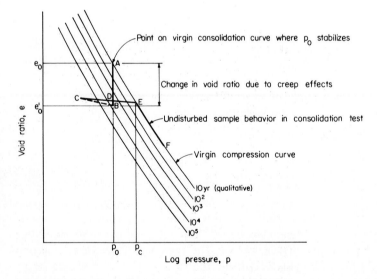

Figure 13-18 Creep (or secondary compression) producing apparent preconsolidation. (*After Bjerrum, 1972.*)

Note that this concept is highly idealized and that at least three factors may act to produce a small p_c:

1. Many clays are postglacial; thus the creep duration is not more than 15 000 years.
2. Creep is not large in inorganic clays.
3. Creep is dependent on p_o, which may not be very large for shallow deposits.

During glacial periods the ice depth produced overconsolidation of some deposits, although clays deposited as a result of ice melt or as glacial till are generally normally consolidated unless desiccation effects have since taken place.

From Fig. 13-17, an overconsolidated soil is defined as having an overconsolidation ratio (OCR) computed as

$$OCR = \frac{p_c}{p_o} > 1$$

An OCR = 1 is a normally consolidated clay. Figure 13-19a illustrates the shear strength envelope ($F'AB$) of an overconsolidated clay which was preconsolidated to p_c due to some geological, chemical, or climate factors and is currently in situ under a present overburden pressure of p_o. The curve ABC of Fig. 13-19b represents the virgin compression curve during preconsolidation (without creep effects)

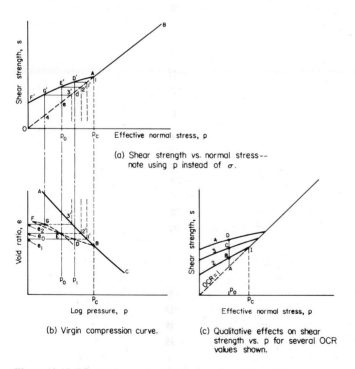

(a) Shear strength vs. normal stress -- note using p instead of σ.

(b) Virgin compression curve.

(c) Qualitative effects on shear strength vs. p for several OCR values shown.

Figure 13-19 Effects of overconsolidation on shear strength.

and later loss of effective pressure along curve branch *BDEFGE*. At each stage along the unload-reload branch we can estimate the shear strength and produce the overconsolidation branch ($F'G'E'D'A$) of the s versus p curve as follows (for points E' and D'):

1. At points D and E on Fig. 13-19b, erect vertical lines from (b) to (a) as shown and locate points d and e on the normal s versus p curve.
2. Project from D and E horizontally and locate points 1′ and 2′ on the virgin compression curve.
3. From points 1′ and 2′, extend vertical lines to (a) and locate corresponding points 1′ and 2′ on the s versus p curve. Note that these two points define the shear stress of the normally consolidated clay under pressures corresponding to the void ratios e_o and e_1.
4. Extend horizontal lines from 1′ and 2′ to intersect the vertical line extensions through d and e to locate points E' and D' as shown. Using these points, point A, and point G' as shown, construct the overconsolidation branch of the curve.

Referring again to Fig. 13-19b, a sample obtained in situ under the present over-burden pressure p_o would have an initial zero pore pressure due to either creep, climatic conditions, or other factors, although there may be a static water pressure in the pore voids based on water table position. When the sample is removed from the site, some small rebound occurs (E to F), with development of negative pore pressures, but the rebound will be much less than for a normally consolidated clay. The sensitivity of the overconsolidated clay will also be much less than for the normally consolidated clay. Now when the cell stress σ_c is applied to simulate p_o, we obtain the curve FGE of Fig. 13-19b. At point E the pore pressure will be again zero.

Note that a preconsolidated laboratory sample can be made by placing a sample in the triaxial cell under a cell pressure $\Delta\sigma_3$ larger than that used for the test for a sufficient time for volume changes to be completed.

If the confining cell pressure is greater or less than p_c, the excess pore pressure is computed as

$$u = \Delta\sigma_3 - p_c$$

and will be positive if $\Delta\sigma_3 > p_c$ and negative if $\Delta\sigma_3 < p_c$. The negative pore pressure is also called *suction* and is caused by the capillary tensions in the water as the soil grains tend to expand due to the reduced consolidation pressure. In partially saturated soils, the capillary tensions produced due to drying also develop negative pore (or suction) pressures.

Application of the deviator stress $\Delta\sigma_1$ in the CU test will produce one of the following:

1. Positive pore pressures if $p_c/p_o < 4^+$; this will produce a reduced shear strength and $\phi < \phi'$.

2. Negative pore pressures if $p_c/p_o > 4^+$ due to a tendency for sample dilation, since the soil with this OCR is approaching a very high density and the shear stresses tend to produce volumetric expansion, which, with no sample drainage, produces suction. The resulting apparent shear strength in this situation will be unrealistically high—higher than the "undrained" shear strength. This condition in the field could produce an unsafe design condition, as the negative pore (suction) pressure will attract water which will:

(a) Reduce the cohesion, i.e., soften the clay.
(b) Increase the negative pore pressure toward zero so that the beneficial effect of the negative pore pressure in increasing the effective stress is lost.

If we perform an additional consolidated-undrained shear test at a cell pressure of $\sigma_c = p_c$ on a perfectly undisturbed sample, the resulting Mohr's circle should be tangent to curve OAB of Fig. 13-19a at point A.

The rupture envelopes for normally consolidated, moderately overconsolidated, and highly overconsolidated clays are qualitatively shown in Fig. 13-19c. Note that the qualitative effect of grain interlocking/density is shown as the differences in shear strength between points A, B, and C for the same *effective normal stress*. The location of point I of the two branches of the Mohr's failure envelope depends on the OCR of p_c/p_o and is not a single point.

The cohesion intercept obtained in CD tests for overconsolidated cohesive soils is probably closer to a true cohesion concept than the analytical parameters obtained from Eq. (13-1). The method of producing this cohesion is not fully understood but may consist of at least the following:

1. Interparticle forces and bonding created by the closer particle spacing resulting both from void ratio reduction and reduced water content as some of the double layer and outer free water is expelled by the consolidation pressure.
2. Interparticle cementation by leaching into the mass of contaminants. This includes the cementation of the byproducts produced by crushing of particle contact points from the consolidation pressure producing yield stresses on these small areas. This crushing of particle points may also produce a limited amount of interparticle cementation.

Shear strength behavior for overconsolidated clays at normal stresses greater than p_c would be expected to be the same as that for normally consolidated clays. Beyond the preconsolidation pressure the soil structure is simply subjected to a stress which tends to cause general particle displacement, with shear resistance developed from interlocking, contact shear stresses, and water content.

Since overconsolidated clays exhibit brittle fracture with a peak deviator stress at a low strain, the residual (also termed ultimate) stress parameters may be of considerable value. Residual shear strength is very nearly

$$s_r = \sigma' \tan \phi_r \qquad (13\text{-}9)$$

since the cohesion intercept is very nearly zero at large strains of 20 to 30 percent or larger. At these strains the soil is essentially a particulate mass with most of the interparticle bonds (cohesion) destroyed.

The residual strength should be investigated if progressive failure may be possible. The residual strength may be difficult to obtain in the standard triaxial test because of the influence of end platen restraint and the confining effect of the rubber membrane as the sample cross section substantially enlarges at the large strains. The direct shear test (Skempton, 1964) or a torsional shear apparatus (de Beer, 1967) may be used to obtain the residual strength.

Fissured Overconsolidated Clays

In many areas, particularly if the clay is heavily overconsolidated, the mass will contain a network of discontinuities. If the discontinuities consist of fine hairline cracks, they may be termed *fissures*. The term *joint* is also used to describe discontinuities in the soil mass. There is no unanimity of agreement at present as to what constitutes a fissure or joint. In any case, these discontinuities may be from a few centimeters to several meters in length. *Slickensides* are the somewhat polished interfaces developed when adjacent blocks of clay slip with respect to each other. Slickensides may also be formed at the interface of a landslide in clay. Figure 13-20 illustrates fissuring, joints, and slickensides using the author's definitions. These discontinuities are caused by cyclic wetting and drying, loss of overburden pressure causing expansion or tension cracks, earthquakes and other tectonic movements, and laminations occurring during sedimentation processes. Crack contamination with dust or organic materials when the fissure is open or exposed

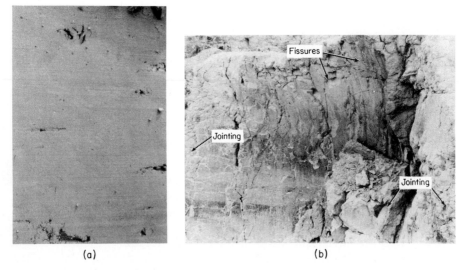

(a)   (b)

Figure 13-20 Slickensides, fissures, and joints. (*a*) Slickenside. Clay has been exposed to the atmosphere and slightly dried. (*b*) Fissures and joints. Fissures are large cracks.

tends to preserve the discontinuity even though later saturation (and swell) or subsequent overburden pressure cause crack closure.

The shear strength of fissured clays will vary widely, depending on whether the test is being made on a small, intact sample or on one that is larger and/or contains one or more discontinuities. The triaxial test produces reasonably reliable results, since the cell pressure will tend to close the fissure as in situ. The major problem with these soils is not in the testing, however, but in the prediction of the in situ shear strength after disturbance. This is particularly important in slopes constructed in this material, since the loss of overburden pressure coupled with crack opening and resultant softening from entering water can result in a failure either shortly after construction or as much as 10 to 20 years later. This phenomenon, with a number of references, is considered in detail by Morgenstern (1967).

Example 13-5

GIVEN Two consolidated-undrained triaxial tests on an unsaturated cohesive soil as follows:

Test	σ_1	σ_3	u, kPa
1	190	65	35
2	340	130	60

REQUIRED
(a) Apparent shear strength parameters c and ϕ.
(b) Effective shear strength parameters c' and ϕ'.

SOLUTION

Step 1 Plot Mohr's circles using total stresses as in Fig. E13-5 (solid lines). Obtain the cohesion intercept as $c = 14$ kPa and $\phi = 23.1°$.

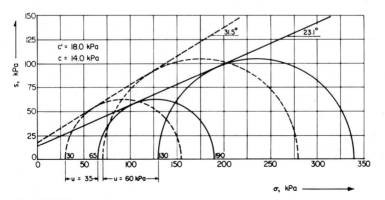

Figure E13-5

Step 2 Subtract u from σ_3 to obtain $\sigma_3' = 30$ for test 1 and 70 for test 2. Using the same radius for each test as in step 1 and the σ_3' starting points, draw the dashed circles of effective stresses as shown and scale $c' = 18.0$ kPa and $\phi' = 31.5°$.

13-7 PORE PRESSURE EFFECTS IN CONSOLIDATED-DRAINED TESTS

Consolidated-drained tests are performed by consolidating the sample as for CU tests. The test is then performed so slowly that "no pore pressures develop." Practically, this is impossible for soils with $S > 80$ to 90 percent, since drainage is necessary for the soil structure to reorient in shear and drainage can only occur with an excess pore pressure. This requires the test to be performed so slowly that small excess pore pressures develop which do not seriously invalidate the test.

Consolidated-drained tests are seldom performed for three major reasons:

1. CU tests with pore pressure measurements will give about the same results at a considerable saving in time.
2. CD tests take too long to be practical.
3. For most in situ loading conditions, U or CU tests adequately describe the shear strength requirements.

The Coulomb shear strength equation for CD tests on normally consolidated clays is (based on observations)

$$s = \sigma' \tan \phi' \tag{13-10}$$

The shear strength for overconsolidated clays for $\sigma' < p_c$ is given by Eq. (13-2). When the effective normal stresses are greater than p_c, for these clays the shear strength is defined by Eq. (13-10), as illustrated in Fig. 13-19c.

13-8 SENSITIVITY OF COHESIVE SOILS

The sensitivity of cohesive soils is defined as

$$S_t = \frac{\text{undisturbed strength}}{\text{remolded strength}}$$

where the strength may be the undrained, consolidated-undrained, or consolidated-drained value. The undrained value is most commonly used. The remolded strength is obtained by remolding the specimen used for the undisturbed strength so that the water content is as nearly the same as practical. The sensitivity description is as follows:

Insensitive	$S_t < 2$
Medium sensitivity	$2 < S_t < 4$
Sensitive	$4 < S_t < 8$
Very sensitive	$8 < S_t < 16$
Quick	$S_t > 16$

Highly overconsolidated clays tend to be insensitive. This is at least partially due to the low natural water contents in these deposits. Most cohesive soils, such as glacial till clays and those found in the B horizon of residual deposits, are of medium sensitivity. A few glacial clays and most fresh-water deposits are very sensitive. A few of the fresh-water and marine deposits are quick. The sensitivity of the large majority of cohesive deposits will range from 2 to 8. Sensitivities greater or less than this are much less commonly encountered. Most quick clays seem to be found (or at least reported) in Canada and Scandinavia.

13-9 EMPIRICAL METHODS FOR SHEAR STRENGTH

Numerous correlations for shear strength or shear strength parameters have been proposed in the literature. Several will be presented here to illustrate some of those available.

One of the earliest correlations is that between the SPT (Sec. 3-9) and the unconfined compression strength, as illustrated in Table 13-1.

Table 13-1 Empirical relationship between SPT and several soil properties†

N	Consistency	Field identification	γ_{sat}, kN/m³	q_u, kPa
< 2	Very soft	Easily penetrated several centimeters by fist	16–19	< 25
2–4	Soft	Easily penetrated several centimeters by thumb	16–19	25–50
4–8	Medium	Moderate effort required to penetrate several centimeters with thumb	17–20	50–100
8–16	Stiff	Readily indented by thumb	19–22	100–200
16–32	Very stiff	Readily indented by thumbnail	19–22	200–400
> 32	Hard	Difficult to ident with thumbnail	19–22	> 400

† Generally determine values locally, as $q_u = CN$, and supplement with visual inspection, as it is not uncommon to get low N values when q_u is large.

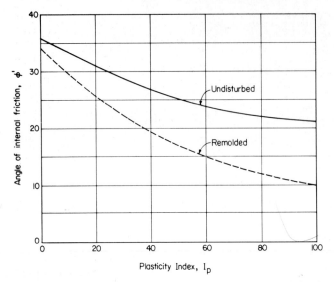

Figure 13-21 Correlation between angle of internal friction ϕ' (true) and plasticity index for both undisturbed and remolded soil. (*After Bjerrum and Simons, 1960.*)

Relationships between ϕ and plasticity index I_p are shown in Fig. 13-21. A relationship between ϕ and percent clay fraction (Skempton, 1964) is shown in Fig. 13-22. Both of these curves should be used cautiously, as there are several major exceptions which can be found in the literature as well as substantial scatter in the data points used to establish the curves. For routine soil work, however,

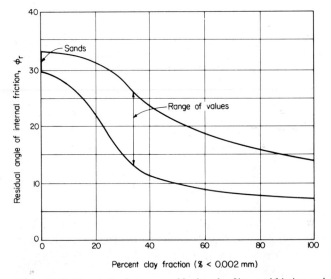

Figure 13-22 Correlation between residual angle of internal friction and percent clay. (*After Skempton, 1964.*)

Figure 13-23 Torvane.

particularly in regions where w_L is on the order of 20 to 45 and I_P on the order of 15 to 30, these curves will be reasonably reliable.

Figure 13-23 illustrates the torvane, which is commonly used to obtain the shear strength of soft to very soft clay. Due to its small size, several tests can be made for statistical determination of s_u from SPT samples or in the sides of test pits.

Figure 13-24 illustrates the pocket penetrometer device, which can be used in test pits or in the base of borings 75 cm or larger in diameter where a person can be lowered into the hole, as in caisson work. This device works well in any fine-grained cohesive soil. The operator simply selects a gravel-free location, pushes the piston rod into the soil to the calibration mark, and simultaneously reads q_u on the graduated scale. Again due to its small size, several tests can be made for a statistical determination of q_u.

The "Rimac" device, shown in Fig. 13-25, was developed for testing automobile valve springs and has been modified for field testing. This device is widely used in field boring operations (see Fig. 3-6e) to determine q_u as the borehole advances. The operator obtains a reasonably intact sample from the SPT test and measures the diameter after carefully trimming a length (determined directly from the scale on machine). The sample is placed between the platens of the Rimac tester and compressed to the maximum load. The load and final sample length are simultaneously read, and the unconfined-compression strength is computed as

$$q_u = \frac{P_{max}(L_o - \Delta L)}{A_o L_o}$$

which is, of course, the same as Eq. (13-4).

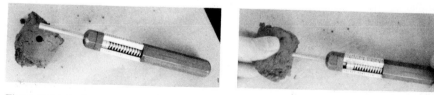

Figure 13-24 Pocket penetrometer.

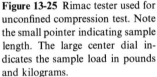

Figure 13-25 Rimac tester used for unconfined compression test. Note the small pointer indicating sample length. The large center dial indicates the sample load in pounds and kilograms.

A borehole shear device has been recently developed, as schematically shown in Fig. 13-26. This device is inserted in the borehole, expanded into the soil, then pulled. The force used to expand the device into the soil is P_v, and the pull to shear is P_h in the direct shear test as used in Eq. (13-5). This device essentially performs the direct shear test in situ. The test should approximate the CU test if a reasonable time is allowed for drainage, as the shear teeth or serrations are not very deep.

A vane shear test, schematically shown in Fig. 13-27, is widely used in soft, cohesive deposits where sample disturbance is critical. This test consists of insert-

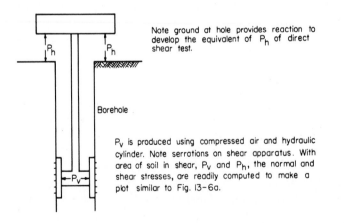

Note ground at hole provides reaction to develop the equivalent of P_h of direct shear test.

Borehole

P_v is produced using compressed air and hydraulic cylinder. Note serrations on shear apparatus. With area of soil in shear, P_v and P_h, the normal and shear stresses, are readily computed to make a plot similar to Fig. 13-6a.

Figure 13-26 Borehole shear device. [For additional details, see Wineland (1975).]

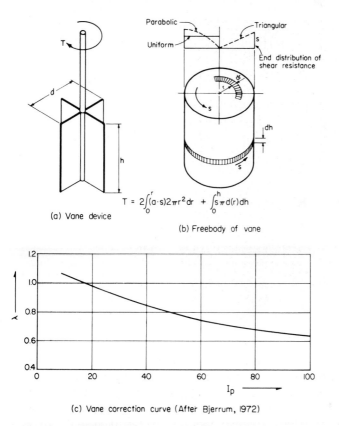

(a) Vane device

(b) Freebody of vane

$$T = 2\int_0^r (a \cdot s)2\pi r^2 dr \; + \; \int_0^h s\pi d(r)dh$$

(c) Vane correction curve (After Bjerrum, 1972)

Figure 13-27 Schematic of vane shear test and strength corrections.

ing a vane into the soil and applying a torque. From a free-body diagram as in Fig. 13-27*b*, the consolidated-undrained shear strength s_u can be computed as

$$s_u = \frac{4T}{\pi(2d^2h + ad^3)} \qquad (13\text{-}11)$$

where T = applied torque

d = diameter of vane (commonly 50 to 150 mm)

h = length of vane (commonly 100 to 225 mm)

$a = \frac{2}{3}$ for uniform end shear

$\frac{3}{5}$ for parabolic end shear

$\frac{1}{2}$ for triangular end shear

In practice a borehole is advanced to less than 1 m from the point of test. The hole is cleaned and the vane inserted and carefully pushed to the desired elevation, then a torque is applied until the cylinder of soil defined by the vane perimeter shears. The CU shear strength is obtained, since the approximate in situ effective stress is acting on the soil at the test point.

Some controversy exists as to how correct the vane s_u value is compared to laboratory tests or for design use. The latest opinion appears to be that the vane s_u should be corrected for design as

$$s_{u(design)} = \lambda s_{u(vane)} \qquad (13\text{-}12)$$

where λ is obtained from the curve shown in Fig. 13-27c as proposed by Bjerrum (1972). This curve was proposed after a study of a large number of embankment failures indicated that the shear strength used in design (and that would be obtained from the vane shear test) was too high.

13-10 FACTORS AFFECTING SHEAR STRENGTH

It should be evident at this point that the shear strength of a soil depends on:

1. Effective or intergranular pressure $\sigma' = \sigma - u$
2. Interlocking of particles; thus, angular particles give greater interlocking and higher shear strength (larger ϕ) than rounded particles such as bank run gravel and sand (see Table 13-2)
3. Packing of particles, or density
4. Cementation of particles, natural or otherwise
5. Particle attraction or cohesion. Note that apparent cohesion due to capillary effects may also be present
6. Water content, particularly for cohesive soil. Cohesion is principally affected by water content as evidenced by behavior of a dry vs. a wet lump of clay
7. Sample quality
8. Test method, as U, CU, or CD
9. Other effects, such as humidity, temperature, testing skill, condition of laboratory equipment, etc.

Table 13-2 Typical range of values of true angle of internal friction ϕ' for several soils

	ϕ'	
Soil	Loose	Dense
Sand, crushed (angular)	32–36°	35–45
Sand, bank run (subangular)	30–34	34–40
Sand, beach (well rounded)	28–32	32–38
Gravel, crushed	36–40	40–50
Gravel, bank run	34–38	38–42
Silty sand	25–35	30–36
Silt, inorganic	25–35	30–35
Clay	See Figs. 13-21 and 13-22	

For cohesive soils the shear strength is also considerably influenced by:

1. The rate of loading or strain rate—higher shear strengths are obtained at higher strain rates
2. Anisotropy of mass—vertical strength not the same as the horizontal value
3. Effects of progressive failure, including the effect of in situ soil not being simultaneously loaded to the same intensity or rate and changes in water content due to strain effects, cracks, and external entering water causing softening

13-11 THE $s_u/\bar{p}$ RATIO

The undrained (either U or CU) shear strength is termed s_u. For the unconfined compression or any $\phi = 0$ shear test, s_u is the cohesion intercept, or

$$s_u = c$$

Figure 13-16d indicates an approximate linear relationship between the undrained shear strength s_u and the effective consolidation pressure $\bar{p}$ for a normally consolidated clay. This relationship is now widely recognized, and several empirical correlations exist between $s_u/\bar{p}$ and I_P, as shown in Fig. 13-28 and in Fig. 13-29

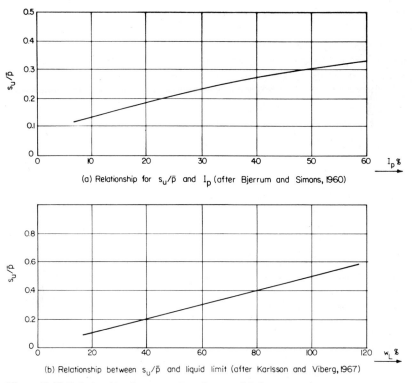

(a) Relationship for $s_u/\bar{p}$ and I_p (after Bjerrum and Simons, 1960)

(b) Relationship between $s_u/\bar{p}$ and liquid limit (after Karlsson and Viberg, 1967)

Figure 13-28 Relationships between $s_u/\bar{p}$ and two soil index properties.

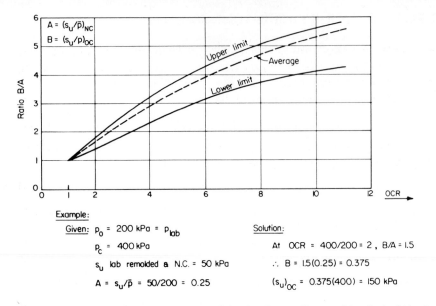

Example:

Given: $P_0 = 200$ kPa $= P_{lab}$

$P_c = 400$ kPa

s_u lab remolded a N.C. = 50 kPa

$A = s_u/\bar{p} = 50/200 = 0.25$

Solution:

At OCR = 400/200 = 2 , B/A = 1.5

∴ B = 1.5(0.25) = 0.375

$(s_u)_{oc} = 0.375(400) = 150$ kPa

Figure 13-29 Relationship between overconsolidated and normally consolidated $s_u/\bar{p}$. (*After Ladd and Foott, 1974.*)

for overconsolidated soils. Figure 13-29 is especially useful in obtaining the normalized soil parameter s_u when the soil exhibits normalized behavior.† According to Ladd and Foott (1974), quick clays and naturally cemented clays with a high degree of structure do not exhibit normalized behavior due to the significant structure alteration at higher consolidation stresses.

The $s_u/\bar{p}$ ratio can be approximately expressed as one of the following:

$$\frac{s_u}{\bar{p}} = 0.11 + 0.0037 I_P \qquad I_P > 10$$

with a scatter on the order of ± 0.05 and using I_p as percent.

$$\frac{s_u}{\bar{p}} = 0.45 w_L \qquad w_L > 40$$

with a scatter on the order of ± 0.10 and using w_L as decimal.

Laboratory shear test values of $s_u/\bar{p}$ for normally consolidated clays tend to range from about 0.09 to 0.35, with the higher ratios for I_P on the order of 50 or more. In situ vane $s_u/\bar{p}$ values tend to be higher, with an approximate range of 0.2 to 0.45, again with the higher values for I_P of 50 or more.

---

† Behavior characterized by similar strength and stress-strain characteristics when the parameter is normalized (divided by) using the effective preconsolidation stress.

13-12 PORE PRESSURE PARAMETERS

A considerable amount of work has been done in attempting to predict pore water pressures in both the saturated and unsaturated soil states. A number of these are cited by Lo (1969) in a paper which has been liberally used here.

One method of expressing the pore water pressure† is

$$u = aJ_1 + bJ_2^{1/2} + cJ_3^{1/3} \tag{13-13}$$

where J_i are the principal stress invariants defined in Eq. (10-4). In an isotropic stress state, $\sigma_1 = \sigma_2 = \sigma_3$ and $a = 1$, so that we obtain

$$u = J_1 + bJ_2^{1/2} + cJ_3^{1/3} \tag{13-14}$$

The difficulty is that the coefficients b and c are dependent on the mechanical properties of the soil mass and are not easily determined.

If the mass were perfectly elastic, no pore water pressures would be caused by J_2 and J_3 and

$$u = \tfrac{1}{3}(\sigma_1 + \sigma_2 + \sigma_3)$$

which has commonly been used for isotropic consolidation and gives $u = \Delta\sigma_3$ as commonly measured in triaxial tests for $S = 100$ percent.

For real soils the coefficients b and c depend on the stress-strain state of the soil, and no unique values are obtained. If we neglect the effect of stress-invariant J_3, and with triaxial conditions of $\sigma_2 = \sigma_3$, for saturated soils Eq. (13-14) may with some algebraic manipulation be transformed to

$$\Delta u = \Delta\sigma_3 + A(\Delta\sigma_1 + \Delta\sigma_3) \tag{13-15}$$

where
$$A = \frac{1}{3} + \frac{m}{3}$$

To give some additional generality to the problem, we may rewrite Eq. (13-15) as

$$u = B[\sigma_3 + A(\sigma_1 - \sigma_3)] \tag{13-16}$$

which is the form suggested by Skempton (1954). Table 13-3 gives some ranges of values for the pore pressure parameter A.

Particular limitations on Eq. (13-16) include:

1. When a stress difference ($\sigma_1 - \sigma_3 = \Delta\sigma_1$) is applied, an induced pore pressure u is set up. When $\Delta\sigma_1$ is removed, a large part of u remains for some time. If $\Delta\sigma_1$ is repeatedly applied and removed, the magnitude of the residual pore pressure will increase.
2. Under sustained application of $\Delta\sigma_1 =$ constant, the pore pressure continues to increase.

---

† Some persons use u while others use Δu for pore pressure. In any case, the pore pressure of interest is that in excess of hydrostatic conditions caused by changes in applied stresses.

Table 13-3 Values of the pore pressure parameter A†

Soil	A
Loose, fine sand	2–3
Sensitive clay	1.5–2.5
Normally consolidated clay	0.7–1.3
Lightly overconsolidated clay	0.3–0.7
Heavily overconsolidated clay	−0.5–0.0
Compacted sandy clay	0.25–0.75

† After Skempton (1954).

3. For sensitive clays the pore pressure continues to increase after failure even though $\Delta\sigma_1$ remains constant or somewhat decreases.
4. Difficulty of reliable prediction of u when S is less than 100 percent.

These several observed phenomena indicate that the pore pressure is strain-dependent rather than stress-dependent as described by Eq. (13-14). Lo (1969) recommended adjusting Eq. (13-14) to

$$u = u_i + u_s$$

where u_i = pore pressure from applying a stress with $\varepsilon_v = \varepsilon_1 = \varepsilon_2 = \varepsilon_3 = 0$, i.e., no volume change as obtained for $S = 100$ percent and isotropic stress conditions. This gives $u_i = \sigma_3$ for an isotropic consolidated triaxial test.

Application of a stress difference $\Delta\sigma_1$ will produce shearing strains (even for constant volume conditions) and additional pore pressures u_s. The pore pressures due to strain effects can be expressed as

$$\frac{u_s}{\bar{p}_o} = b'\varepsilon_1$$

for triaxial stress conditions of $\varepsilon_2 = \varepsilon_3$. For plane strain conditions of $\varepsilon_2 = 0$, the pore pressure is

$$\frac{u_s}{\bar{p}_o} = b''\varepsilon_1$$

where the coefficients b' and b'' would have to be determined somehow. The pore pressure is normalized using $\bar{p}_o$ = effective consolidation pressure in both equations for dimensional analysis. In practice, the values of u_i are combined (Lo, 1969), resulting in the following:

For anisotropic consolidation at $\bar{p}_o$,

$$u = \bar{p}_o\varepsilon_1 + \bar{p}_o(1 - K) \tag{13-17}$$

For isotropic consolidation $(K = 1)$,

$$u = \bar{p}_o \varepsilon_1 + \bar{p}_o \qquad (13\text{-}18)$$

In both of these equations the principal strain ε_1 is a percent. It is, of course, desirable to use stress differences instead of strain, since stresses are more likely to be known in a given pore pressure problem.

Example 13-6

GIVEN Two samples from the same stratum initially subjected to consolidation stresses as follows:

Soil 1	Soil 2
Isotropic: $\sigma_3 = 70$ kPa	$\sigma_2 = \sigma_3 = 55$ kPa
$u = 70$ kPa (measured)	$\sigma_1 = 100$ kPa during test

REQUIRED
(a) Under what circumstances can u be the same for both tests at the stresses indicated?
(b) If the measured u for soil 2 is 110 kPa for the σ_1 given, how much axial strain has occurred?
(c) What are the A and B parameters of Eq. (13-16)?

SOLUTION

(a) For soil 1: $\sigma_{\text{oct}} = \dfrac{\sigma_1 + \sigma_2 + \sigma_3}{3} = \Delta\sigma = 70$ kPa

$$\sigma_1 - \sigma_3 = 0$$

$$u = \sigma_3 + A(\sigma_1 - \sigma_3) = 70 \text{ kPa} \qquad \text{using Eq. (13-15)}$$

For soil 2: $\sigma_{\text{oct}} = \dfrac{100 + 55 + 55}{3} = 70$ kPa

$$u = \sigma_3 + A(\sigma_1 - \sigma_3) = 55 + A(100 - 55) = 55 + 45A$$

For u to be the same, we have

$$45A + 55 = 70$$

$$A = \frac{70 - 55}{45} = \frac{1}{3}$$

(b) Assume

$$K = 1 - \sin \phi = 0.50$$

With anisotropic conditions and $\bar{p}_o = 55$ kPa,

$$u = \bar{p}_o \varepsilon_1 + \bar{p}_o(1 - K)$$

$$110 = 55\varepsilon_1 + 55(1 - 0.5)$$

Solving: $\varepsilon_1 = 1.5$ percent (0.015 m/m)

Thus at $\sigma_1 = 100$ kPa, the deviator stress is $\Delta\sigma_1 = 45$ kPa and the strain is approximately 0.015 m/m.

(c) The B parameter is 1 for the soil, since from soil 1 we have

$$u = B[\sigma_3 + A(\sigma_1 - \sigma_3)] = B\sigma_3 = 70$$

therefore,

$$B = \tfrac{70}{70} = 1.0$$

For soil 2 we have

$$110 = 1[55 + A(100 - 55)]$$

from which

$$A = \tfrac{55}{45} = 1.22$$

13-13 SUMMARY

The concept that shear strength of a soil is not, in general, a unique value has been continually emphasized.

For cohesionless soils the shear strength is

$$s = \sigma_n \tan \phi$$

where σ_n is usually the effective stress, since the permeability is large enough that with a reasonable strain rate, drainage occurs. The principal deficiencies are duplication of the in situ void ratio and cementation.

For cohesive soils we have for the undrained (U) and consolidated-undrained (CU) conditions

$$s_u = c + \sigma \tan \phi$$

In consolidated-drained tests for normally consolidated clays, or if pore pressures are measured in CU tests, we have

$$s = \sigma' \tan \phi'$$

The cohesion in undrained tests will be heavily dependent on water content, while the angle of internal friction depends primarily on the void ratio and particle shape.

We have also considered the concept of peak and residual shear resistance and methods of using Mohr's circle to obtain the corresponding strength parameters. We note that residual parameters may be more appropriate for progressive or large strain failure conditions, whereas peak values are appropriate for small strains. As examples, large strains commonly occur in road and levee embankments, small strains for building foundations.

The effects of overconsolidation on shear strength were considered, and it was noted that long time and/or progressive failure tends to reduce the effects of overconsolidation.

We should note that it is the *effective* stress which has the most influence on shear strength. The effective stress is computed as

$$\sigma' = \sigma - u$$

where σ = total stress and u = pore pressure. Where partially saturated soils exist, u may be negative, temporarily increasing the effective stress. It was shown that pore pressure can be obtained reliably only by measurement, although the computed pore pressure parameters may be used as rough approximations.

Various methods of determining shear strength were presented, including triaxial testing, direct shear testing, unconfined compression, vane shear, borehole shear, using the torvane and pocket penetrometer, and use of the SPT. Correlations using $s_u/\bar{p}$ versus I_P and w_L were also noted.

HOMEWORK PROBLEMS

13-1 Describe the cohesion and friction components for a cohesive soil.

13-2 Why is there no cohesion intercept for a cohesive soil in a CD test?

13-3 Why is there no cohesion intercept for the residual shear strength of a cohesive soil?

13-4 Why can no data be obtained for a undrained test on a sand when $S = 100$ percent?

13-5 Explain why the pore pressure can never be zero in a drained test as long as $S > 0$.

13-6 Compute σ_n and s of Ex. 13-2 and compare them to the scaled values shown.

13-7 Draw stress paths for the following stress conditions:

Test	Initial stresses, kPa		Final stresses, kPa	
	σ_1	σ_3	σ_1	σ_3
1	200	200	700	200
2	200	200	200	700
3	200	200	100	100
4	100	100	300	150
5	300	300	300	100

13-8 A triaxial test was performed and the following data obtained:

Test	σ_3	$\Delta\sigma_1$ failure, kPa
1	70	350
2	140	630
3	210	980

Plot Mohr's circles and find c and ϕ.

   Ans.: $c \cong 0$, $\phi = 45°$.

13-9 An unconfined compression test yielded $q_u = 150$ kPa (average for three tests). What is the percent error, and is it conservative or unconservative to assume $s_u = q_u/2$ when it is known that this soil has a consolidated-undrained angle of internal friction $\phi = 10°$?

13-10 The following data were obtained from a series of drained triaxial tests on a damp sand. Plot the assigned data and determine both ϕ' and, at $\varepsilon > 0.25$, $\phi_{residual}$. Comment on the effect of density both on the slope of the curves and on ϕ. Note that the data were plotted by the author, then the data shown were taken from the smoothed curves for ease in tabulation (and for further plotting).

	Test					
	1	2	3	4	5	6
σ_c, kPa	70	140	210	70	140	210
Void ratio e	0.79	0.78	0.80	0.46	0.47	0.46
Strain ε	Deviator stress $\Delta\sigma_1$, kPa					
0	0	0	0	0	0	0
0.01	50	155	245	140	330	600
0.02	90	220	410	180	420	713
0.04	125	300	460	230	480	710
0.06	140	320	470	236	460	680
0.08	152	316	455	230	440	650
0.10	150	305	450	220	425	640
0.12	145	300	440	215	418	630
0.15	140	290	435	210	408	615
0.20	140	285	430	205	398	600
0.25	140	280	425	200	395	590

13-11 Sketch a qualitative shear envelope for an overconsolidated clay. Show OCR = 1, 2, and 10. Indicate p_c for OCR = 2 and 10. Would a q_u test give a satisfactory answer? Explain.

13-12 In a direct shear test on a sand, $\sigma_n = 180$ kPa and $s = 121.5$ kPa at failure. What is the angle of internal friction ϕ, and what are the principal stresses σ_1 and σ_3?

   Partial ans.: $\sigma_3 = 115.42$ kPa.

13-13 A vane shear test was performed and the following data were obtained:

$$T = 51 \text{ N} \cdot \text{m} \qquad w_L = 57.3$$

$$d = 65 \text{ mm} \qquad w_P = 28.9$$

$$h = 110 \text{ mm} \qquad w_N = 47.5$$

Assume uniform end shear resistance. Is the soil preconsolidated? What value of s_u do you recommend for design? What laboratory test will give this value of s_u?

13-14 In a CU test on a normally consolidated clay with pore pressure measurements, the following data at failure were obtained:

$$\sigma_c = 70 \text{ kPa}$$

$$\Delta\sigma_1 = 74.3$$

$$u = 22.5$$

What is the effective angle of internal friction ϕ'?

    Ans.: $\phi' = 27°$

13-15 Three consolidated-undrained triaxial tests were performed as follows:

Test	σ_3	σ_1	u, kPa
1	0	138.5	? (Unconfined-compression test)
2	70	351.8	56.0
3	140	515.0	105.5

What are the apparent and effective soil parameters c and ϕ? Is the soil saturated?

Chapter 14

Stress-Strain Characteristics and Stresses at a Point

14-1 STRESS-STRAIN DATA REQUIRED

Stress-strain data are obtained as described in Sec. 13-2, using Eq. (13-2) for corrected area and Eq. (13-3) for the deviator stress. The resulting values of stress and strain are plotted to obtain the stress-strain curve as shown in Fig. 13-7a and Fig. 13-11.

From the stress-strain curve one may obtain a value of stress-strain modulus, commonly called the modulus of elasticity E_s, and may compute a value of Poisson's ratio μ.

The stress-strain modulus and Poisson's ratio are widely used in settlement computations (Sec. 15-9)—both in conventional methods, and in finite element analyses, which assume the soil is an elastic continuum.

14-2 THE STRESS-STRAIN MODULUS

The modulus of elasticity, a property of elastic materials, is defined as a constant of proportionality between stress and strain as

$$E = \frac{\Delta\sigma}{\Delta\varepsilon} \tag{14-1}$$

Most engineering materials exhibit this linear behavior over some region of stress (and the corresponding strain). Steel (Fig. 14-1a) exhibits linear behavior over the largest stress range (on the order of 250 MPa); concrete, aluminum, cast iron, wood, etc., are linear over a very limited stress range. Further, even though the stress-strain curve is nonlinear for these materials, if the sample is not stressed beyond the elastic limit, it will return to its original dimensions upon removal of the stress. This is a part of the definition of a homogeneous, isotropic, elastic material.

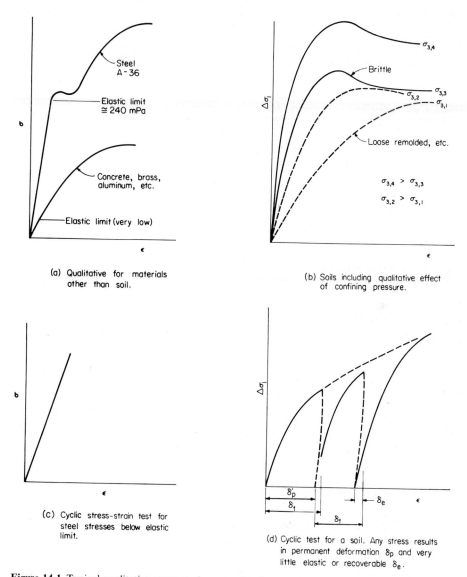

(a) Qualitative for materials other than soil.

(b) Soils including qualitative effect of confining pressure.

(c) Cyclic stress-strain test for steel stresses below elastic limit.

(d) Cyclic test for a soil. Any stress results in permanent deformation δ_p and very little elastic or recoverable δ_e.

Figure 14-1 Typical qualitative stress-strain curves for the materials and conditions indicated.

Soil exhibits linear stress-strain characteristics only at extremely low strain amplitudes—strains generally on an order of magnitude of 10^{-4} and smaller. In conventional triaxial tests no data are obtained at these strain levels, and as a consequence the stress-strain curve has no linear region. Also, unless the test is halted when the sample has strained the low amplitude indicated, the sample will not recover its original shape but will remain permanently deformed, as illustrated in Fig. 14-1d. This is because the principal soil deformations are state changes caused by relative particle motion. Only a very small amount of soil deformation occurs from particle distortion and is elastically recoverable. Because soil deformation is primarily due to relative particle motion, modulus of elasticity is not a correct term to use; rather, one should use the *stress-strain modulus*. This becomes a value describing the relationship between applied stresses and resulting strains. Conventional terminology, however, uses modulus of elasticity, and the two descriptive terms will be used interchangeably in this text. The user should be aware that modulus of elasticity is a term of convenience rather than a true elastic property for soil materials. Some persons use "elastoplastic" or "theory of plasticity" as descriptive terms for soil distortion. Both of these terms are based on an elastic continuum, and thus for soil are also convenience terms, since soil is a particulate mass and not a continuum.

There are two commonly used methods for computing the stress-strain modulus from nonlinear stress-strain curves:

Tangent modulus. Modulus based on the slope of a line which is just tangent to the stress-strain curve at a point. The *initial* tangent modulus is commonly used (a tangent at the origin), since the slope at the origin is not highly subject to environmental factors such as type of test and confining pressure.

Secant modulus. Modulus based on the slope of a secant line. A secant line cuts the stress-strain curve at two points. When used, the two points are commonly equally spaced from the working stress.

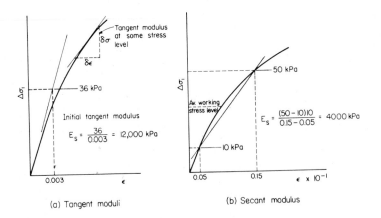

(a) Tangent moduli

(b) Secant modulus

Figure 14-2 Computing stress-strain modulus E_s by several methods.

The tangent and secant moduli are illustrated in Fig. 14-2. In finite element computations, a tangent modulus may be used with the tangent taken at the stress level under consideration. Unfortunately, with the finite element procedure, the stress depends on the modulus of elasticity, and thus iteration is necessary.

Soil is commonly anisotropic rather than isotropic due to formation by sedimentation. This produces a situation where E_h is generally not equal to E_v. The determination of E_s should consider the possibility of anisotropy and orient the test samples to obtain the correct E_s.

14-3 POISSON'S RATIO

Poissons's ratio is a property of elastic materials defined as

$$\mu = \frac{\varepsilon_3}{\varepsilon_1} \tag{14-2}$$

where
$\varepsilon_1 =$ strain colinear with stress of interest

$\varepsilon_3 =$ strain orthogonal to stress of interest

Poisson's ratio is often assumed to be 0.2 to 0.4 in soil mechanics work. A value of 0.5 is usually used for saturated soils, and 0.0 is often used for dry soils and others as a computational convenience. This is because a value of Poisson's ratio is difficult to obtain for soils. The vertical strain is relatively easy; the lateral strain is particularly difficult as compared with, say, any metal where a strain gage may be readily attached. A few soils laboratories have modified triaxial cells to allow diameter measurements to be taken as the test proceeds. We may also take volumetric measurements as the test proceeds, which requires completely filling the cell with a liquid during sample consolidation, and use Eqs. (10-7). For example, in the triaxial test, if we measure the change in sample volume ΔV, the volumetric strain ε_v can be computed as

$$\varepsilon_v = \frac{\Delta V}{V_o} = \varepsilon_1 + \varepsilon_2 + \varepsilon_3$$

Also we have $\sigma_x = \sigma_z = \sigma_c$ and $\sigma_y = \sigma_1$. Substituting into Eqs. (10-7), we obtain

$$\varepsilon_v = \frac{(\sigma_1 + 2\sigma_3)}{E}(1 - 2\mu) \tag{14-3}$$

For plane strain, σ_2 or $\sigma_3 = 0$, and with this substitution into Eqs. (10-7), we obtain

$$\varepsilon_{1p} = \frac{1 - \mu}{E}\left(\sigma_1 - \sigma_3 \frac{\mu}{1 - \mu}\right) \tag{14-4}$$

$$\varepsilon_{1p} = \frac{1}{E_p}(\sigma_1 - \mu_p\sigma_3) \tag{14-4a}$$

From this we have

$$E_p = \frac{E}{1 - \mu^2} \quad \text{and} \quad \mu_p = \frac{\mu}{1 - \mu}$$

Thus, the plane strain values of E_p and μ_p are shown to be larger than the triaxial values. Note also that μ is dependent on E_s so that in anisotropic soils there may be both vertical and lateral values corresponding to E_v and E_h.

If one assumes a sufficiently close strain interval, the stress-strain modulus may be assumed to be linear between any two points, and the general equation defining strain may be used with stress-strain data from a triaxial test to obtain

$$\varepsilon_1 = \frac{1}{E_s}(\Delta\sigma_1 - 2\mu\sigma_3)$$

which can be solved for two adjacent values of ε_1 and $\Delta\sigma_1$ to obtain E_s and μ. This type of computation can be made up to the point where negative values or values of $\mu > 0.5$ are obtained. No real material appears to have $\mu < 0$, and a value of $\mu > 0.5$ indicates plastic behavior of an elastic material. In making this type of computation one finds that $\mu > 0.5$ occurs at relatively low strain levels, as one might intuitively suspect from this type of material. We might note also that most computations involving elastic materials are valid only for small distortions. This computation illustrates that μ is not a constant for a soil but depends on stress level.

Example 14-1

GIVEN The stress-strain curve shown (Fig. E14-1) for a soft clay soil.

REQUIRED Compute a value of E_s and μ.

SOLUTION At $\Delta\sigma_1 = 250$ kPa, obtain $\varepsilon_1 = 0.011$. At $\Delta\sigma_1 = 375$ kPa, obtain $\varepsilon_1 = 0.018$. Substituting,

$$0.011E_s = 250 - 2\mu(100)$$
$$0.018E_s = 375 - 2\mu(100)$$

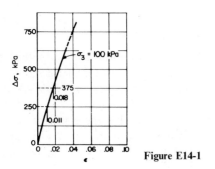

Figure E14-1

By elimination, obtain

$$E_s = \frac{125}{0.007} = 17\,857 \text{ kPa}$$

Back substitution into the first equation gives

$$\mu = \frac{250 - 196}{200} = 0.26$$

14-4 FACTORS AFFECTING THE STRESS-STRAIN MODULUS AND APPROXIMATIONS

Any factor which will modify the slope of the stress-strain curve will affect the stress-strain modulus and Poisson's ratio. These factors include:

Cell (or in situ confining) pressure (qualitatively illustrated in Fig. 14-1*b*)
Soil unit weight
Geological history (producing a brittle or progressive failure)
Grain shape
Sample size
Type of test, as U, CU, or CD
Sample disturbance

For these several reasons, it may be convenient to estimate the value of E_s, at least for preliminary design. For cohesive soils,

Soft clay: $\qquad\qquad\qquad\qquad E_s \cong 250 \text{ to } 750 s_u$

Stiff clay: $\qquad\qquad\qquad\qquad E_s \cong 750 \text{ to } 1500 s_u$

where s_u = undrained shear strength. Sand is so variable that some kind of tests should be performed to estimate a value. One estimate based on the SPT is

$$E_s \cong 30\,000 \text{ to } 50\,000 \log N$$

where N = standard penetration test number as outlined in Sec. 3-9.

14-5 RESILIENT MODULUS

The resilient modulus is defined as the initial tangent modulus of a triaxial test stress-strain curve which has been cycled several times with a deviator stress $\Delta\sigma_1$ of a value of approximately the working stress, as shown in Fig. 14-3. Alternatively, a value of $\Delta\sigma_1 \cong \Delta\sigma_1/2$ (maximum) is sometimes used.

This test was initially used for pavement subgrades, but there is some body of opinion that this value may be more appropriate for many foundation settlement problems. This value is useful partially because the E value increases somewhat in

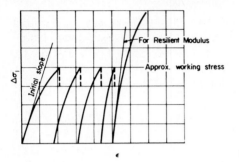

Figure 14-3 Method of obtaining the resilient modulus of deformation.

the test, due to soil particle readjustment, to a more stable structure on successive load-unload cycles. This produces a larger initial tangent modulus on the last cycle and a correspondingly smaller computed settlement which is more in line with observed settlements.

14-6 DYNAMIC SOIL STRESS-STRAIN MODULUS

The dynamic stress-strain modulus is useful in predicting vibration displacements of foundations subjected to cyclic loads. Machine, pump, radar tower, compressor, and turbine foundations are typical foundations which are loaded dynamically. Earthquakes also produce dynamic loadings which require displacement and strength predictions.

Soil dynamic problems fall in two general categories of strain magnitudes:

	Range of strain amplitude ε
Machine foundations	10^{-4} to 10^{-3}
Earthquake	10^{-2} to 10^{-1} and larger

These ranges of strain amplitudes require separate consideration to determine the dynamic properties of stress-strain modulus, damping, and soil strength. This is because the modulus and strength decrease, often markedly, with increasing strain amplitude. The values of stress-strain modulus obtained from the usual triaxial compression test tend to be quite low (perhaps on the order of $\frac{1}{2}$ to $\frac{1}{10}$) compared with dynamic stress-strain moduli determined from low amplitude strain tests. This is because the first sample strain value in the standard triaxial test is often on the order of 0.005 m/m, which may be satisfactory for an earthquake analysis and static analyses in general but is several orders of magnitude too large for machine foundation analysis. With machine foundations, dynamic soil strength is not generally a critical factor, whereas with earthquake studies this will be the principal design consideration, as shown in the next section.

It is preferable to determine the stress-strain properties in situ for machine foundations with low strain amplitudes. Where this is not practical, several laboratory methods are available, including cyclic triaxial testing (Park and Silver, 1975; Weissman and Hart, 1961), cyclic direct shear tests (Kovacs et al., 1971), hollow cylinder torsion tests (Ishihara and Yasuda, 1975), and resonant column testing (Hardin and Music, 1965).

The resonant column test has been used considerably for determining the dynamic modulus for machine foundations. This test utilizes the time of travel of a shear wave through a column of soil which is either hollow or solid. The frequency at resonance is obtained and the stress-strain modulus computed. General details of test details and equipment are beyond the scope of this text, and the reader is referred to Hardin and Music (1965) and Richart et al. (1970) for further details.

In situ determination of the dynamic stress-strain modulus requires measuring either shear, Rayleigh, or compression waves and back computing the modulus. Elastic, homogeneous ground stressed with a vibration produces three elastic waves traveling outward from the source at different speeds. These waves are:

Primary or compression waves, called P waves (have the highest velocity)
Secondary or shear waves, called S waves
Near ground surface waves termed Rayleigh waves, or R waves.

The velocity of the R wave is about 10 percent less than that of the S wave, but the R wave is easier to interpret on an oscilloscope (or from an oscillograph) and is usually used.

The in situ determination of the dynamic stress-strain modulus uses seismic techniques, i.e., a shock source and an electronic pickup unit. The pick-up unit often inputs into an oscilloscope so that the operator may either visually observe the arrival of the R wave (and others) to the pickup or photograph the wave trace for later data reduction. The vibration is commenced at the shock point, a known distance d_o from the pickup unit, and the time of arrival to the pickup unit is observed. The wave velocity is computed as

$$v_c = \frac{d_o}{t_c} \quad \text{or} \quad v_s = \frac{d_o}{t_s} \cong \frac{d_o}{t_R}$$

where t_i is the elapsed time for the compression, shear, or Rayleigh wave, respectively. Relationships between wave velocity and dynamic shear modulus are as follows:

$$v_c = \sqrt{\frac{E_s(1 - \mu)}{\rho(1 + \mu)(1 - 2\mu)}} \qquad (14\text{-}5)$$

$$v_s \cong \sqrt{\frac{G}{\rho}} \qquad (14\text{-}6)$$

where ρ = soil density = γ/g, and other terms are as previously defined, except G. From theory of elasticity, the shear modulus G is related to the modulus of elasticity as

$$G = \frac{E_s}{2(1 + 2\mu)} \qquad (14\text{-}7)$$

The shear modulus is generally required in computations of vibration amplitudes for machine foundations. Equation (14-6) is generally used to determine G directly, as the use of Eqs. (14-5) and (14-7) requires estimating Poisson's ratio, and the small error due to differences in v_s and v_R is less than that in estimating μ.

Empirical equations for G have been proposed by Hardin and Richart (1963) as:

For rounded sand grains,

$$G = \frac{6900(2.17 - e)^2}{1 + e}\bar{\sigma}_o^{0.5} \qquad \text{kPa} \qquad (14\text{-}8)$$

For angular sands and normally consolidated clays of low activity,

$$G = \frac{3230(2.97 - e)^2}{1 + e}\bar{\sigma}_o^{0.5} \qquad \text{kPa} \qquad (14\text{-}9)$$

For overconsolidated clays with overconsolidation ratio OCR (Hardin and Drnevich, 1972),

$$G = \frac{3230(2.97 - e)^2}{1 + e}\text{OCR}^M\bar{\sigma}_o^{0.5} \qquad (14\text{-}10)$$

In all the above equations,

e = in situ (or test) void ratio.

$\bar{\sigma}_o = \dfrac{J_1}{3} = \dfrac{1}{3}(\sigma_1 + \sigma_2 + \sigma_3)$ = effective confining stress, kPa.

M = exponent depending on the plasticity index I_P as follows:

I_P	0	20	40	60	80
M	0	0.18	0.30	0.41	0.48

Example 14-2 Estimate the shear modulus for the following in situ soil tions: $e = 0.76$; $\phi' \cong 36°$; subangular silty sand (fairly dense), $\gamma = 18.1$ kN/m³.

SOLUTION Use Eq. (15-2):

$$K_o = 1 - \sin \phi = 0.41$$

Estimate the effective depth of vibration influence = 3 m.

$$\sigma_v = 3(18.1) = 54.3 = \sigma_1$$

$$\sigma_2 = \sigma_3 = K_o\sigma_1 = 0.41\sigma_1 = 22.3 \text{ kPa}$$

$$\sigma_o = \tfrac{1}{3}(54.3 + 22.3 + 22.3) = 32.9$$

$$G = \frac{3230(2.97 - 0.76)^2}{1 + 0.76}(32.9)^{1/2} = 51\,400 \text{ kPa}$$

14-7 CYCLIC MODULUS OF DEFORMATION AND LIQUEFACTION

The widespread interest in nuclear power has produced a need to investigate potential plant sites for geological faults and possible soil liquefaction during an earthquake. Liquefaction is here defined as a soil state caused by a buildup of pore pressure from a cyclic or dynamic loading until the effective pressure is zero. This is a phenomenon of fine- to medium-fine-grained sands. It has not been observed in gravels and is difficult to develop in fine silty sands or medium to coarse sands. Three types of tests have been used in the last 10 years to study the structural stability (strength) of sands. These include the cyclic triaxial test and direct shear test cited in the previous section and the shaking table test (Seed et al., 1977).

These studies are made to estimate the magnitude and number of cycles of stress (or strain) necessary to cause a pore pressure buildup sufficient to reduce the effective pore pressure to a low enough value that, under the soil loads, the material behaves as a viscous fluid, or liquefies. As was pointed out in Sec. 13-11, researchers have discovered that the residual pore pressure tends to increase on successive stress cycles in a clay. It has also been found that pore pressure increases with increasing strain. This phenomenon would be expected in a fully saturated soil in undrained shear. In natural soils where $S < 100$ percent even with drainage, this phenomenon may also be induced if the pore pressure buildup is larger than the drainage rate, as may occur during earthquakes, where relatively large strains are developed. Note also that "progressive" failure does not require the entire mass to be simultaneously in a state of liquefaction. A local point of liquefaction may result in sufficient load transfer to an adjacent point that the soil is overstressed, resulting in a mass failure.

The cyclic direct shear and cyclic triaxial tests can be used also to determine the dynamic modulus and the damping factor for a soil under the conditions of the test (void ratio, confining pressure, cementation, sample disturbance, strain or stress ratio, etc.). The data from these tests are usually displayed as shown in Fig. 14-4. Generally the strain and stress are electronically monitored, the output is directed to an oscillograph, and the plot shown is made. The plot can also be

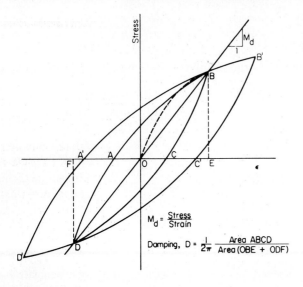

Figure 14-4 Hysteresis loop for a dynamic shear test. Note that this loop will locate about the origin of axis as above. A triaxial test will locate to the right of the axis as in Ex. 14-3. The above are actual data, with curve $ABCD$ for the first cycle and $A'B'C'D'$ for 10 cycles. This soil has a reduced stress-strain modulus with increasing stress cycles.

made by hand by obtaining the stresses and corresponding strains. Generally, since the cycling is on the order of $\frac{1}{6}$ to 10 Hz, it is only practical to use an oscillograph.

From the hysteresis loop shown, the stress-strain modulus is

$$M_d = \frac{\text{stress}}{\text{strain}} \tag{14-11}$$

where $M_d = E_s$ for axial strain and deviator stress
$= G$ for shear strain and shear stress.

Damping is computed as

$$D = \frac{1}{2\pi}\left[\frac{\text{area of loop } ABCD}{\text{area of both triangles } (OBE + ODF)}\right] \tag{14-12}$$

The area of the loop and of the triangles is conveniently obtained using a planimeter.

The cyclic shear test is also used to determine the number of cycles N_c at some deviator stress, generally expressed as a stress ratio

$$R = \frac{\Delta\sigma_1}{2\bar{\sigma}_3}$$

where $\bar{\sigma}_3 =$ effective confining pressure that will cause the measured pore pressure, starting from consolidated-drained conditions ($u = 0$), to equal the confining pressure ($u = \bar{\sigma}_3$). At this pore pressure condition, "liquefaction" is assumed. For the direct shear test, use $\Delta s =$ cyclic shear stress and the applied normal stress σ_v for $\bar{\sigma}_3$.

For the cyclic triaxial test,

1. Build a sample to the desired density (or void ratio). Use a damp sand of known water content and weighed into 5 to 10 separate equal parts to produce a uniformly dense sample by tamping the equal weights into equal sample volumes. Several trial samples may be necessary before a satisfactory sample is obtained. A similar procedure would be used for remolded cohesive soils. Alternatively, where or when practical, undisturbed samples may be used.

2. For liquefaction studies, saturate and consolidate the sample using a back pressure σ_b and check that the B parameter of Sec. 13-11 is 0.98^+. The effective consolidation pressure is obtained as the difference between cell pressure and back pressure,

$$\bar{\sigma}_3 = \sigma_3 - \sigma_b$$

For example, with $\sigma_3 = 650$ kPa and a back pressure on the sample pore pressure fluid reservoir of 500 kPa, the effective consolidation pressure is 150 kPa. A positive back pressure may be maintained during the test to ensure $S \rightarrow 100$ percent by keeping the air dissolved in the pore water.

For determining the dynamic modulus, the in situ water content, or damp sand in a "drained" condition of pore lines open for drainage should be used.

3. Place the cell into a triaxial machine modified for cyclic load application. A modification of existing equipment can be made at modest cost where competent laboratory technicians are available (Cullingford et al., 1972; Chan and Mulilis, 1976). Equipment is available from several commercial laboratory suppliers and/or can be custom built.

4. Apply a strain increment and observe the maximum deviator load (stress). Generally, a strain of 2.5, 5, and/or 10 percent is used in liquefaction studies. For machinery foundations, the strains may be on the order of 0.025 to 0.1 percent or less, and the sample is not cycled to failure; rather, one to four cycles are recorded, then the strain increment is reset and the test repeated.

Adjust the equipment so that a portion of the stabilized stress (or strain) is cycled. It is evident that the cycling must be such that sample contact is maintained on the $(-)$ stress part of the hysterisis loop.

5. Allow several cycles to be superimposed, for a statistical averaging for computation of dynamic moduli. For liquefaction studies, the cycles are counted to liquefaction $(u = \bar{\sigma}_3)$, and a plot of stress ratio versus N_c is made as in Fig. 14-5 for several stress ratios with $\varepsilon = $ constant. For example, with $\bar{\sigma}_3 = 600$ kPa, the initial deviator stress $\Delta \sigma_1$ to produce $\varepsilon = 0.025$ (2.5 percent strain) might be 450 kPa. We may run tests as:

$\Delta\sigma_1$	Cycles N_c to liquefaction	Stress ratio R_c
100 kPa	500	$100/(2 \times 600) = 0.08$
300	30	0.25
400	8	0.33

This is part of the data plotted in Fig. 14-5.

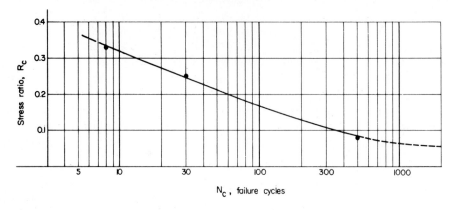

Figure 14-5 Plot of stress ratio R versus number of cycles to liquefaction N_c.

Example 14-3

GIVEN The cyclic triaxial test data shown in Fig. E14-3 for a loose, medium coarse sand $(D_r = 0.42)$ and $G_s = 2.68$. The test is stress-controlled as much as is practical. Cell pressure $\sigma_3 = 140$ kPa and the initial strain is 0.37, producing an initial $\Delta\sigma_1 = 50$ kPa and $\Delta\sigma_1/\sigma_3 = 0.36$.

REQUIRED Initial tangent modulus, dynamic stress-strain modulus, and damping factor D.

SOLUTION Draw lines $O'A'$, AOC, AB, and CD as shown.

$$\text{Static } E_s = \text{slope } O'A' = \frac{0.7(140)}{0.25}$$

$$= 392 \text{ kPa}$$

$$\text{Dynamic } E_s = \text{slope } OA$$

$$= \frac{1.46(140)}{0.68 - 0.27}$$

$$= 498 \text{ kPa}$$

To obtain the damping factor, use a planimeter

$$\text{Area of } AB'CD' = 23 \text{ units} \qquad (3 \text{ trials})$$

$$\text{Area of } OAB = 23.5 \text{ units}$$

$$\text{Area of } OCD = 10.5 \text{ units}$$

$$D = \frac{1}{2\pi}\left(\frac{23}{23.5 + 10.5}\right) = 0.11$$

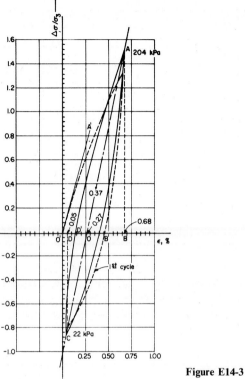

Figure E14-3

Laboratory testing indicates that sample preparation, grain size, void ratio (or relative density), effective confining pressure, and previous strain history all considerably influence the laboratory data. For these reasons, in using laboratory data for design, a correction factor is introduced,

$$\left(\frac{s_e}{\bar{p}_o}\right)_{\text{field}} = C_r\left(\frac{\Delta\sigma_1}{2\bar{\sigma}_3}\right)_{\text{lab}} \tag{14-13}$$

and dividing, we obtain the apparent safety factor as

$$F = \frac{C_r(\Delta\sigma_1/2\bar{\sigma}_3)_{\text{lab}}}{(s_e/\bar{p}_o)_{\text{field}}} \tag{14-14}$$

where s_e = field shear stress caused by earthquake
$\bar{p}_o$ = in situ effective overburden stress
C_r = adjustment factor, ranging from about 0.57 for normally consolidated sands to 1 for sands with OCR $\to$ 8

Note that Eq. (14-14) is compared at the same number of cycles.
The field shear stress can be computed (Seed and Idriss, 1971) as

$$s_e = 0.65(r_d)ma \tag{14-15}$$

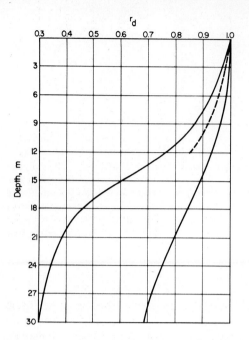

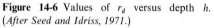

Figure 14-6 Values of r_d versus depth h. (*After Seed and Idriss, 1971.*)

where m = mass of unit column of soil to the depth of interest h, such as location of bedrock (origin of earthquake) = $\gamma h/g$

a = estimated acceleration of earthquake, depending on magnitude and distance from epicenter

r_d = reduction factor to account for the accelerated mass m not being a rigid body; use Fig. 14-6 for values of r_d

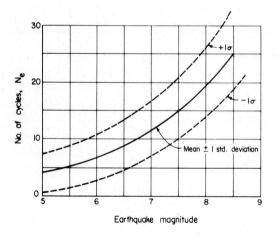

Figure 14-7 Earthquake magnitude (Richter scale) vs. number of resulting stress cycles. (*After Valera and Donovan, 1977.*)

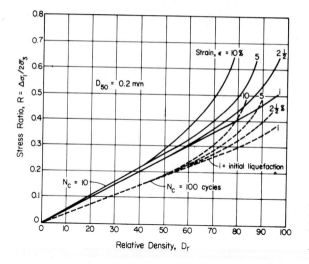

Figure 14-8 Stress ratio vs. relative density for cycles to liquefaction. Initial liquefaction is the stage where the first effects of partial liquefaction are detected and is not "failure," (*After Lee and Seed, 1967.*)

The factor 0.65 is used to statistically average the shear stress over the cycle duration.

The number of field cycles for comparison with the laboratory curve may be estimated from Fig. 14-7. This curve is based on a scatter of at least ± 1 standard deviation.

Figure 14-8 illustrates a summary of data for a clean, isotropically consolidated sand with the mean (D_{50}) grain size $= 0.2$ mm. This may be used to indicate the general range of stress ratios R to be expected. It also indicates variation of R with relative density and number of cycles to cause liquefaction for either stress ratio or relative density.

Figure 14-9 illustrates the effect of grain size on the strength ratio and may be used to adjust data from Fig. 14-8 for grain sizes other than 0.2 mm. These data were obtained from test samples formed by sedimentation and light, wet compaction. Moist tamping gives cyclic strengths as much as 140 percent higher. Confining pressures significantly higher than 100 to 200 kPa may give lower cyclic strengths. Naturally cemented soils may possess higher cyclic strengths.

Clay soils generally do not liquefy in cyclic loading; instead, the strain progressively increases. An exception is noted with some naturally cemented, sensitive clays, which develop sharply defined shear planes on which the soil remolds to the much lower residual strength. Figure 14-10 illustrates the stress ratio of clay, defined as

$$R_c = \frac{s_c}{s_u}$$

where s_c = cyclic shear strength (also the deviator stress applied)
s_u = undrained shear strength

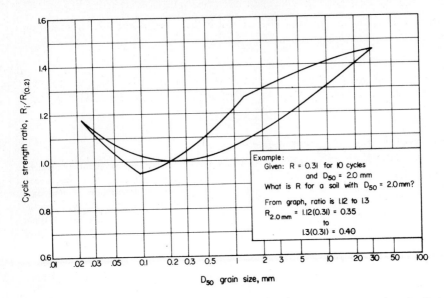

Figure 14-9 Relationship between mean (D_{50}) grain size and the cyclic strength ratio. Data from several sources.

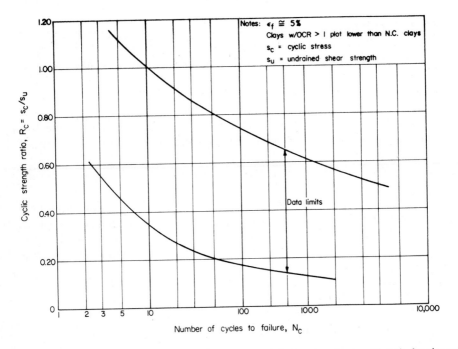

Figure 14-10 Relationship between number of cycles to failure and cyclic strength ratio for clay soils. (*After Lee and Focht, 1975.*)

This figure is based on a wide range of clay soils (Lee and Focht, 1975). Equation (14-3) may be used for clay soils with $C_r = 1$.

Example 14-4 Estimate the safety factor of a nuclear power plant site for the conditions shown in Fig. E14-4 and an earthquake of magnitude 8 on the Richter scale.

SOLUTION Preliminary computations:
Estimate $G_s = 2.68$. From the gravimetric-volumetric relationships in Chap. 2, derive for γ, in grams per cubic centimeter,

$$\gamma_{dry} = \frac{G_s(\gamma_{sat} - 1)}{G_s - 1} = \frac{2.68(1.937 - 1)}{2.68 - 1}$$

$$= 1.495 \text{ g/cm}^3$$

The volume of solids is

$$V_s = \frac{1.495}{2.68} = 0.558$$

The volume of voids is

$$1 - 0.558 = 0.442$$

The void ratio is

$$e = \frac{0.442}{0.558} = 0.792$$

Estimate $e_{min} = 0.4$ and $e_{max} = 1.2$, and compute the relative density [Eq. (6-1)] as

$$D_r = \frac{e_{max} - e_n}{e_{max} - e_{min}} = \frac{1.2 - 0.79}{1.2 - 0.4} = 0.51$$

$$D_r = 0.51 \times 100 = 51 \text{ percent}$$

Compute

$$\text{Average } \bar{p}_o = 2(18.0) + 7.5(19.0 - 9.81) = 105 \text{ kPa}$$

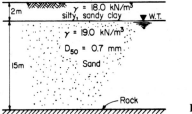

Figure E14-4

Note that this is between 100 and 200 kPa of the test data ranges used to develop Figs. 14-8 and 14-9. Compute

$$s_e = 0.65(r_d)ma$$

$$m = \frac{[2(18.0) + 7.5(19.0)]1000}{9.807} = 18\,201 \text{ kg}$$

Estimate the acceleration as

$$a = 0.1\,g = 0.1(9.807) = 0.981 \text{ m/s}^2$$

r_d is estimated as 0.92 at $h = 9$ m from Fig. 14-6.

$$s_e = 0.65(0.92)(18\,201)(0.981) = 10\,667 \text{ N} = 10.7 \text{ kN}$$

$$\frac{s_e}{\bar{p}_o} = \frac{10.7}{105} = 0.10$$

For earthquake magnitude $= 8$, obtain $N_e \cong 20$ cycles from Fig. 14-7. From Fig. 14-8, for $N_c = 20$ cycles at $D_r = 51.0$, interpolate at $\varepsilon = 2.5$ percent to obtain $R_c = 0.25$. From Fig. 14-9 at $D_{50} = 0.7$ mm, interpolate midway between range lines for factor $= 1.06$.

$$R_c = 0.25 \times 1.06 = 0.27 \qquad \text{(design value)}$$

Compute F using Eq. (14-14).

$$F = \frac{0.57(0.27)}{0.10} = 1.54 \qquad \text{(which appears adequate)}$$

If F were on the order of 1.1 to 1.2, it would be necessary to reevaluate assumptions and/or perform cyclic tests on undisturbed soils. If this is not possible, consider a grout wall around the site to maintain the exterior water table to satisfy environmental concerns and dewater the interior so that liquefaction is not a problem.

14-8 SUMMARY

This chapter has introduced means of computing stress-strain moduli and Poisson's ratio for soils.

Both static, as in conventional triaxial tests, and dynamic stress-strain moduli are considered, and the reader should be aware that "static" values at low strains may be less than 50 percent of the dynamic values. At larger static or dynamic strains (as "residual" strength or liquefaction is developed) $E_s \to 0$.

The concept of liquefaction as used in earthquake studies was introduced.

Cyclic shear tests as used to evaluate earthquake resistance were presented. This chapter has introduced the reader to considerable empirical data, compared with earlier chapters. This is due to the highly uncertain nature of earthquake intensity, the number of stress cycles likely to be introduced, and the mass, or quantity, of soil involved. These are coupled with the problems of relating laboratory prepared samples to field conditions with regard to methods of sample preparation, degree of saturation, confining pressure, type of soil, equipment limitations, accuracy of test simulation, and interpretation of data.

In general, however, it has been found that

1. Cyclic stresses that will induce liquefaction are much smaller than those required under static loading conditions.
2. The frequency is not particularly critical in that frequencies of $\frac{1}{6}$ to 10 Hz produce essentially the same results.
3. Cyclic stresses will produce liquefaction over a considerable range of unit weight γ.
4. When sands liquefy under cyclic stresses, the deformations immediately become very large. This causes some problems in the laboratory in determining "failure."
5. Partial liquefaction of dense sand produces a condition of near zero stress at low strains, but if the strain is increased, an appreciable resistance is recovered.
6. The higher the stress ratio, the smaller the number of cycles to liquefaction.
7. The lower the confining pressure, the lower the number of cycles required to cause liquefaction.

HOMEWORK PROBLEMS

14-1 What is the approximate initial tangent modulus of Ex. 14-1?
   Ans.: $E_s \cong 25\,000$ kPa.

14-2 What is the secant modulus using the two points of Ex. 14-1?
   Ans.: $E_s \cong 17\,800$ kPa.

14-3 Plot the stress-strain data of Prob. 13-10, tests 1 and 3, and compare the initial tangent moduli.

14-4 Plot the stress-strain data of Prob. 13-10, tests 2 and 5, and compare the initial tangent moduli. For this problem, prepare a table of G versus μ.

14-5 Plot the assigned test data of Prob. 13-10, and find the strain at which Poisson's ratio μ is either $(-)$ or > 0.50. Comment on the strain level at which this occurs. Be sure to use a large enough scale or enlarge the initial part of the graph. Also, use the smooth curve and not point-to-point of data.

14-6 Redo Ex. 14-5 for N_e at $+1$ standard deviation $(+1\alpha)$.

14-7 Redo Ex. 14-5 for N_e at -1 standard deviation (-1α).

14-8 Redo Ex. 14-5 if the sand density is increased by vibroflotation to 20.5 kN/m³, but all other data are the same.
   Ans.: $F \cong 4.0^+$.

14-9 Redo Ex. 14-5 and plot F versus ground acceleration a. Use values of $a = 0.05$ to 1.0. If the earthquake intensity and site is in your area, what is the F value?

14-10 For the hysteresis loop shown in Fig. P14-10, compute the dynamic stress strain modulus and indicate whether it is G or E and the damping factor D.

 Ans.: Modulus = 2300 kPa, $D = 0.17$.

14-11 For the hysteresis loop shown in Fig. P14-11, compute the dynamic stress-strain modulus and the damping factor.

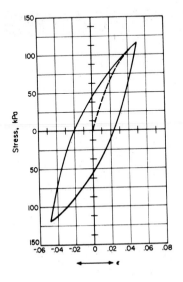

Figure P14-10

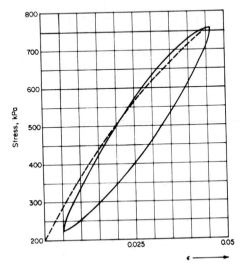

Figure P14-11

Chapter 15
Soil Stresses
and Soil Pressures

15-1 SOIL STRESSES AT A POINT

Soil formed into a residual or sedimentary deposit produces a column of soil over any element, as in Fig. 15-1a. The vertical pressure is $\sigma_v = p_o = \gamma h$, as shown. During the formation of the deposit, the element will be consolidating under the pressure σ_v. The vertical stress produces a lateral flow into the surrounding soil due to the Poisson's ratio effect. The surrounding soil resists the lateral flow effect with a developed lateral stress σ_h. Over geological periods, consolidation, and both vertical and lateral creep strains, will become zero. At this time a stable stress state will develop in which σ_h and σ_v will become principal effective stresses, since zero displacements will produce zero shear stresses on the vertical and horizontal planes defining the soil element. The equilibrium in situ condition produced at this stress state is commonly termed the K_o condition.

The ratio of the lateral and vertical in situ soil pressures may be defined by a factor K as

$$K = \frac{\sigma_h}{\sigma_v}$$

The K_o condition, in particular, is the ratio of the effective equilibrium pressures,

$$K_o = \frac{\sigma_h}{\gamma h} = \frac{\sigma'_3}{\sigma'_1} \tag{15-1}$$

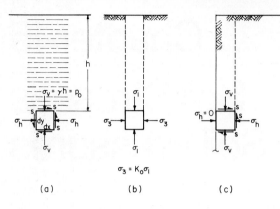

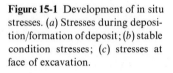

Figure 15-1 Development of in situ stresses. (*a*) Stresses during deposition/formation of deposit; (*b*) stable condition stresses; (*c*) stresses at face of excavation.

The range of K_o is qualitatively

$K_o < 1$ for *normally* consolidated soil

$K_o < 1$ for *overconsolidated* soil (OCR < 3 approximately)

$K_o > 1$ for *overconsolidated* soil (OCR > 3 approximately)

The overconsolidation ratio OCR, defined in Chap. 11 as OCR $= p_c/\gamma h$, is not the same ratio as K_o, which is always as given by Eq. (15-1).

The K_o condition can be illustrated by the several Mohr's circles, as in Fig. 15-2, for conditions of both normally and overconsolidated (OC) soils. The in situ stress conditions represent some elastic equilibrium state, since with no displacements, the soil may be considered an elastic continuum. The application of additional stresses will produce changes in the size of Mohr's circle, and when the shear stresses are of sufficient magnitude, the soil fails. Figure 15-3 illustrates the failure condition as sufficient additional stresses are imposed on the soil mass.

The determination of K_o by measuring σ_h in situ is nearly impossible, since it is irrecoverably lost when a cavity is excavated alongside the element. It is altered even when one is close, as in Fig. 15-1c, where $\sigma_h = 0$ on the left side. This is true no matter how carefully one were to place a measuring device into the cavity.

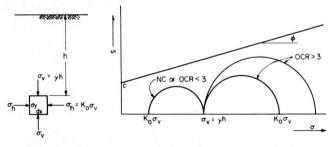

(a) Stressed point in situ. (b) Mohr's circles for in situ stress conditions.

Figure 15-2 Qualitative representation of in situ stresses at a point.

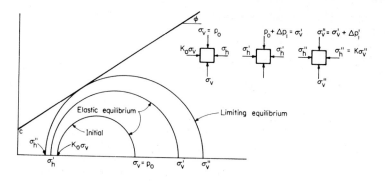

Figure 15-3 Mohr's circle representation of soil stress changes from K_o to the limiting equilibrium state as the vertical stress is increased from p_o to $p_o + \Delta p_1$. In design, Δp_1 should produce σ_v', not σ_v''. Note that there is both a decrease in lateral pressure and an increase in vertical pressure.

Wroth (1975) reports on a device which removes the soil and simultaneously self-inserts into the cavity; however, this may not completely recover the in situ stress σ_h due to excavation disturbance. The next section gives an analytical reason for loss of σ_h and not being able to "push" the soil back into place. On a nonanalytical basis, however, the reader should be aware that due to the particulate nature of the material, any displacement produces a new structure (material) and interparticle stresses.

Based on observations of grain pressure in silos, Jaky (1948), and later Brooker and Ireland (1965), using a large series of laboratory tests on five clay soils, suggested the following equation for the in situ lateral earth pressure:

$$K_o = M - \sin \phi' \qquad (15\text{-}2)$$

where $M = 1$ for normally consolidated, cohesionless and cohesive soils; also for the lateral pressures developed in piles of grain, such as corn, barley, wheat, etc.

$\quad = 0.95$ for overconsolidated clays on the order of OCR > 2.

$\phi' =$ effective angle of internal friction.

Use of this equation allows a reasonable estimate of the in situ lateral earth pressure. Available evidence indicates that the estimate is reliable enough for most engineering purposes.

15-2 ACTIVE AND PASSIVE EARTH PRESSURES

The concept of active and passive earth pressure is of particular importance in soil stability problems, bracing of excavations, design of retaining walls, and development of pullout resistances using various types of anchorage devices. Consider first the statics of a block of material on a plane at a slope angle of ρ, as in Fig. 15-4. The friction coefficient between the block and plane is v, defined on the figure.

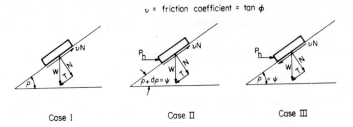

Figure 15-4 Block sliding down a plane to develop the concept of active, at rest, and passive earth pressure.

Consider the three cases shown as follows:

Case I $P_h = 0$

From summing forces parallel to the plane, obtain

$$T - vN = 0$$

and substituting $v = \tan \phi$ and the W components for T and N,

$$W \sin \phi - W \cos \rho \tan \phi = 0$$

from which obtain

$$\tan \rho = \tan \phi$$

or the block is on the verge of slip when the slope of the plane is $\rho = \phi$.

Case II When the slope angle is $\psi = \rho + \Delta\rho$.

Again summing forces parallel to the plane and with the block just restrained against sliding down the plane with the external force P_h, we have

$$P_h \cos \psi + W \cos \psi \tan \phi - W \sin \psi = 0$$

from which obtain

$$P_h = W(\tan \phi - \tan \psi)$$

In this case P_h will be some minimum value, since the friction resistance is aiding P_h to restrain the block.

Case III The slope angle is the same as in Case II, but we wish to push the block up the plane.

Summing forces parallel to the plane, obtain

$$P_h \cos \psi - W \cos \psi \tan \phi - W \sin \psi = 0$$

from which

$$P_h = W(\tan \psi + \tan \phi)$$

In this case P_h is a maximum, since friction resistance as well as the tangential component of the weight vector must be overcome to have a movement up the plane.

These three cases in a soil represent approximately:

Case I. K_o conditions
Case II. Active earth pressure conditions—note that to develop the limiting friction, there must be a slight movement down the plane
Case III. Passive earth pressure conditions—note that to develop the limiting resisting friction there must be a slight movement up the plane. In soil the angle ψ changes in value from the active to the passive case

Now let us investigate an impossible condition of the insertion of a frictionless wall of substantial rigidity and zero volume into a cohesionless soil mass along the vertical plane *AB* of Fig. 15-5a, then excavate the soil from the left side of the wall as shown in Fig. 15-5b while simultaneously fixing the wall against any movement

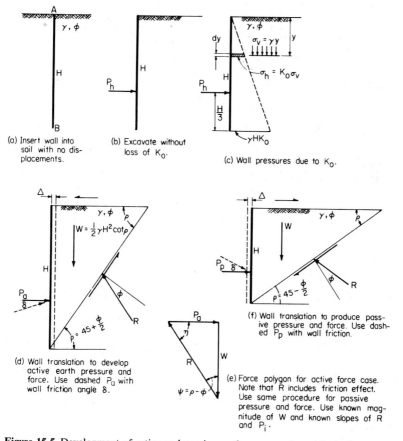

(a) Insert wall into soil with no displacements.

(b) Excavate without loss of K_o.

(c) Wall pressures due to K_o.

(d) Wall translation to develop active earth pressure and force. Use dashed P_a with wall friction angle δ.

(e) Force polygon for active force case. Note that R includes friction effect. Use same procedure for passive pressure and force. Use known magnitude of W and known slopes of R and P_i.

(f) Wall translation to produce passive pressure and force. Use dashed P_p with wall friction.

Figure 15-5 Development of active and passive earth pressures in a cohesionless soil mass.

whatsoever. The stresses acting against the wall at this point would be K_o stresses, since the conditions imposed have not produced any strains in the soil mass behind the wall. The wall force necessary to hold the soil/wall system in place is a K_o force, and from Fig. 15-5c, obtain

$$\sigma_h = K_o \gamma y$$

The wall force is the summation of the unit stresses over the differential of area, to obtain

$$P_h = \int_0^H \sigma_h \, dy = \int_0^H K_o \gamma y \, dy$$

$$= \tfrac{1}{2}\gamma H^2 K_o \qquad (a)$$

This is the force necessary to hold the wall in place against any lateral movement. Since the pressure diagram is triangular, the force P_h acts through the centroid of the pressure area or at the $H/3$ location from the bottom of the wall, as shown.

If the wall is allowed to translate into the excavation with sufficient soil strains that the full frictional resistance is mobilized and without destroying the soil structure, the force P_h will become some minimum value (somewhat as Case II earlier). The approximate failure zone will develop behind the wall, as shown in Fig. 15-5d. Friction will be mobilized along the plane OB and in the direction shown and will be a limiting value, since some slip has occurred.

The forces on the sliding wedge of Fig. 15-5d produce the force polygon of Fig. 15-5e. Solving for $P_a = P_h$, which must be obtained from a maximum value of the possible force combinations,

$$P_a = \tfrac{1}{2}\gamma H^2 \cot \rho \tan (\rho - \phi) \qquad (b)$$

The maximum value of P_a is obtained from $dP_a/d\rho = 0$, from which we can obtain

$$\rho = 45° + \frac{\phi}{2}$$

Substituting this value of ρ in Eq. (b), and substitution of $\cot (45 + \phi/2) = \tan (45 - \phi/2)$, obtain

$$P_a = \frac{1}{2}\gamma H^2 \tan^2 \left(45 - \frac{\phi}{2}\right) \qquad (15\text{-}3)$$

which may be rewritten as

$$P_a = \tfrac{1}{2}\gamma H^2 K_a \qquad (15\text{-}3a)$$

Since $P_a = \int_0^H \sigma_a \, dy$ as in Eq. (a), the *active* earth pressure is

$$\sigma_a = \gamma y \tan^2 \left(45 - \frac{\phi}{2}\right) = \gamma y K_a \qquad (15\text{-}4)$$

A similar analysis for passive pressure, as when the wall is forced into the soil (Fig. 15-5f), gives

$$\rho = 45° - \frac{\phi}{2}$$

and for P_p,

$$P_p = \frac{1}{2}\gamma H^2 \tan^2\left(45 + \frac{\phi}{2}\right) = \frac{1}{2}\gamma H^2 K_p \qquad (15\text{-}5)$$

and the passive earth pressure is

$$\sigma_p = \gamma y \tan^2\left(45 + \frac{\phi}{2}\right) = \gamma y K_p \qquad (15\text{-}6)$$

Note that the active earth pressure is a condition of loosening strains where the friction resistance is mobilized to reduce the force necessary to hold the soil in position. Passive earth pressure is a condition of densifying the soil by a lateral movement into the soil mass, with the friction mobilized to increase the force necessary to cause strain. Note also that the failure slope angle decreases by ϕ over the active case, producing a much larger failure wedge. For $\phi = 30°$, the possible range of earth pressures is as follows:

Earth pressure	Symbol	Computed as	K coefficient
Active	K_a	$\tan^2\left(45 - \dfrac{\phi}{2}\right)$	0.333
At rest	K_o	$1 - \sin\phi$	0.50
Passive	K_p	$\tan^2\left(45 + \dfrac{\phi}{2}\right)$	3.000

Thus, passive earth pressures are on the order of $K_p \rightarrow 10 K_a$.

Observations and model wall tests indicate that the computations are valid for K_o and K_a pressures and that the failure slope is nearly planar, as in Fig. 15-5. The K_p values tend to be too large, especially when wall geometry differs from that used in Fig. 15-5 when ϕ is larger than, say, about 38°. Available evidence indicates that for a sloping backfill and/or a rough wall where significant friction resistance develops, the failure surface is partly curved near the wall base. In general, the ground surface being a principal plane, the failure surface angle is $45 + \phi/2$ for the active and $45 - \phi/2$ for the passive condition (see Fig. 15-6a).

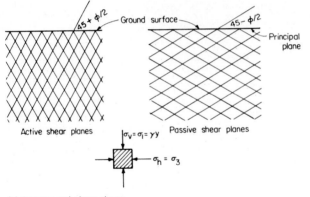

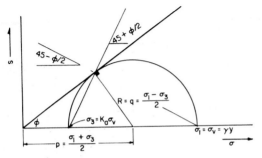

(a) Stresses and shear planes.

(b) Mohr's circle to orient shear planes.

Figure 15-6 Rankine stress state using Mohr's circle.

15-3 PRESSURES AGAINST WALLS

There are two common methods of computing lateral earth pressures against walls. The earliest analytical solution was the Coulomb method, developed by C. A. Coulomb in 1776. Later Rankine (ca. 1857) proposed a procedure for cohesionless soils which is actually a simplification of the Coulomb method. Equations (15-4) and (15-6) describe both the Rankine and Coulomb methods for horizontal ground surfaces; dry, cohesionless soils; and smooth walls. No wall friction is developed between a smooth wall and soil during wall movement; thus, the wall pressure becomes a principal stress. Only the Rankine method will be considered further in this text due to its simplicity and widespread use.

The Rankine lateral earth pressure can be studied using Mohr's circle. Consider Fig. 15-6a, where an element of soil is acted on by $p_o = \gamma y$ and is in a state of incipient failure. The "active state" shear planes are as shown (also shown are shear planes for the passive state for reference). This orientation can be obtained from Mohr's circle at an angle of $45 + \phi/2$ with the horizontal as in Fig. 15-6b.

The major principal stress is $p_o = \gamma y$ and the minor principal stress is $\sigma_3 = K_a \gamma y$, and the active earth pressure coefficient is

$$K_a = \frac{\sigma_3}{\sigma_1} = \frac{\sigma_3}{\gamma y}$$

Also using p and q as defined in Sec. 13-3, obtain

$$\sin \phi = \frac{q}{p} = \frac{\frac{1}{2}(\sigma_1 - \sigma_3)}{\frac{1}{2}(\sigma_1 + \sigma_3)}$$

With ϕ known and some rearranging, obtain

$$\sigma_a = \sigma_3 = \gamma y \frac{1 - \sin \phi}{1 + \sin \phi} = \gamma y \tan^2 \left(45 - \frac{\phi}{2} \right)$$

With passive pressure, the values of ϕ and $\sigma_3 = \gamma y$ are known, and rearranging, obtain

$$\sigma_p = \sigma_1 = \gamma y \tan^2 \left(45 + \frac{\phi}{2} \right)$$

Note again that the shear planes intercept the ground surface (a principal plane) at $45 + \phi/2$ for the active earth pressure condition and $45 - \phi/2$ for the passive pressure condition.

Alternative values for K_p are widely used, including those of Caquot and Kerisel (1948), Sokolovski (1965), and more recently Rosenfarb and Chen (1972), which are somewhat smaller than the values here when $\beta > 15°$, wall friction δ is included, and $\delta > \phi/2$ and $\phi > 36$ to $38°$. Generally, unless all these conditions are met the Rankine value for K_p is satisfactory to use. The Rosenfarb and Chen values and a computer program for generating the values are readily available in Bowles (1977).

15-4 INCLINED COHESIONLESS GROUND

With inclined ground the Rankine method considers static equilibrium of an element at a depth y. The soil weight acts vertically, and the lateral earth pressure is conjugate to the weight as in Fig. 15-7a. Thus, the lateral earth pressure acts parallel to the ground surface. Note that the Rankine method assumes a frictionless wall; therefore, the stresses on the vertical face of the element are principal stresses. Rankine made an analytical solution of this case to obtain

Active pressure:

$$K_a = \cos \beta \, \frac{\cos \beta - \sqrt{\cos^2 \beta - \cos^2 \phi}}{\cos \beta + \sqrt{\cos^2 \beta - \cos^2 \phi}} \qquad (15\text{-}7)$$

Passive pressure:

$$K_p = \cos \beta \, \frac{\cos \beta + \sqrt{\cos^2 \beta - \cos^2 \phi}}{\cos \beta - \sqrt{\cos^2 \beta - \cos^2 \phi}} \qquad (15\text{-}7a)$$

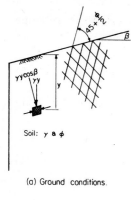

(a) Ground conditions.

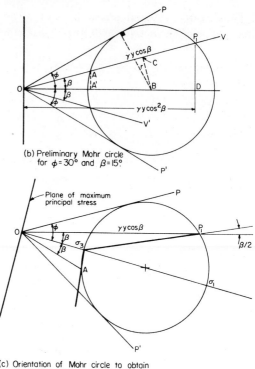

(b) Preliminary Mohr circle
for $\phi = 30°$ and $\beta = 15°$

(c) Orientation of Mohr circle to obtain
plane of maximum principal stress.

Figure 15-7 Using Mohr's circle to obtain Rankine earth pressure for an inclined ground surface.

Inspection of Eq. (15-7) indicates that the maximum stable slope occurs when $\beta = \phi$ for a cohesionless soil.

The active earth pressure can be obtained directly from Mohr's circle as follows:

1. Draw a set of orthogonal axes and locate OV and OV' as shown in Fig. 15-7*b*.
2. Compute $\gamma y \cos \beta$ and locate point P_1 along OV as shown.
3. Lay off lines OP and OP' at $\pm \phi$ as shown. At this point we do not know enough data to directly construct a Mohr's circle. However, since we recognize that the circle must be tangent to OP and OP_1 and pass through point P_1 to satisfy the stress condition and shear strength simultaneously, we may continue.
4. By trial find a circle tangent to OP and OP_1 and through point P_1.
5. Construct a new set of axes as in Fig. 15-7*c*. Line OV is horizontal, with point P_1 located using $\gamma y \cos \beta$ as in step 2. Lay off lines $O\sigma_1$ at β to OV and locate OP and OP' at $\pm \phi$ from $O\sigma_3$. Also locate OA at β from $O\sigma_3$ as shown.
6. Scale the circle center from step 4 and locate it on $O\sigma_3$. Use radius from step 4 and draw circle as shown.

7. Scale σ_a as the intersection of circle and OA. The slope of $\sigma_3 P_1$ is $\beta/2$ and slope of a line through σ_3 and σ_a deviates from the vertical by $\beta/2$.

The potential shear (or slip) planes are inclined to the ground surface at $45 \pm \phi/2$, as shown in Fig. 15-7a.

An analytical solution can be obtained for the Rankine equations using Mohr's circle as follows:

Referring to Fig. 15-7b, K_a is

$$K_a = OA' = OA \cos \beta \tag{c}$$

The conjugate stresses are

$$K = \frac{OA}{OP_1} = \frac{OC - CA}{OD + CP_1} \tag{d}$$

But the following relations also can be obtained from Fig. 15-7b:

$$OC = OB \cos \beta$$
$$CA = CP_1 = \sqrt{r^2 - (BC)^2}$$
$$BC = OB \sin \beta$$
$$r = OB \sin \phi$$

Substituting these values into Eqs. (d) and (c), and using $\cos^2 \phi = 1 - \sin^2 \phi$, we obtain Eq. (15-7). Notice, however, that we have ignored the shear stress AA' and $P'D$, introducing some approximation, or inconsistency, into the solution.

Example 15-1

GIVEN A cohesionless soil, $\phi = 32°$; $\gamma = 17.5$ kN/m^3.

REQUIRED What is the safe slope angle for an excavation?

SOLUTION From Eq. (15-7), we note that the $\sqrt{}$ term is negative for values of $\beta = \rho > 32°$. From Fig. 15-4 we note that in Case I the maximum value of $\rho = \phi$. Therefore the maximum safe slope is

$$\rho = 32°$$

Example 15-2

GIVEN A wall in cohesionless soil; $\phi = 34°$, $\beta = 10°$, $\gamma = 17.9$ kN/m^3, $H = 4$ m.

REQUIRED What are the active and passive wall forces?

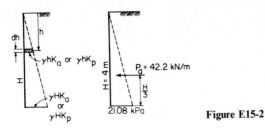

Figure E15-2

SOLUTION Use the Rankine equations [Eqs. (15-7), (15-7a)].

$$K_a = \cos 10 \, \frac{\cos 10 - (\cos^2 10 - \cos^2 34)^{1/2}}{\cos 10 + (\cos^2 10 - \cos^2 34)^{1/2}} = 0.2944$$

$$K_p = \cos 10 \, \frac{\cos 10 + (\cos^2 10 - \cos^2 34)^{1/2}}{\cos 10 - (\cos^2 10 - \cos^2 34)^{1/2}} = 3.2946$$

The active wall force is

$$P_a = \int_0^H \gamma h K_a \, dh = \tfrac{1}{2}\gamma H^2 K_a = 42.2 \text{ kN/m of wall width}$$

Alternatively from Fig. E15-2, P_a = area of pressure diagram = $\tfrac{1}{2}H(\gamma H K_a)$.
The passive pressure is

$$P_p = \int_0^H \gamma h K_p \, dh = \tfrac{1}{2}\gamma H^2 K_p = 471.8 \text{ kN/m}$$

The location is at the one-third height from base, since the pressure diagram is triangular as in Fig. E15-2. The location from the top of the wall is:

$$P_a \bar{y} = \int_0^H \gamma h K_a h \, dh = \frac{\gamma H^3}{3} K_a$$

Substituting for $P_a = \tfrac{1}{2}\gamma H^2 K_a$ we obtain

$$\bar{y} = \frac{\gamma H^3 K_a/3}{\gamma H^2 K_a/2} = \frac{2H}{3} \text{ from top of wall}$$

15-5 LATERAL EARTH PRESSURE FOR COHESIVE SOILS

Bell (1915), working with clay soils and the Rankine and Coulomb equations, recognized that Mohr's circle could be used to obtain Eqs. (15-4) and (15-6), which are either the Rankine or Coulomb equations for a smooth wall, cohesionless soil, and horizontal backfill.

Using Mohr's circle for a cohesive soil, one can, through some trigonometric manipulations, obtain in a similar manner:

Active case:

$$\sigma_a = \gamma h \tan^2 \left(45 - \frac{\phi}{2}\right) - 2c \tan \left(45 - \frac{\phi}{2}\right) \qquad (15\text{-}8)$$

Passive case:

$$\sigma_p = \gamma h \tan^2 \left(45 + \frac{\phi}{2}\right) + 2c \tan \left(45 + \frac{\phi}{2}\right) \qquad (15\text{-}8a)$$

Inspection of Eq. (15-8) indicates that a vertical excavation can be made in a cohesive soil (it is impossible in cohesionless soil) as in Fig. 15-8 according to the following:

At the ground surface, $\sigma_v = \gamma h = 0$, and

$$\sigma_a = -2c \tan \left(45 - \frac{\phi}{2}\right) = -2c\sqrt{K_a}$$

At the point h_t, we have $\sigma_a = 0$, and

$$h_t = \frac{2c}{\gamma\sqrt{K_a}}$$

The theoretical maximum excavation depth H_c is obtained where $P_a = 0$ as

$$P_a = 0 = \sigma_a \, dh = \frac{4c}{\gamma\sqrt{K_a}} = 2h_t$$

Since the soil will be in tension through the depth h_t, this depth may, and often does, form tension cracks which can be readily observed along and near the edges of vertical cuts (see Fig. 8-13a).

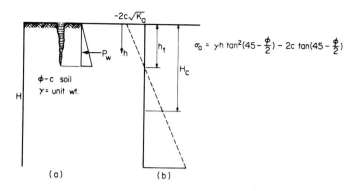

Figure 15-8 Excavation in a cohesive soil. (a) Effect of a tension crack filling with surface water; (b) theoretical pressure diagram for active pressure conditions.

The theoretical depth H_c is not used for design due to the possibility of tension cracks into which rainwater can enter, reducing the cohesion as well as causing an additional lateral force from hydrostatic pressure. There is usually a reduction in cohesion when clay is exposed due to various environmental factors which also reduces the theoretical depth H_c.

The lateral pressure against a wall in cohesive soil is computed as

$$P_i = \int_0^H \sigma_i \, dh$$

For the active earth pressure case, this produces

$$P_a = \tfrac{1}{2}\gamma H^2 K_a - 2cH\sqrt{K_a}$$

The lateral force may be computed directly from the pressure diagram using the appropriate equations for pressure areas. A typical active earth pressure diagram is shown in Fig. 15-8b.

Example 15-3

GIVEN Cohesive soil, $\gamma = 18.0$ kN/m^3; undrained shear strength $s_u = c = 30$ kPa; wall height $= 4$ m.

REQUIRED Wall pressure profile and the location and magnitude of active force.

SOLUTION For undrained shear conditions, $\phi = 0$ and

$$K_a = \tan^2(45 - 0) = 1.00$$

At top of wall,

$$\sigma_a = -2c = -60 \text{ kPa}$$

At base of wall,

$$\sigma_a = \gamma H - 2c = (18)(4) - 2(30) = 12 \text{ kPa}$$

$$P_a = \text{area of pressure diagram}$$

$$= \frac{-60(3.333)}{2} + 12\left(\frac{0.667}{2}\right)$$

$$= -96 \text{ kN/m}$$

Alternatively, integrate Eq. (15-8) to obtain

$$P_a = \tfrac{1}{2}\gamma H^2 - 2cH$$

$$= \tfrac{1}{2}(18)(4) - 60(4)$$

$$= -96 \text{ kN/m}$$

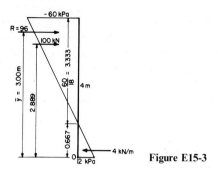

Figure E15-3

h_t is shown on Fig. E15-3. The location of P_a is found by summing moments about point O of the figure.

$$-96\bar{y} = 4\left(\frac{0.667}{3}\right) - 100(2.889) = -288.0$$

$$\bar{y} = \frac{288}{96} = 3.0 \text{ m above point } O$$

Example 15-4

GIVEN Soil of Ex. 15-3.

REQUIRED What is the depth of tension crack? What is the theoretical depth of unbraced excavation?

SOLUTION As shown on Fig. E15-3,

$$h_t = \frac{2c}{\gamma} = 3.333 \text{ m}$$

Note that $\sigma_a = 0$ at this point.

$$H_c = 2h_t = 2(3.333) = 6.67 \text{ m}$$

It may be possible (but OSHA regulations may disallow this) to excavate, say,

$$H_e = \frac{H_c}{F} = \frac{H_c}{2} = \frac{6.7}{2} = 3 \text{ m}$$

without bracing if the cut is very temporary and if no heavy material is stored along the bank.

15-6 THE TRIAL WEDGE SOLUTION

The force polygon of Fig. 15-5 is a graphical solution for the *trial wedge* of Fig. 15-5d if the weight vector is drawn to scale and the angles η and ψ are measured. The maximum value of P_a or P_p can be obtained from several trial failure wedges defined by the wall, backfill slope, and trial failure slope angle ρ.

The more general trial wedge for either cohesionless or cohesive soil, including backfill slope, tension crack, wall friction, and cohesion effect on the wall (termed adhesion), is shown in Fig. 15-9. It is necessary to make several trials to obtain the maximum value of P_i, as in Fig. 15-9c for P_a. This general case may be used to obtain a solution for irregular backfill surfaces.

The wall adhesion c_a is some value intermediate between c and, say, $0.6c$, depending on an engineering assessment of how well the soil adheres to the wall. The soil-to-soil interface uses the values of ϕ and c. The wall angle δ may be obtained from tables found in Bowles (1977) or elsewhere. For rough concrete or masonry walls, $\delta \to \phi$; for wood or steel sheet pile walls, δ ranges from about 15 to 26°. It may be measured in the laboratory by a direct shear type test, i.e., pulling a piece of the material across the soil and simultaneously applying a normal pressure.

The trial wedge may be programmed for the digital computer to compute the weight vector and values of C_s, C_w, and h_t. The R and P_i vectors can be solved by

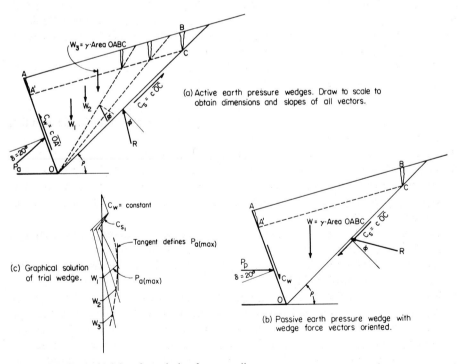

(a) Active earth pressure wedges. Draw to scale to obtain dimensions and slopes of all vectors.

(c) Graphical solution of trial wedge.

(b) Passive earth pressure wedge with wedge force vectors oriented.

Figure 15-9 General trial wedge solution for any soil.

a pair of simultaneous equations, since $\sum F_h$ and $\sum F_v$ must be satisfied for closure of the force polygon. By incrementing ρ in intervals of, say, 1°, a satisfactory maximum value of P_i can be readily found.

The trial wedge for cohesionless soils is considerably simpler than that for cohesive materials. It is evident that C_s and C_w are both zero and that no tension crack needs to be accounted for with cohesionless materials.

15-7 LOGARITHMIC SPIRAL AND ϕ-CIRCLE METHODS FOR PASSIVE PRESSURE IN COHESIONLESS SOIL

It has been shown by several investigators that the slope of at least a part of the slip surface for *passive pressure* is part of a logarithmic spiral for an ideal plastic material. This may be true also for active pressure, but the difference—if any—is so small that no significant error is introduced by considering a plane slip surface. The equation of the logarithmic spiral as used in soil mechanics work is

$$r = r_o e^{\theta \tan \phi}$$

where terms are identified in Fig. 15-10a and ϕ = angle of internal friction of the soil. For $\phi = 0$ soils, the spiral becomes a circle. A circle is sometimes assumed as a computational convenience for other soils. Regardless of the assumed shape of the lower part of the failure surface, the upper part is an approximate Rankine solution with a ground surface (principal plane) intersection of $45 - \phi/2$ and with the maximum principal stress as shown in Fig. 15-10b.

The logarithmic spiral method is illustrated in Fig. 15-10b. The steps to obtain the passive pressure include:

1. Select a trial origin O with $OB = r_o$. At A lay off a line AD at an angle with the ground surface of $45 - \phi/2$ as shown.
2. Extend a trial spiral from B to intersect line AD at some point X.
3. At X and tangent to the spiral, lay off XC such that the ground surface is intersected with an angle of $45 - \phi/2$, as shown. Readjust point O and the spiral orientation until these conditions are met. Draw vertical XY to ground surface.
4. Compute the passive resistance of wedge XYC on the curved block using the Rankine method and apply at the one-third point as shown.
5. Subdivide $ABXY$ into four to six strips which can be treated as trapezoids, as shown for element 4 in Fig. 15-10b.
6. For each trapezoidal element, compute W_i and scale the moment arm $\bar{x}_i$ and r_i. Also compute the friction resistance as $N \tan \phi$. Note that the normal force N acts at $90 - \phi$ to the tangent of the spiral at any point (including X). Since this direction passes through the spiral origin, no rotational moment is caused.
7. Scale the moment arms of P_p and P'_p as $\bar{y}_2$ and $\bar{y}_1$ with respect to point O.

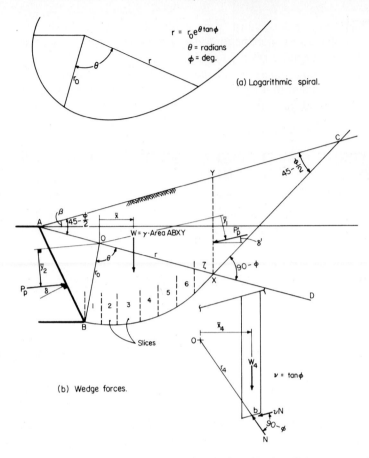

Figure 15-10 Passive earth pressure using the logarithmic spiral as a part of the assumed failure surface.

8. Find $\sum M_o = O$ and obtain

$$P_p \bar{y}_2 - \sum [W_i \bar{x}_i + (vN)r_i] - P'_p \bar{y}_1 = 0$$

from which P_p may be readily obtained.

Alternatively, one may use a circular arc for the part BX with little loss of accuracy, as shown in Fig. 15-11a. The use of a circular arc greatly simplifies the work as follows:

1. Draw wall system AB and ground surface AC. Note that AC may slope as in Fig. 15-10 but is horizontal here for simplicity.
2. Lay off line AD at $45 - \phi/2$ to ground surface. Also select a trial failure surface XC which intersects the ground surface at $45 - \phi/2$ as shown.

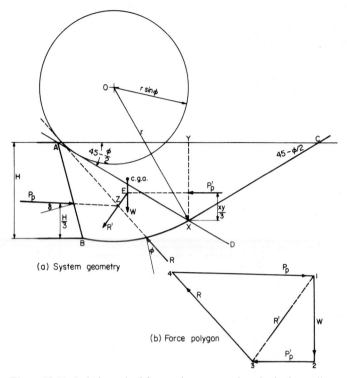

Figure 15-11 ϕ-circle method for passive pressure in cohesionless soil.

3. At point X, which is tangent to the circular part BX, erect a perpendicular to XC and, by trial, locate a point O which produces a radius through X and B and draw arc BX.

4. Erect vertical XY as shown and compute the passive earth force acting on this vertical plane as P'_p using the Rankine method. Locate P'_p at $XY/3$ as shown.

5. By some means find the center of area (c.g.a) of area $ABXY$. A cardboard cutout suspended by a thread at several points is recommended. Use a planimeter to measure the area of $ABXY$ to compute the weight W.

6. Compute the radius of the ϕ circle as $r \sin \phi$ where $r = OX$, and draw the circle about point O as shown.

7. Extend the line of action of P'_p to intersect the W vector at point E. Start the force polygon for the system of forces as shown in Fig. 15-11b by drawing W from point 1 to 2 and P'_p from 2 to 3, and obtain the vector R'.

8. Through E and using the slope of R', extend the vector, and from P_p extend the vector to intersect at point Z.

9. Through Z and tangent to the ϕ circle, draw the R vector.

10. Now complete the force polygon by transferring the slope of R and P_p. Note that R continues from point 3 and P_p must terminate on point 1 to close the polygon. The intersection of R and P_p at point 4 allows scaling of P_p for magnitude.

It will be necessary to make several trials to obtain the maximum value of P_p for either the log spiral or the circular arc method. In all trials, the arc will intersect the line AD at some point X, and the slopes of both AD and XC are constant for the given soil conditions and ground geometry.

In cohesive soils there will be an additional resistance due to wall adhesion and due to cohesion along the slip surface BX. The effect of cohesion along XC will be included in P'_p. It will be easier and sufficiently accurate to always use BX = circular arc in cohesive soil.

15-8 SHEAR FAILURE AND BEARING CAPACITY

The ultimate bearing capacity of soil beneath a foundation load depends primarily on the shear strength. The working, or allowable, value for design will take into consideration both strength and deformation characteristics.

Most of the currently used bearing capacity theories are based on plasticity theory. Prandtl (ca. 1920) developed expressions from analysis of the assumed flow conditions of Fig. 15-12. The curved part of the arc ed or ce is assumed to be a part of a log spiral. For foundations on saturated clay, it is usual to assume undrained ($\phi = 0$) conditions; from the preceding section, we have a circular arc, and the ultimate bearing capacity by Prandtl's method is

$$q_{ult} = (\pi + 2)c = 5.14c$$

Others have found values of 5.64 to $5.74c$ for surface footings.

Terzaghi (1943) investigated the problem and obtained for strip footings

$$q_{ult} = cN_c + \gamma D N_q + \tfrac{1}{2}\gamma B N_\gamma \tag{15-10}$$

for square footings

$$q_{ult} = 1.3cN_c + \gamma D N_q + 0.4\gamma B N_\gamma \tag{15-10a}$$

for round footings

$$q_{ult} = 1.3cN_c + \gamma D N_q + 0.6\gamma R N_\gamma \tag{15-10b}$$

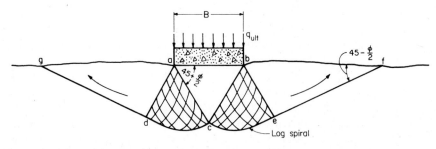

Figure 15-12 Assumed failure surface for bearing capacity failure for a footing on the ground surface. Note the upward assumed soil movement adjacent to the footing.

where D = footing depth

B = footing width (least dimension)

R = footing radius

γ = *effective* unit weight of soil

N_i = bearing capacity factors, shown in Fig. 15-13

In general, the Terzaghi equations are applicable for shallow foundations ($D \leq B$). Using Terzaghi's equations and $\phi = 0$, we obtain for the N_i terms

$$N_c = 5.74 \qquad N_q = 1.00 \qquad N_\gamma = 0.00$$

The Terzaghi value of 5.74 instead of 5.14 for N_c is an increase due to friction between soil and footing. The N_q term is for the surcharge effect due to overburden, which tends to confine the soil and avoid the upward movement along plane *bf* or *ag* of Fig. 15-12. This term produces a very significant contribution to bearing capacity, especially for large values of ϕ. Since bearing capacity does not increase without bound with increase in depth, either the value of D is limited to not more than 1 or $2B$ or a reduced N_q factor is used for greater depths. The N_γ term reflects footing width for frictional soils but is seldom a large contribution unless the footing is very wide.

Numerous researchers have proposed bearing capacity equations similar to the Terzaghi equations. Of these, those of Hansen (1970) appear to be the most widely used because they include shape and depth factors in a more rational

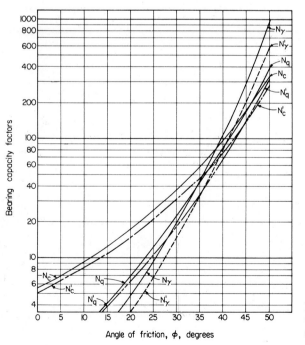

Figure 15-3 Bearing capacity factors for Terzaghi's equations (N_i) and the Hansen values (N_i').

manner than any other proposal and they are relatively easy to use. The Hansen equation is

$$q_{\text{ult}} = cN_c s_c d_c + \gamma D N_q s_q d_q + \tfrac{1}{2}\gamma B N_\gamma s_\gamma d_\gamma \qquad (15\text{-}11)$$

where $N_q = \tan(45 + \phi/2) \exp(\pi \tan \phi)$
$\quad N_c = (N_q - 1) \cot \phi$
$\quad N = 1.5(N_q - 1) \tan \phi$

and the shape and depth factors are approximately

$$s_c = 1 + \frac{N_q B}{N_c L} \qquad d_c = 1 + \frac{0.4D}{B}$$

$$s_q = 1 + \frac{B}{L} \tan \phi \qquad d_q = 1 + 2 \tan \phi (1 - \sin \phi)^2 \frac{D}{B}$$

$$s_\gamma = 1 - \frac{0.4B}{L} \qquad d_\gamma = 1.00$$

Other terms are the same as in the Terzaghi equations, and $L =$ footing length. For a round footing, use $B =$ diameter. The N_i factors are shown on Fig. 15-13 with the Terzaghi values for comparison.

Skempton (1951) made a comprehensive study of foundations in London clay and proposed the N_c bearing capacity factor as shown in Fig. 15-14. Using Fig. 15-14, the bearing capacity is

$$q_{\text{ult}} = cN_c + \gamma D N_q$$

Later work in the United States and elsewhere indicates that $N_c = 9$ is satisfactory for deep foundations $(D/B \geq 5)$ in clay for such round footings as bases of caissons or belled piers.

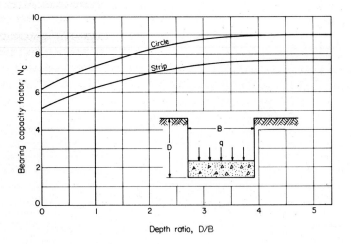

Figure 15-14 Bearing capacity factor N_c for a deep foundation in clay. (After Skempton, 1951.)

In all cases the allowable bearing capacity is reduced from the ultimate value by a suitable factor of safety F to obtain

$$q_a = \frac{q_{ult}}{F}$$

where $F = 2.0$ for cohesionless soils (usually)

$\qquad = 3.0$ for cohesive soils

It should be noted that if soil is permanently removed, the net increase in soil pressure is of particular design interest. The net pressure increase is

$$q_{net} = q_{loads} + q_{footing\ weight} - q_{soil\ excavated}$$

It is evident that the net pressure increase may be made near zero by an appropriate type of foundation at an adequate depth. A soil pressure increase of zero would not create either a shear failure or any settlement.

Example 15-5

GIVEN Square $(B/L = 1)$ footing with $B = 1$ m, $D = 0.5$ m $(D/B = 0.5)$, $\phi = 20°$, $c = 30$ kPa, $\gamma = 17.6$ kN/m^3.

REQUIRED Estimate the bearing capacity using Hansen's equation [Eq. (15-11)].

SOLUTION From Fig. 15-13, obtain $N_c = 15$, $N_q = 6.2$, and $N_\gamma = 3.5$.

Compute $s_c = 1 + \dfrac{6.2}{15} = 1.4$ $\qquad s_q = 1 + \tan \phi = 1.4$

$$s_\gamma = 1 - \frac{0.4B}{L} = 0.6 \qquad d_c = 1 + 0.4(0.5) = 1.2$$

$$d_q = 1 + 2 \tan 20(1 - \sin 20)^2 \frac{D}{B} = 1 + 0.315(0.5) = 1.2$$

$$d_\gamma = 1.0$$

Substituting into Eq. (15-11),

$$q_{ult} = 30(15)(1.4)(1.2) + 0.5(17.6)(6.2)(1.4)(1.2)$$
$$+ \tfrac{1}{2}(17.6)(1)(3.5)(0.6)(1)$$
$$= 756 + 92 + 18 = 866 \text{ kPa}$$

$$q_a = \frac{q_{ult}}{3} = \frac{866}{3} = 289 \text{ kPa}$$

15-9 IMMEDIATE ELASTIC SETTLEMENTS

Loads applied to a soil mass cause an increase in the existing pressures and produce settlements. We considered $\Delta H = f(\text{time})$ in Chap. 11. Elastic settlements can be computed as

$$\Delta H = \frac{qL}{E_s}$$

where E_s = stress-strain modulus
L = depth of stress influence
q = increase in soil pressure above equilibrium.

The Boussinesq equation in Sec. 10-7 indicates that q varies over the depth of stress influence L. From Sec. 14-4 we note that E_s depends on confinement (or cell pressure), soil density, and other factors; however, these two in particular indicate that E_s increases with depth. The depth of stress influence is somewhat uncertain but may be estimated from the Boussinesq equation at the point where the computed stress is on the order of 0.01 to $0.05q_o$. With these several variables some kind of integration over the depth of interest is necessary to compute the settlement using this equation.

An equation using theory of elasticity and based on a load on the surface of a semi-infinite, elastic, weightless half space gives

$$\Delta H = \frac{qB(1 - \mu^2)}{E_s} I_w \tag{15-12}$$

where terms not previously defined include:
B = width of loaded foundation area (use least lateral dimension)
I_w = shape factor (see Bowles, 1977), depending on whether the foundation is rigid or flexible. For rigid foundations,
$\quad I_w = \pi/4$ (round, B = diameter)
$\quad\quad = 0.82$ (square)
$\quad\quad = 1.20$ (rectangle length/width = 2)

Example 15-6

GIVEN A square rigid footing 3×3 m with a load of 1300 kN. Laboratory tests give $E_s = 16\,000$ kPa. The footing is located as shown in Fig. E15-6.

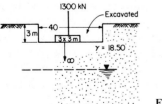

Figure E15-6

REQUIRED Estimate the elastic settlement.

SOLUTION

Step 1   The contact pressure is $q = P/A$,

$$q = \frac{1300}{9} = 144.4 \text{ kPa}$$

Step 2   The net pressure is

$$144.4 - 3(18.50) = 88.9 \text{ kPa}$$

Step 3   Estimate Poisson's ratio $= 0.3$ and substitute this into Eq. (15-12):

$$\Delta H = \frac{88.94(3)(1 - 0.3^2)}{16\,000} 0.82 = 0.012 \text{ m}$$

15-10 SUMMARY

This chapter has introduced the concept of earth pressure as

Pressure	Type
Active	Minimum earth pressure due to expansive relative movements away from soil mass
K_o	In situ equilibrium earth pressure
Passive	Maximum earth pressure due to compressive relative movements into soil mass

We note that the soil resistance aids in active earth pressure and that the volume of soil is a minimum. The soil resistance must be overcome for a larger volume of soil in developing passive earth pressure. This change in the volume of soil in the failure wedge plus the particulate nature of soil and the associated soil flow with loss of lateral pressure creates a situation where K_o conditions are difficult to impossible to measure.

The trial wedge method of determining passive earth pressure was introduced, along with use of a log spiral and the ϕ-circle methods. It was pointed out that the Rankine and Coulomb earth pressure methods are commonly used to determine wall pressures. The Rankine method was extensively covered both by use of Mohr's circle and by use of the trial wedge method. The use of a plane failure surface gives little loss of accuracy for the active earth pressure case, and for small slope angles, small angles of wall friction, and relatively small angles of internal friction gives reasonably good values for passive earth pressure.

A brief introduction to bearing capacity theories and elastic settlements has been made and the two most common methods of computing ultimate bearing capacity have been presented. For refinements in bearing capacity and settlement computations, the reader should refer to a text on foundation engineering, e.g., Bowles (1977).

HOMEWORK PROBLEMS

15-1 What is the lateral pressure at the base of a 5-m bin of wheat which has a unit weight of 7.5 kN/m³ and $\phi = 27°$? What is the wall force per meter?

Ans.: Pressure at base = 14.1 kPa.

15-2 For the conditions of Fig. 15-5d and e, make several trials to obtain $P_{a, \max}$. Take $\phi = 30°$; $\gamma = 18.00$ kN/m³; $H = 5$ m. Make trials for $\rho = 50, 55, 60,$ and 65°. Plot ρ versus P_a and obtain a graphical value of P_a and ρ.

Ans.: $P_a = 75$ kN/m.

15-3 For $H = 6$ m, $\phi = 36°$, $\gamma = 18$ kN/m³ and the method shown in Fig. 15-5d and e, make trials for $\rho = 55, 60, 65, 70,$ and 75°. Plot ρ versus P_a, and obtain P_a and ρ.

15-4 Redo Prob. 15-2 for P_p. This time try $\rho = 20, 25, 30, 35,$ and 40°.

Ans.: $P_p = 675$ kN/m.

15-5 Compute the active earth force of Prob. 15-3 using the Rankine equation. Draw a neat sketch, and show the pressure profile and the location and magnitude of the resultant wall force.

15-6 Make a plot of shear planes based on Mohr's circle as in Fig. 15-6 for $\phi = 36°$.

15-7 Compute the active and passive wall forces for the following conditions: $H = 6$ m; $\gamma = 19.15$ kN/m³; $\phi = 32°$; $\beta = 10°$.

15-8 Verify the procedure of Fig. 15-7 for Prob. 15-7 for the active earth pressure at the base of the wall.

15-9 Use the ϕ-circle procedure to determine the approximate passive earth force of Prob. 15-7. Limit the number of trials to not over three.

15-10 What is the active earth force and point of application for the following conditions: $\phi = 5°$; $c = 18$ kPa; $\gamma = 17.00$ kN/m³; $H = 6$ m.

Ans.: $\bar{y} = -1.35$ m.

15-11 Redo Prob. 15-10 for the passive earth force.

Ans.: $P_p = 1749.4$ kN/m at $\bar{y} = 2.39$ m above base.

15-12 What are the depth of tension crack and the theoretical depth of unbraced excavation for a soil with $\phi = 0°$, $q_u = 80$ kPa, and $\gamma = 17.50$ kN/m³? What unbraced depth of excavation do you recommend and why?

15-13 What is the maximum depth of tension crack in Prob. 15-10, and what is the resulting active wall force if the crack fills with water?

Ans.: $P_a = 119.9$ kN/m.

15-14 Compare the ultimate bearing capacity by the Terzaghi and Hansen bearing capacity equations for a footing in a cohesionless soil. General data: $\phi = 32°$, $\gamma = 18.00$ kN/m³, footing is 1.5 m square, and $D = 1.5$ m ($D/B = 1$).

Ans.: Percent increase of Hansen equation = 30.

15-15 Redo Prob. 15-14 if the soil also has cohesion, $c = 20$ kPa. Comment on the effect of this small amount of cohesion on the bearing capacity.

15-16 Using procedures outlined in Sec. 15-4, show the derivation of Eq. (15-7a).

15-17 Redo Ex. 15-6 if the footing is located on the ground surface and not at a depth of 3 m.

Ans.: $\Delta H = 0.020$ m.

15-18 Redo Ex. 15-6 if E_s is to be determined from SPT data and the average value of N in the depth $3B$ is $N = 9$.

Chapter 16
Stability of Slopes

16-1 GENERAL CONSIDERATIONS IN STABILITY OF SLOPES

Slopes may be man-made, as in:

Cuts and fills for highways and railroads
Earth dams
River levees
Dikes for containment of water, including leachates, industrial wastes, and
    sewerage
Landscaping operations for industrial or other development
Banks of canals and other water conduits
Temporary excavations

Slopes may also be naturally formed as hillsides or stream banks.

In any case, the ground not being level results in gravity components of the weight tending to move the soil from the high point to a lower level. Seepage may be a very important consideration in moving the soil where water is present. Earthquake forces may also be important in stability analysis on occasion.

These several forces produce shear stresses throughout the soil mass, and a movement will occur unless the shearing resistance on every possible failure surface throughout the mass is sufficiently larger than the shearing stress. The shearing resistance depends on the shear strength of the soil and other natural factors, such as instant presence of water from seepage and/or rainfall infiltration as well as roots, ice lenses, frozen ground, or rocks which must be severed along the slip surface. Animal, worm, and reptile burrows or decayed roots may produce a progressive failure mechanism which initiates a slope failure.

The major factor, however, is the shear strength of the soil, which may:

1. Be undrained (s_u) for some cases of loading
2. Be effective (ϕ', c') for some cases of loading
3. Increase with time (as consolidation) or with depth
4. Decrease with time due to later saturation, development of excess pore pressure, as when a downward sloping pervious stratum has the exit blocked and fills with water, or loss of negative pore pressure (see Sec. 13-6)

A stability analysis involves making an estimate of both the failure model and the shear strength. The failure model will require prediction of the weights (or loads) to be resisted and the effect of water. The water estimate requires consideration of seepage forces and saturated and effective unit weights. The shape of the failure model can usually be reasonably well defined; however, for the center of rotation, it may require numerous trials to find the worst case.

The solution is highly sensitive to the shear strength, and the shear strength is generally the most difficult parameter in the analysis to predict. This is because of its variation with depth and the difficulty of deciding whether to use the effective or undrained strength. It has already been pointed out in earlier chapters that the shear strength is sensitive to disturbance and testing procedures. It is also difficult to predict changed soil water conditions.

Care must be exercised to produce an economical solution, since one could always insert overly conservative values of shear strength into the problem.

Some solutions entail careful monitoring of the construction (fill) using piezometers to ensure that pore pressures do not build sufficiently to lower the shear stress to a failure. This procedure is routinely done at present on large earth dam fills as a result of at least two major construction failures involving several million cubic meters of soil (see Casagrande, 1965). It is important to have the piezometers located where they can measure critical pore pressures, as in the Fort Peck slide (in 1938) piezometers were installed, but not into the zone where the pore pressures caused failure.

In general, slope stability is a plane strain problem, i.e., the length compared to cross section is very large. It is usual to investigate a typical cross section which is 1 unit thick with plane strain, ignoring the perpendicular strains (and stresses). Many small slope failures, as can be readily observed along road cuts in mountainous areas, are nearly equal in lateral and vertical dimensions; nevertheless, plane strain conditions are assumed in current analyses.

16-2 INFINITE SLOPES

Figure 16-1a illustrates the cross section of an infinite slope in a cohesionless soil without seepage. The thickness perpendicular to the plane of the paper is 1 unit (1 m). The soil is often assumed to be homogeneous; however, in real situations,

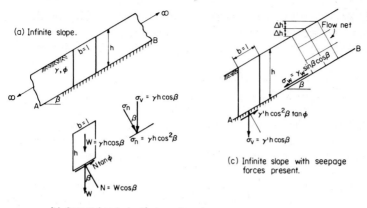

(b) General free-body of slope element.

Figure 16-1 Infinite slope with cohesionless soil.

the soil may be highly stratified with widely varying shear strengths. If we isolate an element, as in Fig. 16-1b, and examine the forces for stability, $\sum F$ parallel to the slope gives

$$W \tan \beta - W \cos \beta \tan \phi = 0$$

and solving we obtain

$$\beta = \phi$$

for stability. This is the same as the Rankine solution of Sec. 15-4.

With a constant seepage and with the water table at the ground surface, we have from Fig. 16-1c

$$\sigma_n = \gamma'h \cos \beta \cos \beta = \gamma'h \cos^2 \beta$$
$$\sigma_t = \gamma'h \sin \beta \cos \beta$$

The pore pressure stress is

$$\sigma_w = \gamma_w h \sin \beta \cos \beta$$

and summing stresses parallel to the plane AB, obtain

$$\gamma_w h \sin \beta \cos \beta + \gamma'h \sin \beta \cos \beta - \gamma'h \cos^2 \beta \tan \phi = 0$$

Solving for the critical slope angle β, obtain

$$\tan \beta = \frac{\gamma'}{\gamma' + \gamma_w} \tan \phi$$

or

$$\beta = \tan^{-1} \left(\frac{\gamma'}{\gamma' + \gamma_w} \tan \phi \right)$$

For example, with no water, $\gamma' = \gamma_t$ and $\beta = \phi$; with water and $\gamma' = 17.8 - 9.8 = 8.0 \, \text{kN/m}^3$ and $\phi = 32°$, the critical slope angle is

$$\beta = \tan^{-1}\left(\frac{8}{17.8} \tan 32\right) = 15.7°$$

Thus with water the theoretical safe slope angle β is only about half that without water.

16-3 STABILITY OF INFINITE COHESIVE SLOPES

Referring to Fig. 16-2, and with no water, we have

$$\sigma_t = \gamma h \sin \beta \cos \beta \qquad \text{(stress)} \qquad (a)$$

$$\sigma_n = \gamma h \cos^2 \beta \qquad \text{(stress)} \qquad (b)$$

The resisting stress is

$$s = c_d + \sigma \tan \phi_d \qquad (c)$$

where c_d and ϕ_d are design shear strength parameters and not the soil values unless the safety factor $F = 1$. At $F = 1$, the shear strength $s = \sigma_t$, and by substitution of Eqs. (a) and (b) into (c), obtain

$$\gamma h \sin \beta \cos \beta = c_d + \gamma h \cos^2 \beta \tan \phi_d$$

or the design cohesion is

$$c_d = \gamma h \cos^2 \beta (\tan \beta - \tan \phi_d)$$

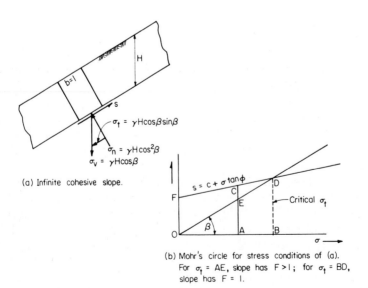

(a) Infinite cohesive slope.

(b) Mohr's circle for stress conditions of (a). For $\sigma_t = AE$, slope has $F > 1$; for $\sigma_t = BD$, slope has $F = 1$.

Figure 16-2 Infinite slope in cohesive soil.

The critical value of clay thickness is

$$H = \frac{c_d}{\gamma} \frac{\sec^2 \beta}{\tan \beta - \tan \phi_d}$$

This can be illustrated by Mohr's circle as in Fig. 16-2b, where OA represents the normal stress at some height less than critical and OB represents the normal stress at the critical height, at which time $\sigma_t = BD$. We note that if $\beta < \phi$ the rupture line is never intersected (theoretically safe). Alternatively, we may write

$$N_s = \frac{c_d}{\gamma H} = \cos^2 \beta \, (\tan \beta - \tan \phi_d) \tag{16-2}$$

where N_s = stability number as commonly used in the literature (some writers have used N_s as $\gamma H/c$, so that the reader should be careful to see how the term is used or strange results may be obtained). The stability number, being dimensionless, allows the combining of three problem parameters into a single value and allows the use of simple charts for representing stability relationships.

With seepage the full depth of interest, the stability number becomes

$$N_s = \cos^2 \beta \left(\tan \beta - \frac{\gamma'}{\gamma} \tan \phi_d \right) \tag{16-2a}$$

If the top flow line is a distance h_1 below and parallel to the ground surface and the soil weight in zone h_1 is γ_1, we obtain

$$N_s = \cos^2 \beta \left[\left(1 - \frac{h_1}{H} \frac{\gamma - \gamma_1}{\gamma} \right) \tan \beta - \left(\frac{\gamma'}{\gamma} + \frac{h_1}{H} \frac{\gamma_1 - \gamma'}{\gamma} \right) \tan \phi_d \right] \tag{16-3}$$

16-4 CIRCULAR ARC ANALYSIS

Where slopes are finite in extent, as most man-made embankments and roadway cuts are, the failure surface is curved. Various workers have suggested that the curved surface is a part of a circular arc or a log spiral. Observed slip surfaces tend to be a combination of a circular arc and a log spiral, somewhat oval with relative flat arcs on each end and a sharper arc on the interior. There may be plane discontinuities if the surface intersects a hard zone such as very stiff clay, dense sand, or a rock surface.

The circular arc is the simplest solution and the only one considered here. In the author's opinion, the errors in a slope analysis are not so much in the shape of the assumed failure surface but in the soil properties and the search for the critical failure location. If one can find the critical location (which depends on both geometry and soil properties) from among the infinite number possible, it will be a happy coincidence.

When the slope is a homogeneous clay and an undrained strength analysis is used, the stability analysis is relatively straightforward. Figure 16-3 illustrates trial

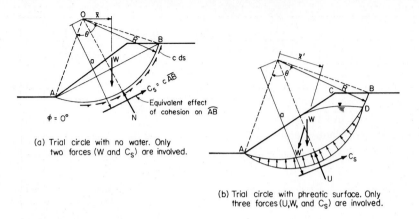

(a) Trial circle with no water. Only
two forces (W and C_s) are involved.

(b) Trial circle with phreatic surface. Only
three forces (U,W, and C_s) are involved.

Figure 16-3 Trial failure circles with and without water forces.

failure circles and the forces involved. Figure 16-3a is for a case with no water, and
Fig. 16-3b includes seepage forces.

When seepage forces are present, it is necessary to locate the phreatic line and
sketch in a flow net. The equipotential lines intersect the trial arc, and with the
head known, the pressure at these points can be computed to give the pressure
profile as shown in Fig. 16-3b. A numerical integration of this area can be made to
obtain the total water force U, which has a line of action through the circle center
O. This value of U can be added vectorially to the weight vector W to obtain the
new force vector W' with a new line of action and scaled moment arm $\bar{x}'$. The
resulting factor of safety is

$$F = \frac{R(c\widehat{AB})}{W'\bar{x}'} \tag{16-4}$$

Taylor (1937) solved Eq. (16-4) for $F = 1$ in terms of β and N_s ($= c/\gamma H$)
and presented the results in chart form. Figure 16-4 is a replot by the author of the
Taylor charts, which can be used for $\phi = 0°$ and the slope conditions shown on
the figure. The dashed curves, along with the curve labeled $\phi = 0°$, are particularly
applicable to Eq. (16-4). These curves can be developed as follows:

1. Obtain the weight of the failure mass W and its moment arm with respect to
   point O. This may be done by using a planimeter for the area (and weight) and
   making a cardboard cutout which is suspended by a thread at two or more
   points to find the center of area.
2. Measure angle θ and compute the arc length as $AB = R\theta$ (θ in radians).
3. Compute the factor of safety as

$$R = \frac{\sum \text{ resisting moments}}{\sum \text{ overturning moments}} = \frac{R(c\widehat{AB})}{W'\bar{x}'}$$

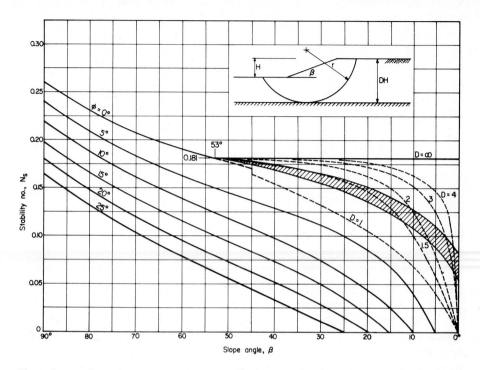

Figure 16-4 Taylor's curves for stability number replotted by the author. For $\phi = 0°$ and $\beta \leq 53°$, use dashed curves. The shaded zone is approximate for toe circles: above for base circles as shown in the sketch, and below for slope circles.

It is necessary to make several trial circle analyses with the safety factor plotted on the center point so that contours of F can be made which will, hopefully, give the minimum value. The values obtained from Fig. 16-4 are supposed to give the minimum value of safety factor F directly.

16-5 THE ϕ-CIRCLE METHOD

The friction circle concept may be used for the particular slope condition of a homogeneous soil with a shear strength of

$$s = c_d + \bar{\sigma} \tan \phi_d$$

where c_d, ϕ_d = design shear strength parameters

Figure 16-5 illustrates the general concept. Figure 16-5a illustrates the trial circle and all the forces involved in a soil where no water is present. The shear resistance is obtained from integrating the cohesion and normal forces along the arc to obtain

$$F = \int c_d \, ds + \int dN \tan \phi_d$$

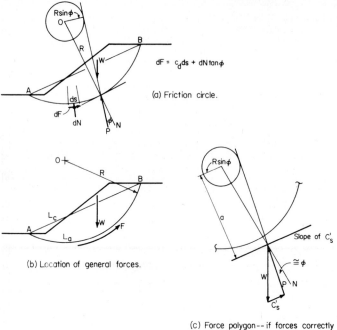

$$dF = c_d ds + dN \tan \phi$$

(a) Friction circle.

(b) Location of general forces.

Slope of C_s'

(c) Force polygon -- if forces correctly computed and located, the resultant P is nearly tangent to the ϕ-circle.

Figure 16-5 The ϕ-circle force system.

or, alternatively,

$$F = C_s + F_f$$
$$= c_d AB + N \tan \phi$$
$$= c_d AB + W \cos \phi \tan \phi$$

as shown in Fig. 16-5b. Note, as in Chap. 15, that the friction resistance combines with N to give a vector P which acts at the angle ϕ to the tangent to the arc. The line of action of P extends and gives a moment arm with respect to point O of $\bar{x}$, computed as

$$\bar{x} = R \sin \phi$$

For all the vectors of dP, the moment arms $\bar{x}$ trace a portion of a small circle with radius $R \sin \phi$. This small circle about point O is called the ϕ circle.

The cohesion term is independent of ϕ, and noting that a general tangent can be obtained for arc AB which is parallel to the chord, and further noting that the normal component of C_s cancels, by equating moments along the arc L_a and with respect to the chord length L_c, we obtain (refer to Figs. 16-5a and 16-3a)

$$cL_c a = cL_a R$$

which produces an equivalent moment arm of

$$a = \frac{L_a}{L_c} R$$

With this value of moment arm, the equivalent cohesion force is computed as

$$C_s' = cL_c$$

For a force system in moment and static equilibrium, the system must be concurrent, with both $\sum F_H$ and $\sum F_v = 0$. A force polygon can be used for the forces as shown in Fig. 16-5c, with the slope of P obtained from the intersections of the line of action of W and C_s' and approximately tangent to the ϕ circle. Usually we assume that P is exactly tangent to the ϕ circle, as the maximum difference is generally less than 7 percent. The body force W is obtained by planimetering the area or by analysis for simple geometric sections. The line of action is obtained by cutting a cardboard model and suspending by a thread at two or more points. Several trial circles are necessary to obtain the minimum F, which is computed as before, i.e.,

$$F = \frac{\sum \text{ resisting moments}}{\sum \text{ overturning moments}}$$

Inspection of the force polygon in Fig. 16-5c indicates that

$$W = f(\beta, H, \phi)$$

$$C_s' = f(c)$$

$$P = f(\phi)$$

thus there are five variables. We may combine H, γ, and c into the single variable N_s as earlier and reduce the variables to three. In practice, four of the variables must be given; however, by use of the dimensionless variable N_s, a parametric study for a particular slope can be made rather easily.

Taylor (1937) also presented a table of N_s versus ϕ which has been put into chart form by the author as Fig. 16-4. This chart represents the results of several trials to obtain the critical circle; thus, the user merely has to enter with ϕ and either β or N_s.

Where slope geometry and a stratum of very stiff material are located as in Fig. 16-6, we have the possibility of

Toe circle—all circles for soils with $\phi > 3°$ and all $\beta > 53°$
Slope circle—always for $D \to 0$ and $\beta < 53°$
Midpoint circle—circle center on a vertical line bisecting the slope and depends on D and for $\beta < 53°$; always for $D > 4$

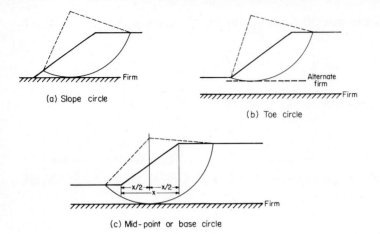

(a) Slope circle

(b) Toe circle

(c) Mid-point or base circle

Figure 16-6 Trial circles controlled by soil geometry.

Here the factor of safety F is slightly redefined to be a comparison of the required, or developed, shear strength compared to the actual shear strength, or

$$F = \frac{\text{actual soil shear strength } s}{\text{developed shear resistance}}$$

It appears from numerous analyses of slope failures and designs that if:

F	Event
$F <$ about 1.07	Failures are common
$1.07 < F < 1.25$	Failures do occur
$F > 1.25$	Failures almost never occur

In determining the factor of safety F, we can use Fig. 16-4 and obtain for β and ϕ the corresponding N_s. Since we know γ and H, the actual cohesion required can be computed as

$$c_{\text{required}} = N_s \gamma H$$

and the factor of safety is

$$F = \frac{c_{\text{actual of soil}}}{c_{\text{required}}}$$

Similarly, we could use H or γ instead of the soil cohesion.

When the shear strength is $s = c + \sigma \tan \phi$, it is necessary to apply F to both c and ϕ as

$$s_{\text{required}} = \frac{c}{F} + \frac{\sigma \tan \phi}{F}$$

This is illustrated in the following example.

Example 16-1

GIVEN A simple slope with $1V$ to $0.75H$ $(0.75:1)$; $H = 12$ m, $\gamma = 18.0$ kN/m³, $\phi = 15°$, $c = 30$ kPa.

REQUIRED What is F when applied to both ϕ and c?

SOLUTION Compute

$$\beta = \tan^{-1} \frac{1}{0.75} = 53°$$

From Fig. 16-4, obtain $N_s = 0.1$ at $\phi = 15°$ and $\beta = 53°$. Compute

$$\gamma' = \frac{30}{(0.1)(12)} = 25 \text{ kN/m}^3$$

$$H' = \frac{30}{(0.1)(18)} = 16.7 \text{ m}$$

$$c' = 0.1(12)(18) = 21.6 \text{ kPa}$$

The several factors of safety are

For γ:
$$F = \frac{25}{18} = 1.39$$

For H:
$$F = \frac{16.7}{12} = 1.39$$

For c:
$$F = \frac{30}{21.6} = 1.39$$

Since we used a value of $\phi = 15°$, the F on ϕ is 1. Re-enter Fig. 16-4 with an estimated $F = 1.3$, giving $\phi = 15/1.3 = 11.5°$, and obtain N_s of approximately 0.12. Compute

$$c' = 0.12(12)(18) = 25.9$$

$$F = \frac{30}{25.9} = 1.16$$

Re-enter again, using $F = 1.2$ for $\phi = 12.5°$, and obtain N_s approximately 0.115

$$c' = 0.115(12)(18) = 24.8$$

$$F = \frac{30}{24.8} = 1.21 \text{ (which is sufficiently close to the used value of 1.20)}$$

Therefore the safety factor is 1.20 for this problem when applied on both ϕ and c.

16-6 SLOPE ANALYSIS BY METHOD OF SLICES

Most natural slopes and many man-made slopes consist of more than one soil, or the soil properties vary so much that some type of finite element solution is mandated. The finite element method generally used is to divide the failure section into a series of vertical slices as illustrated in Fig. 16-7a.

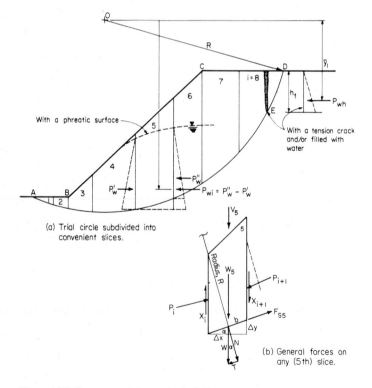

(a) Trial circle subdivided into convenient slices.

(b) General forces on any (5th) slice.

Figure 16-7 Geometry of the method of slices.

The slice width is sufficiently small that the actual shape can be replaced with a trapezoid, as shown in Fig. 16-7b. It is assumed that the slice weight W_i acts through the midpoint of the area as shown. With this assumption the following relationships are developed:

$$N_i = (W_i + V_i) \cos \alpha$$

$$T_i = (W_i + V_i) \sin \alpha$$

$$F_s = N_i \tan \phi + cb$$

$$= (W_i + V_i) \cos \alpha \tan \phi + c \frac{\Delta x}{\cos \alpha}$$

and

$$\alpha = \tan^{-1} \frac{\Delta y}{\Delta x}$$

It is usual practice to neglect the interelement forces of X_i and P_i. Some persons have used these forces, but the point of application and line of action of the P forces are indeterminate in stratified soils or where the soil properties (ϕ, c, γ) vary with depth. In these cases, about all that is known for certain is that the line of action of P is inside the failure surface. The vertical force depends on both P and the soil properties. Several researchers have shown that little error is introduced by neglecting the X and P forces. Note, too, that at slip the soil properties at the trial circle boundary are all that are valid—those inside the failure zone are those of a highly remolded soil and are unknown.

Moment equilibrium about point O, using a summation for all the slices in the failure circle with attention to signs, gives

$$\sum RF_s - \sum R(W_i + V_i) \sin \alpha = 0$$

The resisting moment is $\sum RF_s$, and the safety factor F is

$$F = \frac{\sum \text{resisting moments}}{\sum \text{overturning moments}} = \frac{\sum RF_s}{\sum R \sin (W_i + V_i)}$$

Cancelling R and inserting for the shear strength s, obtain

$$F = \frac{\sum (cb + (W_i + V_i) \cos \alpha \tan \phi}{\sum (W_i + V_i) \sin \alpha} \tag{16-5}$$

We may use either effective or total stresses and with the appropriate c and ϕ parameters in evaluating Eq. (16-5). The effective stress is most conveniently developed by using γ and γ' as appropriate in computing the weight vector W.

It is considered more correct to apply F to the soil parameters in Eq. (16-5) as was done earlier. If we do this, we obtain

$$F = \frac{\sum [cb/F + (W_i + V_i) \cos \alpha (\tan \phi)/F]}{\sum (W_i + V_i) \sin \alpha} \qquad (16\text{-}6)$$

Since $b = \Delta x/\cos \alpha$, we have the α angle producing a significant role in the problem. Bishop (1955) suggested that the effect of α could be reduced by an alternative method of determination of the normal force. Referring to Fig. 16-7b, $\sum F_v$ on the element (and neglecting the X's) is

$$N_i \cos \alpha - (W_i + V_i) + F_{si} \sin \alpha = 0 \qquad (a)$$

The friction resistance due to ϕ is

$$F_{si} = \frac{N_i \tan \phi}{F} \qquad (b)$$

Substituting Eq. (b) into (a) and dividing by $\cos \alpha$, obtain

$$N_i + \frac{N_i \tan \phi \tan \alpha}{F} = \frac{W_i + V_i}{\cos \alpha}$$

or

$$N = \frac{W_i + V_i}{\cos \alpha} \frac{1}{1 + (\tan \phi \tan \alpha)/F} \qquad (c)$$

Since $(W_i + V_i) \cos \alpha$ in Eq. (16-5) is N of Eq. (c), substitution for N gives:

$$F = \frac{\sum [c \, \Delta x + (W_i + V_i) \tan \phi] \dfrac{\sec \alpha}{[1 + (\tan \phi \tan \alpha)/F]}}{\sum (W_i + V_i) \sin \alpha} \qquad (16\text{-}7)$$

This form of the equation has been programmed by the author (Bowles, 1974). Note again that when effective stresses are used, the weight vector W_i uses the appropriate unit weight (γ or γ') in each slice.

It may be necessary to consider tension cracks, water in the tension crack, and/or unbalanced water pressure (as a seepage force) on each side of the slices. Noting that the radius R effect has been factored both top and bottom and that the numerator represents the resistance in Eq. (16-7), we may readily adjust the denominator with reference to Fig. 16-7a as follows:

Let $\qquad\qquad\qquad D = \sum (W_i + V_i) \sin \alpha$

Also, let the unbalanced water pressure on any slice be P_{wi} and the water force in the tension crack be P_{wh}. Moment arms are $\bar{y}$ and $\bar{y}'$, respectively, and a new denominator D' can be written as

$$D' = \sum \left[(W_i + V_i) \sin \alpha + \frac{P_w \bar{y}}{R} \right] + \frac{P_{wh} \bar{y}_1}{R}$$

The value of D' is now used instead of D in Eq. (16-7). The failure surface is now defined by sector $ABCEF$ of Fig. 16-7a.

An iterative analysis is necessary to obtain F in Eq. (16-7), since it is on both sides of the equality sign. Programming on the computer allows a rapid solution after only a few cycles (usually 2 to 3) by assuming $F = 1$ initially for the right-hand side and computing F on the left. This value of F is compared with the assumed value; if it is not sufficiently close, the just computed F is used in the next iteration and the cycle repeated. Inspection of Eq. (16-7) shows that for $\phi = 0$ soils an iteration is not necessary, as the equation reduces to Eq. (16-5).

A computer program must develop the arc based on initializing point O and the area entrance coordinates. The arc is divided into i slices and α, b, and $\bar{y}$ (if necessary) are computed for each slice. The weight is computed based on the soil(s) in the slice and/or making allowance for the phreatic surface. The quantities cb and $(W_i + V_i) \tan \phi$ are computed using soil parameters for the soil on the base of the slice. Once F is computed, either the arc entrance or circle center coordinates are incremented and additional analyses made for obtaining a minimum F.

An approximate analysis can be made by hand. The work is generally too prohibitive to iterate or to make an extensive critical circle search. By hand, the trial circle is drawn and subdivided into convenient slices, as shown in Ex. 16-2. The weight W_i (and V_i) is computed and plotted to scale, assuming that the vector acts through the midpoint of each slice as shown. The normal and tangent components can be directly obtained graphically from a radius through the projection of W_i onto the failure arc. The normal acts through point O and can be neglected. Compute the overturning moment, using the scaled values of T_i and the correct sign depending on which side of O the slice is located, as

$$M_{\text{overturning}} = R \sum T$$

Compute the resisting moment as

$$M_{\text{resisting}} = R \sum (cb + N \tan \phi)$$

The factor of safety is computed as

$$F = \frac{M_{\text{resisting}}}{M_{\text{overturning}}} = \frac{\sum (cb + N \tan \phi)}{\sum T} \tag{16-8}$$

Example 16-2 Hand solution of method of slices.

Soil properties: $\gamma = 17.6 \text{ kN/m}^3$; $c = 40 \text{ kPa}$; $\phi = 24°$

Slice	Area	Scaled				N tan ϕ	cb
		W	b	T	N		
1	$(0 + 2.8)(4.5/2) = 6.3$	110.9	5.3	−60	95	42	212
2	$(2.8 + 16.1)(15.3/2) = 144.6$	2544.7	16.0	−800	2440	1086	640
3	$(16.1 + 23.3)(12.6/2) = 248.2$	4368.7	12.8	0	4369	1945	512
4	$(23.3 + 27.0)(12.6/2) = 316.9$	5577.3	13.3	1600	5250	2337	532
5	$(27.0 + 25.5)(12.6/2) = 330.8$	5821.2	15.5	3260	4850	2159	620
6	$(25.5 + 0)(13.6/2) = 173.4$		28.9	2710	1800	801	1156
		$3051.8 + 200 = 3251.8$			6710	8370	3762

$$F = \frac{8370 + 3762}{6710} = 1.81$$

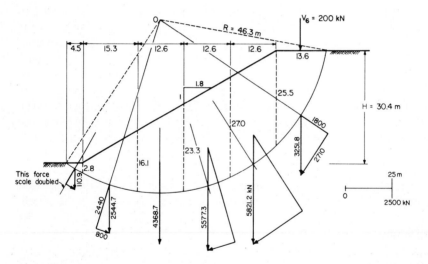

Figure E16-2 Hand solution of method of slices.

16-7 WEDGE BLOCK ANALYSIS

A sliding block or wedge may be a more appropriate cross section for many stability problems (Fig. 16-8) where the failure surfaces may be defined by a series of broken lines. This method may be extended to analysis of rock slopes, particularly in layered strata.

Figure 16-9 illustrates how to apply the wedge analysis method. The reader should make the necessary adjustments in active and passive pressures and the weight of the block for phreatic surfaces, stratification, etc.

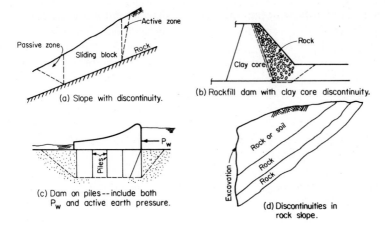

(a) Slope with discontinuity.

(b) Rockfill dam with clay core discontinuity.

(c) Dam on piles -- include both
P_w and active earth pressure.

(d) Discontinuities in
rock slope.

Figure 16-8 Typical situations where a wedge analysis may be appropriate.

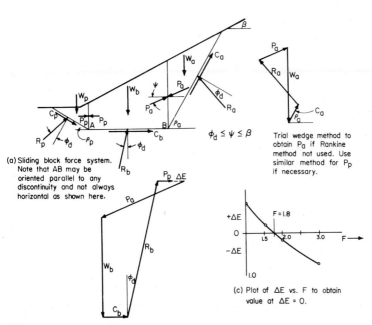

(a) Sliding block force system.
Note that AB may be
oriented parallel to any
discontinuity and not always
horizontal as shown here.

$\phi_d \leq \psi \leq \beta$

Trial wedge method to
obtain P_a if Rankine
method not used. Use
similar method for P_p
if necessary.

(b) Force polygon for central block to obtain ΔE.

(c) Plot of ΔE vs. F to obtain
value at $\Delta E = 0$.

Figure 16-9 The wedge block analysis. Note that ΔE may be oriented other than horizontally, with either the value or the horizontal component used in (c) to obtain F.

The active and passive earth forces are computed using either the Rankine earth pressure or the trial wedge method. The Rankine method, with $\rho = 45 \pm \phi/2$, can be used where the soil is homogeneous in these zones, but the trial wedge method is necessary where a fault or discontinuity locates the failure surface and may be used for stratified soils. An assumed F is used to compute the values of

$$c_d = \frac{c}{F} \qquad \phi_d = \frac{\phi}{F}$$

for computing the active and passive Rankine earth pressure forces, and the base resistance of the block of C_s and $F_f = N \tan \phi_d$. The problem is conveniently solved using a force polygon for each assumed value of F, as in Fig. 16-9b for the sliding block, with the closure error scaled as ΔE. A plot of ΔE versus F can be made as in Fig. 16-9c; the value of F at $\Delta E = 0$ is the desired value. This method of analysis is illustrated in Ex. 16-3.

Example 16-3

GIVEN The sliding block geometry of Fig. E16-3a and the following soil data: $\gamma = 17.0$ kN/m^3; $c = 60$ kPa; $\phi = 30°$.

REQUIRED Make a sliding block analysis and find the factor of safety.

SOLUTION Make the following assumptions:

1. No tension crack
2. For slope of P_a, take $\psi = \beta$
3. Use Rankine K_p for passive earth force P_p

Compute

$$\beta = \tan^{-1} \tfrac{1}{2} = 26.6°$$

$$K_p = \tan^2 \left(45 + \frac{\phi}{2} \right) \qquad P_p = 0.5 \gamma H^2 K_p + 2cH \sqrt{K_p}$$

Scale selected dimensions from Fig. E16-3a, including the dimensions of base and height of the active pressure wedges. It will be necessary to use the trial wedge method to obtain P_a, since $\beta > \phi$ when F is applied to ϕ. Set up the following table of values:

					Scaled					
F	ϕ_d	K_p	c_d	P_p, kN	b, m	h, m	$W_a = \tfrac{1}{2}\gamma bh$	C_a	P_a	C_b
1	30°	3.00	60	2758	57.5	17.5	8 553	3450	1400	3000
1.5	20	2.04	40	1848	65.6	20.0	11 152	2624	3950	2000
2.0	15	1.70	30	1523	70.8	21.5	12 939	2124	5900	1500
3.0	10	1.42	20	1255	78.2	22.8	15 155	1564	8450	1000

In table: $C_a = c_d b$, $C_b = 50c_d$.

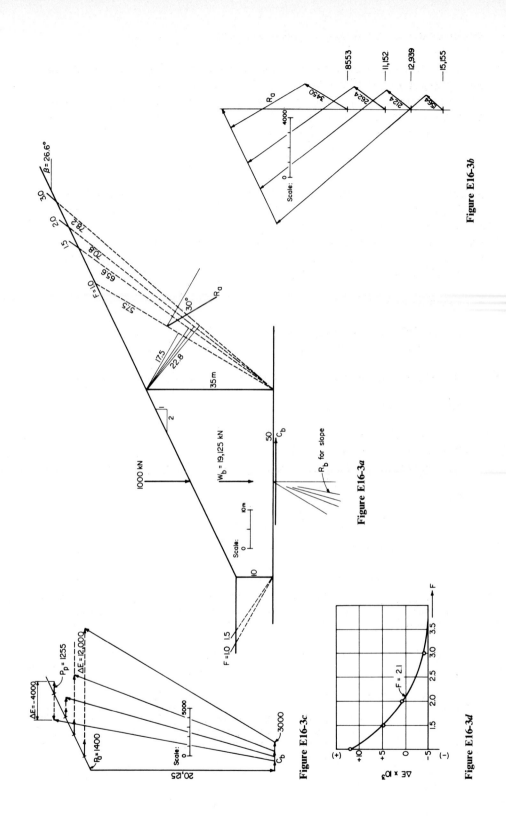

β = 26.6°

30
20
1.5
782
708
656
575
F = 1.0

175
22.8

30°

R_a

35 m

2
1

1000 kN

W_b = 19,125 kN

50

C_b

R_b for slope

Figure E16-3a

Scale:
0
10m

10

F = 1.0   1.5

R_a

4000

Scale:
0

3450
2624
2122
1654

8553
11,152
12,939
15,155

Figure E16-3b

P_p = 1255
ΔE = 4000
ΔE = 12,000

R_b = 1400

Scale:
0
5000

20,125

C_b   3000

Figure E16-3c

(+) +10
     +5

ΔE × 10³   0

     -5   (-)

F = 2.1

1.5   2.0   2.5   3.0   3.5   F

Figure E16-3d

P_a is obtained graphically from Fig. E16-3b using W_a, C_a, and the slopes of P_a and R_a obtained from Fig. E16-3a.

The values of P_p, P_a, C_b, and W_b are used, together with the slope of R_b from Fig. E16-3a, to plot the force polygon of Fig. E16-3c to obtain values of ΔE for the corresponding F used.

A plot of ΔE versus F is made in Fig. E16-3d to obtain $F = 2.1$.

16-8 SUMMARY

The mechanics of slope stability have been considered in some detail. We note that in any stability analysis two factors are of paramount importance:

1. Soil properties
2. Shape, and the instant center, of the potential failure mass

Where the slope is isotropic and homogeneous, solutions may be tabulated as in Fig. 16-4 but are relatively easy to obtain analytically.

Where the soil is stratified, contains a phreatic surface, is nonisotropic, or has a discontinuity, some type of finite element solution is necessary.

The method of slices is commonly used where no clearly defined discontinuities are present. The sliding wedge solution is used where a discontinuity forces the location of a part of the failure surface.

If the safety factor is $F > 1.25$, we may have considerable confidence that the slope is safe. If F is less than about 1.07, we may expect a slope failure.

HOMEWORK PROBLEMS

16-1 Derive Eq. (16-2a).

16-2 For Fig. P16-2, what is the maximum value of H for $F = 1$?

16-3 For Fig. P16-2, what is the maximum value of H for $F = 1.25$?
  Ans.: $H = 21$ m.

16-4 What is F for the conditions shown in Fig. P16-4, using the circle center coordinates as assigned and all entrance points at B?

No.	(a)	(b)	(c)	(d)	(e)	(f)	(g)	(h)	
X	150	150	150	150	155	155	155	155	
Y	140	145	150	155	140	145	150	155	
	—	—	1.569	1.411	—	—	1.507	1.389	Ans. for 16-4
	1.569	1.412	—	1.397	—	—	1.389	1.331	Ans. for 16-6

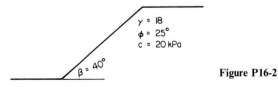

$\gamma = 18$
$\phi = 25°$
$c = 20$ kPa

$\beta = 40°$

Figure P16-2

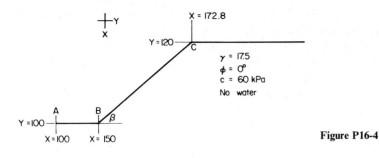

Figure P16-4

16-5 Compute minimum F using the assigned problem number of Prob. 16-4.

Ans.: $x_c = 50$, $y_c = 33$, $x_e = 85$, $y_e = 30$, $F = 1.761$.

16-6 Compute F for Prob. 16-4 as assigned if the soil properties include $\phi = 25°$.

16-7 Compute F for the assigned problem number of 16-4 with $\gamma = 17.5$, $\phi = 25°$, and $c = 60$ kPa, and including a tension crack at or to the right of point C such that the largest part of crack is included to intersect the trial failure circle. Do not include water in the tension crack unless specifically assigned.

16-8 Verify Ex. 16-2 using a suitably large scale for the drawing.

16-9 Verify Ex. 16-3 using a suitably large drawing scale.

16-10 Redo Ex. 16-3 if the base of the sliding block has an angle $\beta = 15°$, no surcharge, and soil properties of $\gamma = 18$ kN/m^3, $\phi = 32°$, and $c = 80$ kPa.

Bibliography

The following abbreviations are used to simplify the reference list:

AASHTO	American Association of State Highway and Transportation Officials
HRB	Highway Research Board [presently Transportation Research Board (TRB), Washington, D.C.]. Publications available from TRB if in print
HRR	Highway Research Record. Publications available from TRB if in print
ASCE	American Society of Civil Engineers
JGED	Journal of Geotechnical Engineering Division, ASCE (1974–)
JSMFD	Journal of Soil Mechanics and Foundations Division ASCE (1956–1973)
PSC	Proceedings ASCE Specialty Conferences
	1st PSC: Shear Strength of Cohesive Soils (1960)
	3d PSC: Placement and Improvement of Soil to Support Structures (1968)
	5th PSC: Performance of Earth and Earth Supported Structures (1972)
	6th PSC: In Situ Measurement of Soil Properties (1975)
ASTM	American Society for Testing and Materials, Philadelphia, Pa.
STP	Special Technical Publication published by ASTM.

CGJ   Canadian Geotechnical Journal, Ottawa, Canada.

ICSMFE   Proceedings of International Conference on Soil Mechanics and Foundation Engineering

ISSS   Proceedings of International Symposium on Soil Structure, Gothenburg, Sweden

Geotechnique is published by the Institution of Civil Engineers, London.

AASHTO (1971): "Highway Materials, Part 2," AASHTO, Washington, D.C.

Abbot, M. B. (1960): One-Dimensional Consolidation of Multi-layered Soils, *Geotechnique*, vol. 10, no. 4, pp. 151–165.

Aboshi, H., and H. Monden (1961): Three-Dimensional Consolidation of Clay, *5th ICSMFE*, vol. 1, pp. 559–562.

ASTM (1978): "Annual Book of Standards: Soil and Rock; Building Stones; Peats," Part 19, ASTM, Philadelphia, Pa.

ASTM (1973): "Evaluation of Relative Density and its Role in Geotechnical Projects Involving Cohesionless Soils," *STP* no. 523, 510 pages.

Azzouz, A. S., R. J. Krizek, and R. B. Corotis (1976): Regression Analysis of Soil Compressibility, *Soils and Foundations*, Tokyo, vol. 16, no. 2, pp. 19–29.

Barden, L., and R. J. McDermott (1965): Use of Free Ends in Triaxial Testing of Clays, *JSMFD*, ASCE, vol. 91, SM 6, pp. 1–23.

Bell, A. L. (1915): The Lateral Pressure and Resistance of Clay and the Supporting Power of Clay Foundations, in "A Century of Soil Mechanics," Institution of Civil Engineers, London, pp. 93–134.

Berry, P. L., and W. B. Wilkinson (1969): The Radial Consolidation of Clay Soils, *Geotechnique*, vol. 19, no. 2, pp. 253–284.

Bertram, G. E. (1940): "An Experimental Investigation of Protective Filters," publication of the Graduate School of Engineering, Harvard University, no. 267.

Bishop, A. W. (1955): The Use of the Slip Circle in the Stability Analysis of Slopes, *Geotechnique*, vol. 5, no. 1, pp. 7–17.

Bjerrum, L. (1972): Embankments on Soft Ground, *Proc. 5th PSC*, vol. 2, pp. 1–54.

Bjerrum, L., and N. E. Simons (1960): Comparison of Shear Strength Characteristics of Normally Consolidated Clays, *Proc. 1st PSC*, pp. 711–726.

Bjerrum, L., A. Casagrande, R. B. Peck, and A. W. Skempton (1960): "From Theory to Practice in Soil Mechanics: Selections from the writings of K. Terzaghi," John Wiley and Sons, Inc., New York, 425 pages.

Blessey, W. E. (1970): Pile Foundations Loads and Deformations—Mississippi River Deltaic Plain, *Proc. Conf. Design and Installation of Pile Foundations and Cellular Structures, Lehigh University*, pp. 1–26.

Bowles, J. E. (1978): "Engineering Properties of Soils and Their Measurement," 2d ed., McGraw-Hill Book Company, New York, 213 pages.

Bowles, J. E. (1977): "Foundation Analysis and Design," 2d ed., McGraw-Hill Book Company, New York, 750 pages.

Bowles, J. E. (1974): "Analytical and Computer Methods in Foundation Engineering," McGraw-Hill Book Company, New York, 519 pages.

Brooker, E. W., and H. O. Ireland (1965): Earth Pressures at Rest Related to Stress History, *CGJ*, vol. 2, no. 1, pp. 1–15.

Caquot, A., and J. Kerisel (1948): "Tables for the Calculation of Passive Pressure, Active Pressure, and Bearing Capacity of Foundations," trans. M. A. Bec, Gauthier-Villars, Paris.

Caroll, D. (1970): "Rock Weathering," Plenum Publishing Corporation, New York, 203 pages.

Casagrande, A. (1965): Role of the "Calculated Risk" in Earthwork and Foundation Engineering, *JSMFD, ASCE*, vol. 91, SM 4, pp. 1–40.

Casagrande, A. (1958): Notes on the Design of the Liquid Limit Device, *Geotechnique*, vol. 8, no. 2, pp. 84–91.

Casagrande, A. (1948): Classification and Identification of Soils, *Trans. ASCE*, vol. 113, pp. 901–930.

Casagrande, A. (1937): Seepage through Dams, in "Contributions to Soil Mechanics 1925–1940," Boston Society of Civil Engineers, pp. 295–336.

Casagrande, A. (1932): Research on the Atterberg Limits of Soils, *Public Roads*, vol. 13, no. 8, pp. 121–136.

Cedergren, H. (1977): "Seepage, Drainage and Flow Nets," 2d ed., John Wiley and Sons, Inc., New York, 534 pages.

Cedergren, H. (1948): Use of Flow-Net in Earth Dam and Levee Design, *2d ICSMFE*, vol. 5, pp. 293–298.

Chan, C. K., and J. P. Mulilis (1976): Pneumatic Sinusoidal Loading System, *JGED, ASCE*, vol. 102, GT 3, pp. 277–282.

Collins, K., and A. McGown (1974): The Form and Function of Microfabric Features in a Variety of Natural Soils, *Geotechnique*, vol. 24, no. 2, pp. 223–254.

Cullingford, G., A. K. Lashine, and G. B. Parr (1972), "Servo-Controlled Equipment for Dynamic Testing of Soils," *Geotechnique*, vol. 22, no. 3, pp. 526–529.

de Beer, E. (1967): Shear Strength Characteristics of the "Boom" Clay, *Proc. Geotechnical Conf.*, Oslo, Norway, vol. 1, pp. 83–88.

De Leeuw, E. H. (1965): The Theory of Three-Dimensional Consolidation Applied to Cylindrical Bodies, *6th ICSMFE*, vol. 1, pp. 287–290.

DOT (1976): "Flyash: A Construction Material," U.S. Department of Transportation, Washington, D.C., 198 pages.

Fadum, R. E. (1948): Influence Values for Estimating Stress in Elastic Foundations, *2d ICSMFE*, vol. 3, pp. 77–84.

Fang, H. Y., and W. F. Chen (1971): New Method for Determining the Tensile Strength of Soils, *HRR* no. 345, pp. 62–68.

Flint, R. F. (1970): "Glacial and Quaternary Geology," John Wiley and Sons, Inc., New York, 892 pages.

Gibson, R. E., and P. Lumb (1952): Numerical Solution of Some Problems in Consolidation of Clay, *Proc. Inst. Civil Engineers*, London, pt. 1, no. 2, March, pp. 182–198.

Gilboy, G. (1933): Soil Mechanics Research, *Trans. ASCE*, vol. 98, p. 239.

Grimm, R. E. (1968): "Clay Mineralogy," 2d ed., McGraw-Hill Book Company, New York, 596 pages.

Gromko, G. J. (1974): Review of Expansive Soils, *JGED, ASCE*, vol. 100, GT 6, pp. 667–687.

Hansen, J. B. (1970): A Revised and Extended Formula for Bearing Capacity, *Danish Geotechnical Inst. Bull.* 28, Copenhagen, 21 pages.

Hardin, B. O., and V. P. Drnevich (1972): Shear Modulus and Damping in Soils: Design Equations and Curves, *JSMFD, ASCE*, vol. 98, SM 7, pp. 667–692.

Hardin, B. O., and J. Music (1965): Apparatus for Vibration of Soil Specimens During a Triaxial Test, *ASTM STP* no. 392, pp. 55–73.

Hardin, B. O., and F. E. Richart Jr. (1963): Elastic Wave Velocities in Granular Soils, *JSMFD, ASCE*, vol. 89, SM 1, pp. 33–65.

Hazen, A. (1911): Discussion: Dams on Sand Foundations, *Trans. ASCE*, vol. 73, p. 199.

Henkel, D. J. (1960): The Shear Strength of Saturated Remolded Clays, *1st PSC, ASCE*, pp. 533–553.

Hiriart, F., and R. J. Marsal (1969): The Subsidence of Mexico City, in Nabor Carrillo, "The Subsidence of Mexico City and Texcoco Project," Secretaria de Hacienda Credito Publico Fiduciacia: Nacional Financiera, S.A., Mexico, pp. 109–147.

Holtz, W. G., and H. J. Gibbs (1956): Engineering Properties of Expansive Clays, *Trans. ASCE*, vol. 121, pp. 641–677.

HRR (1968): "Soil Cement," *HRR* no. 255 (see also *HRR* no. 235, "Compaction and Lime Stabilization").

Hvorslev, M. J. (1949): "Subsurface Exploration and Sampling of Soils for Civil Engineering Purposes," Waterways Experiment Station, Vicksburg, Miss. (available from Engineering Foundation, New York), 521 pages.

Hvorslev, M. J. (1937): "Physical Properties of Remolded Cohesive Soils," Ph.D. thesis translated and published (1969) as Translation No. 69-5 by Waterways Experiment Station, Vicksburg, Miss., 165 pages.

Ishihara, K., and S. Yasuda (1975): Sand Liquefaction in Hollow Cylinder Torsion under Irregular Excitation, *Soils and Foundations*, Tokyo, vol. 15, no. 1, pp. 45–59.

Jaky, J. (1948): Pressure in Silos, *2d ICSMFE*, vol. 1, pp. 103–107.

Janbu, N. (1965): Consolidation of Clay Layers Based on Non-Linear Stress-Strain, *6th ICSMFE*, vol. 2, pp. 83–87.

Johnson, A. W., and J. R. Sallberg (1962): Factors Influencing Compaction Test Results, *HRB Bull.* no. 319, 148 pages.

Johnson, A. W., and J. R. Sallberg (1960): Factors That Influence Field Compaction of Soils, *HRB Bull.* no. 272, 206 pages.

Jones, D. E., and W. G. Holtz (1973): Expansive Soils—The Hidden Disaster, *Civil Engineering*, ASCE, pp. 49–51.

Karlsson, R., and L. Viberg (1967): Ratio c/p' in Relation to Liquid Limit and Plasticity Index, with Special Reference to Swedish Clays, *Proc. Geotechnical Conf.*, Oslo, Norway, vol. 1, pp. 43–47.

Kovacs, W. D., H. B. Seed, and C. K. Chan (1971): Dynamic Moduli and Damping Ratios for Soft Clay, *JSMFD, ASCE*, vol. 97, SM 1, pp. 59–75.

Ladd, C. C., and R. Foott (1974): New Design Procedure for Stability of Soft Clays, *JGED, ASCE*, vol. 100, GT 7, pp. 763–786.

Lambe, T. W. (1958): The Structure of Compacted Clay, *JSMFD, ASCE*, vol. 84, SM 2, Proc. no. 1654.

Lambe, T. W., and R. V. Whitman (1969): "Soil Mechanics," John Wiley and Sons, Inc., New York, 553 pages.

Langfelder, L. J., and V. R. Nivargikar (1967): "Some Factors Influencing Shear Strength and Compressibility of Compacted Soils," *HRR* no. 177, pp. 4–21.

Lee, K. L., and J. A. Focht, Jr. (1975): Strength of Clay Subjected to Cyclic Loading, *Marine Geotechnology*, New York, vol. 1, no. 3.

Lee, K. L., and H. B. Seed (1967): Cyclic Stresses Causing Liquefaction of Sand, *JSMFD, ASCE*, vol. 93, SM 1, pp. 47–70.

Lee, K. L., and A. Singh (1971): Relative Density and Relative Compaction, *JSMFD, ASCE*, vol. 97, SM 7, pp. 1049–1052.

Leggett, R. F. (1962): "Geology and Engineering," 2d ed., McGraw-Hill Book Company, New York, 884 pages.

Leonards, G. A. (1962): Engineering Properties of Soils, chap. 2 in "Foundation Engineering," McGraw-Hill Book Company, New York.

Lo, K. Y. (1969): The Pore Pressure–Strain Relationship of Normally Consolidated Undisturbed Clays, Parts I and II, *CGJ*, vol. 6, no. 4, pp. 383–412.

Lo, K. Y. (1961): Secondary Compression of Clays, *JSMFD, ASCE*, vol. 87, SM 4, pp. 61–87.

Mansur, C. I., and R. I. Kaufman (1962): Dewatering, chap. 3 in "Foundation Engineering," McGraw-Hill Book Company, New York.

Martins, J. B. (1965): Consolidation of a Clay Layer of Non-Uniform Coefficient of Permeability, *6th ICSMFE*, vol. 1, pp. 308–312.

Matsuo, S., and M. Kamon (1973): Microscopic Research on the Consolidated Samples of Clayey Soils, *Proc. ISSS*, pp. 194–203.

Meehan, R. L. (1967): The Uselessness of Elephants in Compacting Till, *CGJ*, vol. 4, no. 3, pp. 358–360.

Menard, L., and Y. Broise (1975): Theoretical and Practical Aspects of Dynamic Consolidation, *Geotechnique*, vol. 25, no. 1, pp. 3–18.

Morgenstern, N. R. (1967): Shear Strength of Stiff Clay, *Proc. Geotechnical Conf.*, Oslo, Norway, vol. 2, pp. 59–69.

Middlebrooks, T. A. (1942): Fort Peck Slide, *Trans. ASCE*, vol. 107, pp. 723–742.

Miller, R. M. (1938): Soil Reactions in Relation to Foundations on Piles, *Trans. ASCE*, vol. 103, pp. 1193–1216.

Mills, John S. (1913): "The Panama Canal," Thomas Nelson and Sons, London, 344 pages.

Mitchell, J. K. (1976): "Fundamentals of Soil Behavior," John Wiley and Sons, Inc., New York, 422 pages.

Mitchell, J. K. (1968): In-Placement Treatment of Foundation Soils, *3d PSC, ASCE*, pp. 93–134.

Mitchell, J. K., V. Vivatrat, and T. W. Lambe (1977): Foundation Performance of Tower of Pisa, *JGED, ASCE*, vol. 103, GT 3, pp. 227–249.

NAFAC (1971), "Design Manual: Soil Mechanics, Foundations and Earth Structures," NAFAC DM-7, Department of the Navy, Washington, D.C.

Newmark, N. M. (1942): Influence Charts for Computation of Stresses in Elastic Foundations, *University of Illinois Engineering Experiment Station Bull.* no. 338 (reprinted as vol. 61, no. 92, June 1964).

Park, T. K., and M. L. Silver (1975): Dynamic Triaxial and Simple Shear Behavior of Sand, *JGED, ASCE*, vol. 101, GT 6, pp. 513–529.

PCA (1958): "Soil-Cement Laboratory Handbook," Portland Cement Association, Chicago, Ill.

Peck, R. G., and F. G. Bryant (1952–53): The Bearing Capacity Failure of the Transcona Elevator, *Geotechnique*, vol. 3, pp. 201–208.

Poskitt, T. J. (1969): The Consolidation of Clay with Variable Permeability and Compressibility, *Geotechnique*, vol. 19, no. 2, pp. 234–252.

Proctor, R. R. (1933): Fundamental Principles of Soil Compaction, *Engineering News-Record*, Aug. 31, Sept. 7, Sept. 21, and Sept. 28.

Richart, F. E., Jr., J. R. Hall, and R. D. Woods (1970): "Vibrations of Soils and Foundations," Prentice-Hall, Inc., Englewood Cliffs, N.J., 414 pages.

Rosenfarb, J. L., and W. F. Chen (1972): Limit Analysis Solution of Earth Pressure Problems, *Fritz Engineering Laboratory Report* 355.14, Lehigh University, 53 pages.

Rowe, P. W. (1964): The Calculation of the Consolidation Rates of Laminated, Varved or Layered Clays with Particular Reference to Sand Drains, *Geotechnique*, vol. 14, no. 4, pp. 321–339.

Rumer, R. R. (1964): Discussion: Laminar and Turbulent Flow through Sand, *JSMFD, ASCE* vol. 90, SM 2, pp. 205–207.

Schmertmann, J. H. (1955): The Undisturbed Consolidation Behavior of Clay, *Trans. ASCE*, vol. 120, pp. 1201–1233.

Seed, H. B., and C. K. Chan (1959): Structure and Strength Characteristics of Compacted Clays, *JSMFD, ASCE*, vol. 85, SM 5, pp. 87–128.

Seed, H. B., K. Mori, and C. K. Chan (1977): Influence of Seismic History on Liquefaction of Sands, *JGED, ASCE*, vol. 103, GT 4, pp. 257–270.

Seed, H. B., and I. M. Idriss (1971): Simplified Procedure for Evaluating Soil Liquefaction Potential, *JSMFD, ASCE*, vol. 97, SM 9, pp. 1249–1273.

Sherard, J. L., R. J. Woodward, S. F. Gizienski, and W. A. Clevenger (1963): "Earth and Earth-rock Dams," John Wiley and Sons, Inc, New York, 725 pages.

Sherman, G. B., R. O. Watkins, and R. H. Prysock (1967): A Statistical Analysis of Embankment Compaction, *HRR* no. 177, pp. 157–185.

Simons, N. E. (1960): Comprehensive Investigation of the Shear Strength of an Undisturbed Drammen Clay, *1st PSC, ASCE*, pp. 727–745.

Skempton, A. W. (1964): Long Term Stability of Clay Slopes, *Geotechnique*, vol. 14, no. 2, pp. 77–101.

Skempton, A. W. (1961): Effective Stress in Soils, Concrete and Rocks, *Proc. Conf. Pore Pressure and Suction in Soils*, Butterworths, London, pp. 4–16.

Skempton, A. W. (1954): The Pore Pressure Parameters *A* and *B*, *Geotechnique*, vol. 4, no. 4, pp. 143–147.

Skempton, A. W. (1951): "The Bearing Capacity of Clays," The Building Research Congress, London.

Sokolovski, V. V. (1965): "Statics of Granular Media," Pergamon Press, London, 270 pages.

Sowers, G. F. (1963): Strength Testing of Soils, *ASTM STP* no. 361, pp. 3–21.

Taylor, D. W. (1948): "Fundamentals of Soil Mechanics," John Wiley and Sons, Inc., New York, pp. 229–239.

Taylor, D. W. (1937): Stability of Earth Slopes, *Contributions to Soil Mechanics 1925–1940*, Boston Society of Civil Engineers, pp. 337–386.

Terzaghi, K. (1943): "Theoretical Soil Mechanics," John Wiley and Sons, Inc., New York, 510 pages.

Terzaghi, K. (1925): Principles of Soil Mechanics II—Compressive Strength of Clay, *Engineering News-Record*, vol. 95, no. 20, p. 799.

Terzaghi, K., and R. B. Peck (1967): "Soil Mechanics in Engineering Practice," 2d ed., John Wiley and Sons, Inc., New York, p. 72.

Tovey, N. K., and W. K. Yan (1973): The Preparation of Soils and Other Geological Materials for the SEM, *ISSS*, pp. 59–69.

Tschebotarioff, G. P. (1936): Comparison between Consolidation, Elastic and Other Properties ..., *1st ICSMFE*, vol. 1, pp. 33–36.

Turnbull, W. J. (1950): Compaction and Strength Tests on Compacted Soil, paper presented at annual ASCE meeting.

Turnbull, W. J., J. R. Compton, and R. G. Ahlvin (1966): Quality Control of Compacted Earth Work, *JSMFD, ASCE*, vol. 92, SM 1, pp. 93–103.

USBR (1968), "Earth Manual," U.S. Bureau of Reclamation, Denver, Colo., 783 pages.

Valera, J. E., and N. C. Donovan (1977): Soil Liquefaction Procedures—A Review, *JGED, ASCE*, vol. 103, GT 6, pp. 607–625.

Weber, W. G., Jr. (1969): Performance of Embankments Constructed over Peat, *JSMFD, ASCE*, vol. 95, SM 1, pp. 53–76.

Weissmann, G. F., and R. R. Hart (1961): The Damping Capacity of Some Granular Soils, *ASTM STP* no. 305, pp. 45–54.

White, L. S. (1952–53): Transcona Elevator Failure: Eye Witness Account, *Geotechnique*, vol. 3, pp. 209–214.

Wilson, S. D. (1950): Small Soil Compaction Apparatus Duplicates Field Results Closely, *Engineering News-Record*, Nov. 2, pp. 34–36.

Wineland, J. D. (1975): Borehole Shear Device, *6th PSC, ASCE*, vol. 1, pp. 511–522.

Wroth, C. P. (1975): In Situ Measurement of Initial Stresses and Deformation Characteristics, *6th PSC, ASCE*, vol. 2, pp. 181–230.

Yong, R. N., and D. E. Sheeran (1973): Fabric Unit Interaction and Soil Behavior, *Proc. ISSS*, pp. 176–183.

Indexes

Name Index

Subject Index